Information Economics

Information is a magic commodity: easy to spread, hard to control but crucial to economic decisions. Important economic phenomena, like markets, rating agencies, banking or other forms of financial intermediation can be understood as the resourceful use of information. The failure of information transmission can cause severe problems such as market breakdowns or financial bubbles. This new text book by Urs Birchler and Monika Bütler is an introduction to the study of how information affects economic relations. The authors provide a narrative treatment of the more formal concepts of Information Economics, using easy to understand and lively illustrations from film and literature and nutshell examples.

Birchler and Bütler adopt three separate approaches for explaining the concepts. The book first covers the economics of information in a 'man versus nature' context, explaining basic concepts like rational updating or the value of information. Then in a 'man versus man' setting, Birchler and Bütler describe strategic issues in the use of information: the make-buy-or-copy decision, the working and failure of markets and the important role of outguessing each other in a macroeconomic context. The book also looks at the classical problems of asymmetrical information, optimal contracts, and incentives. It closes with a 'man versus himself' perspective, focusing on information management within the individual.

The concepts covered in this book cast light on many issues from genetic testing to life insurance and pensions to banking and finance and would be of great interest for both undergraduate and postgraduate students interested in information and its role in individual decision making, markets, financial disturbances, and macroeconomics. This is an ideal textbook for students seeking a way in to understanding the key concepts in this field.

Urs Birchler is a Director at the Swiss National Bank and a former member of the Basel Committee on Banking Supervision. He has taught at the universities of Zurich, Berne, St. Gallen and Leipzig.

Monika Bütler is Professor of Economics and Public Policy at St. Gallen University, CESifo Fellow and CEPR affiliate.

Routledge advanced texts in economics and finance

Financial Econometrics
Peijie Wang

Macroeconomics for Developing Countries (2nd edition)
Raghbendra Jha

Advanced Mathematical Economics
Rakesh Vohra

Advanced Econometric Theory
John S. Chipman

Understanding Macroeconomic Theory
John M. Barron, Bradley T. Ewing and Gerald J. Lynch

Regional Economics
Roberta Capello

Mathematical Finance
Core theory, problems and statistical algorithms
Nikolai Dokuchaev

Applied Health Economics
Andrew M. Jones, Nigel Rice, Teresa Bago d'Uva and Silvia Balia

Information Economics
Urs Birchler and Monika Bütler

Information Economics

Urs Birchler and Monika Bütler

LONDON AND NEW YORK

First published 2007
by Routledge
2 Park Square, Milton Park, Abingdon, Oxon OX14 4RN

Simultaneously published in the USA and Canada
by Routledge
270 Madison Ave, New York, NY 10016

Routledge is an imprint of the Taylor & Francis Group, an informa business

Cover Design: Sarah Eva Birchler

Typeset in Times New Roman by
Newgen Imaging Systems (P) Ltd, Chennai, India
Printed and bound in Great Britain by
Antony Rowe Ltd, Chippenham, Wiltshire

British Library Cataloguing in Publication Data
A catalogue record for this book is available
from the British Library

Library of Congress Cataloging in Publication Data
Birchler, Urs W.
Information economics / Urs Birchler and Monika Bütler.
p.cm. – (Routledge advanced texts in economics and finance)
Includes bibliographical references and index.
1. Economics–Decision making. 2. Information technology–Economic aspects. 3. Information resources–Economic aspects. I. Bütler, Monika. II. Title.

HB74.2.B57 2007
303.48′33–dc22 2006102041

ISBN: 978–0–415–37346–3 (hbk)
ISBN: 978–0–415–37345–6 (pbk)
ISBN: 978–0–203–94655–8 (ebk)

To Peter and Eugen

Contents

PART III
Asymmetric information 253

Figures

Tables

Boxes

Problem sets

Preface

It may appear arrogant, if not downright misguided, to write a book on a subject as wide and demanding as the economics of information. Such a book would hardly withstand comparison to *The Analytics of Uncertainty and Information* by Hirshleifer and Riley, published 15 years ago. To make matters worse, since the publication of that seminal work, the field of information economics has flourished, and even a summary of only the most important findings would seem to go beyond the scope of a single monograph.

Of course, we were aware of the difficult task facing us, and resolved not to produce a mere "updated" version of the Hirshleifer and Riley work. In fact, the origins of our book were rather more modest, namely a desire to put down on paper the accompanying lecture notes to a course in *Asymmetric Information* which Urs had taught at the University of St. Gallen for several years. However, before long this endeavor began to mushroom, given that information economics is much more than just the economics of asymmetric information. For instance, new models of information acquisition and processing had come to play an increasingly important role in public policy issues and in the macroeconomic models Monika was teaching. In addition, there were some valuable older ideas Urs had first encountered in a course on information theory given by Gerold Hauser at the University of Zürich in the late 1970s. Given that the *The Analytics of Uncertainty and Information* still seemed to be the most recent textbook on the main concepts in information economics, we began to wonder if it would be possible to write a new textbook, which would present the present state of the field in a readable, intuitive, but still consistent way. After much consideration, we decided to rise to this challenge.

It is up to the readers to judge how successful we have been. At a personal level, the book soon proved to be an exciting experience for us for completely different reasons. Gradually, family and friends became directly involved in the project. First, Monika joined Urs as co-author, coming to the rescue of her husband at a time when the task threatened to overwhelm him. Long before the text was anywhere near completion, Sarah Birchler came up with a cover design which we fell in love with at first sight. If Sarah made the book look well, Elaine Sheerin made it read well. Her role was to make the text, written by two non-native speakers, sound more "English". Being the last on the production line, Elaine also absorbed the time constraints accumulated over the previous stages. She did so heroically without ever losing her sense of humor. Silently in the background, our computers, including our LaTeX systems, ran smoothly, thanks to our most diligent and circumspect Chief IT Officer, Andreas Lang.

Monika's crew at the University of St. Gallen's Research Institute for Empirical Economics and Economic Policy did a great job at various stages of the endeavor. Sharon Bochsler not only provided wonderful LaTeX support, but also provided her pedagogical skills and advice. The influence of many of her ideas and suggestions can be observed throughout the book. Sabine Müller translated many of our hand-scribbled drawings into stunning graphs. Stefan Staubli was our graphics' supremo, juggling around with all kinds of file formats. Martin Ruesch, Jonathan Schulz and, again, Stefan Staubli acted as test readers and provided us with a number of extremely helpful comments and suggestions. Jonathan also helped us to compose problem sets and proposed solutions.

A considerable number of students at the University of St. Gallen and the University of Zürich were exposed to earlier versions of the text and directly or indirectly helped to improve it. Another valuable source of inspiration and advice were our colleagues at the University of St. Gallen and at the Swiss National Bank.

We are grateful to both of our employers for their support during this project. The University of St. Gallen provided research assistance as well as the necessary infrastructure. We owe special thanks to Franz Jaeger, managing director of the Research Institute for Empirical Economics and Economic Policy, for his encouragement and for generously picking up the bill for external expenses. The Swiss National Bank provided an ideal working environment to write the book. Particularly during the decisive last few months prior to publication, Daniel Heller effectively shielded Urs from many daily obligations that might have prevented us from finishing the book on time. None of the institutions mentioned interfered in any way with the content of the book. The views expressed are those of the authors alone.

This book might never have been written were it not for the initiative of Rob Langham, our partner on the publishing side. It was Rob's unshakeable confidence which first seduced us to write the book. Not having much of a stick with which to beat us, Rob used his gentle but steady hand to guide us along every step of the project. We would also like to thank Rob's colleagues for their excellent technical support. Finally, we are indebted to three anonymous referees who advised the publisher on an early draft. Not only were they generous enough not to rubbish the draft, but they also provided many valuable and pertinent suggestions which helped to shape the book as you see it today.

Although we are happy to acknowledge support and inspiration from many sides, we claim responsibility for any outstanding shortcomings.

Everyone mentioned above supported us more or less voluntarily. However, we did not seek approval from our two sons, Peter and Eugen. Despite the loving help of Alexandra Bürgi-Maldonado, they suffered most from the project over the last two years, having to endure busy, distracted and sometimes jittery parents.

Nevertheless, they managed to remain good-natured and obliging against all the odds. Even if this had not been the case, dedicating this book to Peter and Eugen is the least we could do to show our gratitude.

October 2006
Urs Birchler and Monika Bütler

1 Why study information economics?

"There is no doubt about it: we are living in the Information Age," Hal Varian, an eminent economist, writes.[1] According to a study by Hal Varian and Peter Lyman,[2] the amount of new information stored on print, film, magnetic, and optical storage media in 2002 was about $5 * 10^{18}$ bytes, or 5 exabytes (at 1 byte = 8 bit). Five exabytes are equivalent to all words ever spoken by human beings, or to the information contained in 37,000 new libraries the size of the Library of Congress book collections. And the volume of additional information continues to grow by one-third every year.

Information is produced, stored, copied and traded. Information is an economic good and, as we will argue, a strange one. Which other good can be kept and passed on at the same time? Which other good often cannot be destroyed? Which other good cannot be discovered in a person even if the person is scanned with the most advanced technology? Consequently, the first reason for studying the economics of information is:

- **Information is an interesting economic good.**

For many people outside of the economics profession, economics is about money, or about getting rich. For most economists (not all of whom have become rich), economics is about understanding "what, how, and for whom" the economy produces—as the 1970 Nobel laureate in economics, Paul Samuelson (born 1915), put it. That is why economists have intensely studied the market, the very mechanism that brings about the what, how, and for whom. The task of the market is far from trivial: "the knowledge of the circumstances of which we must make use never exists in concentrated or integrated form, but solely as the dispersed bits of incomplete and frequently contradictory knowledge which all the separate individuals possess" as Friedrich A. Von Hayek (1945), another Nobel laureate, remarked 60 years ago. The true "economic problem", therefore, is how to aggregate dispersed information. Our second reason for studying the economics of information then is:

- **Economics is about information.**

As we said before, information is a good, and the market brings all the tiny individual parts of this good together to express information in the price system. It is the scent of profit with which the market seduces us to collect and convey information. An illustration is the Waterloo legend concerning the origins of Nathan Rothschild's fortune (see Box 1.1).

Box 1.1 The Nathan Rothschild Waterloo legend

A famous (and fictitious) legend has it that Baron Nathan de Rothschild made a fortune on the London stock exchange, as he was the first to learn about Napoléon Bonaparte's defeat in the final battle of Waterloo (1815) against the united British and Prussian armies, under Wellington and Blücher respectively.[3] Rothschild had established a courier system by which he learnt the outcome of the battle a full day earlier than the British war office. Upon hearing of Napoléon's defeat, Rothschild allegedly went to the London exchange, put on a sombre face and started to sell British stock and government debt. Other traders, assuming that Rothschild had bad news and trying to free-ride on it, also began to sell. Prices tumbled, and Rothschild's straw men could buy cheaply, leaving the cunning baron with a fortune.

Assume for a moment that the story is true. Imagine you are a trader and you see Rothschild sell. You know that Rothschild knows how the battle ended. Should you follow his lead and sell too? Is that not what Rothschild had expected to happen? After all, he knew that other traders knew about his information advantage. Thinking just one step further, a cunning trader might have bought. And if all traders had, Rothschild would have lost, rather than have made his fortune.

As the example shows, information has highly strategic aspects. It is not only important to know about facts, but also about what others know, and about what they know about what we know. Information strategy is a tricky business, as Lady Kunigund (see Box 1.2) shows.

Box 1.2 The Glove (Friedrich Schiller)

The ballad *Der Handschuh* (The Glove; quoted from an anonymous translation) was written in 1797 by the German dramatist Friedrich Schiller. In this ballad the king and his court are seated around an arena filled with lions, tigers and leopards. Lady Kunigund tosses her glove among the beasts and turns to Knight Delorges:

> "Sir Knight, if the love that thou feel'st in thy breast
> Is as warm as thou'rt wont at each moment to swear,
> Pick up, I pray thee, the glove that lies there!"

Delorges casually descends among the animals, picks up the glove, and returns amid shouts of praise and wonder.

> The praise of his courage each mouth employs;
> Meanwhile, with a tender look of love,
> The promise to him of coming joys,
> Fair Kunigund welcomes him back to his place.
> But he threw the glove point-blank in her face:
> "Lady, no thanks from thee I'll receive!"
> And that selfsame hour he took his leave.

Lady Kunigund discovered another crucial aspect of information: You cannot *obtain* information from others without *giving* some information. By asking for proof of love, Lady Kunigund proved the absence of true love on her part. The act of observing alters the reality being observed; this can be seen as the equivalent of Heisenberg's uncertainty principle in information economics.

Baron Rothschild and Lady Kunigund are just two examples of the very special and important role information plays in economic and strategic situations. They also highlight that information is an unruly kind of economic good and that its aggregation by individuals or by the market may not always run so smoothly. This bring us to the third reason for studying information economics:

- **Information is of strategic importance.**

Information is at the root of many real world phenomena and problems. In this book we will encounter a number of examples: Why doctors do not always tell the truth; why examiners hate mediocre papers; why entrepreneurs are constructive destroyers; why health insurance may not work; why genetic testing may destroy welfare; why workers subsidize professors; why prices are sticky; why winning an auction may be bad news; why companies should not have an incentive to lie; why banks are supervised and why they sometimes fall victims to runs; and why we sometimes prefer to turn a blind eye to the information at our disposal.

Over the last two decades, economists have devoted a great deal of time and effort to improving their understanding of the role of information, thus building on seminal contributions to the field from the early 1970s (see Chapter 2). Information economics is no longer in its infancy; it has now reached its adolescent phase. In conjunction with the above argument the fourth reason for studying information economics is therefore:

- **Information economics is a young field with practical relevance in many contexts.**

Information economics may even be somewhat *en vogue*. This should not come as a surprise. Already many years ago, the fashion industry, known for its infallible sense for cultural undercurrents, gave us two fragrances called *Knowing* (1988), and *Guess* (1990). Linking scent to information, the fashion leaders also expressed a deep understanding of the importance of information in human relations. This importance could be a covert fifth reason for studying information economics.

In a music shop, a young man is leisurely browsing through the racks of CDs. A woman, evidently in a hurry, brusquely pushes past him. “Sorry!” he says, “Maybe I could help you—if you don’t mind telling me what you’re looking for.” “Don’t You Know Who I Think I Was?”, she replies. “Pardon?” “Don’t You Know Who I Think I Was?” Looking rather perplexed, he answers, “No, I have to say I don’t!” “Sorry, I should explain. You asked me what I was looking for. So I told you the title of the album. Some songs from the 80s.”

Unfortunately they cannot find the album. “Fancy a coffee?” the woman asks. “Great idea”, the man replies, “I’m Bob, by the way.” “Pleased to meet you, Bob. I’m Alice.”

2 How to use this book

2 A The purpose of the book

This textbook is aimed at students of economics, business and related fields, particularly those in their third or fourth year of undergraduate study. However, we believe that the book could also be beneficial to Masters and even PhD students. Many parts were written as to be comprehensible also to a second-year undergraduate student equipped with some rudimentary knowledge of game theory and basic microeconomics, not to mention a healthy dose of curiosity.

The goal of the book is to introduce students to the concepts of information economics which have developed during the so-called "information revolution" of the last thirty years (see Box 2.1). Presenting a new field is like showing a foreign visitor around one's home town or country. Should we go by taxi, allowing us to take the visitor anywhere? Or, should we recommend the sight-seeing bus, which does not give the traveller much flexibility (except falling asleep, perhaps), but guarantees stop-offs at a few important sites?

In this book we opted for the sight-seeing approach. We try to give readers an idea of some of the central ideas behind information economics, the equivalents of the Eiffel Tower, Notre Dame, or the Louvre. Once readers are familiar with the structure of the field and its most important features, they can strike out on their own: either taking a taxi and looking at some places of private interest or embarking on the lone walk of research . . .

Box 2.1 The information revolution in economics

Economists have long been interested in the role of information in the economy. Yet, much of economic theory was developed under the assumption that individuals make their decisions based on perfect knowledge. This assumption enables us to model the economy in terms of a general equilibrium framework, known as the Arrow–Debreu model. Such a model is able to cope with uncertainty, insofar as it is of the well-behaved kind, under which all agents believe in the same probabilities for the same events.

In the second half of the twentieth century, advances in telecommunications led to a growing interest in "information theory", first among physicists, and later among economists. Physicists and engineers tackled questions such as how information could be described or measured, often with a view to its optimal transmission through a given channel, like a cable. At the same time, mathematicians developed the basics of strategic behavior theory, better known as game theory. If one single event should symbolize this development, it would be the formulation of the equilibrium in a non-cooperative game by John Nash in 1950.

In the early 1970s a number of economists started to analyze a new type of problem, namely situations in which some agents had better information than others. George Akerlof (1970), in his famous "market for lemons" paper, showed that even slight informational asymmetries could lead to complete market failure.

This "new information economics" was very successful in explaining real world phenomena and institutions like auctions, banks, brand names, stock market bubbles, credit crunches, or incentive systems, to name but a few. The Economics Nobel Prize (Bank of Sweden Prize in Economic Sciences in Memory of Alfred Nobel) was twice awarded to representatives from the field of information economics. In 1996, James A. Mirrlees and William Vickrey were honored for their fundamental contributions to the economic theory of incentives under asymmetric information. In 2001, George A. Akerlof, A. Michael Spence and Joseph E. Stiglitz received the prize for their analyses of markets with asymmetric information. In addition, several other Nobel laureates had contributed to information economics and to its game theoretic roots: Kenneth J. Arrow (1972), George J. Stigler (1982), John F. Nash (1994), as well as Robert J. Aumann and Thomas C. Schelling (2005). One of the leading figures in the "information revolution", Jack Hirshleifer (1925–2005), did not receive a Nobel prize, although many believe that he would have been a worthy winner (Hausken, 2006).

2 B Ways of reading the book

We have tried to make this book readable and easy to use. To begin with, the sections in each chapter follow the *same structure*:

- **Introduction:** Chapter overview and how it relates to other chapters.
- **Main ideas:** A presentation of the basic ideas in an intuitive, non-technical fashion, using day-to-day language and aided by a wealth of examples.
- **Theory:** An introduction to the relevant model(s) using as little formal apparatus as possible.

- **Applications:** "Real world" examples in which the models developed in the theory sections play a role, or extensions and topics complementary to the theory section.
- **Conclusions and further reading:** A short look at what has been learnt and what questions remain unanswered.
- **Problem sets:** A selection of problems for self-training.

Second, we have placed some material in *boxes*. Boxes are "plug-ins" which differ in content from the main body of the text: (i) illustrations (example: Solomon's judgement), (ii) repeatedly used concepts (example: The Elementary Game), (iii) background information from theory or statistics (example: The Uniform distribution). Third, at the end of each chapter we include a checklist of the concepts a reader should by then be familiar with.

And, last but not least, Alice and Bob will guide readers through the text.

Alice and Bob are two young individuals who have already met at the beginning of the book (Chapter 1) and will share several experiences, which illustrate some of the situations analyzed in the book. Their adventures highlight analytical concepts, but often illustrate the limits of a mechanical application.

Of course, one could read all Alice and Bob episodes first, like nibbling off all the icing before eating the cake. A similar strategy of cutting "horizontally" through the book would be to read the "Main Ideas" sections first, before deciding whether any of these ideas deserves more in-depth study, and thus turning to the relevant "Theory" section. Some students may find they have no time to lose and will go directly to the Theory sections. The most ambitious (but by no means an inefficient) way to learn information economics would be to start with the problem sets and go back to the theory, whenever one fails at a problem set.

Solving *only* the problem sets is a better strategy than reading everything *except* the problem sets. There is no substitute for practice, as Professor Myrna Wooders notes on her homepage,[4] offering her students the following advice:

> Watching your instructor solve problems and thinking that you are learning how to solve problems is like watching an aerobics class and thinking that you are becoming fit. Watching helps to learn the 'moves', mental and physical, but cannot substitute for doing the workouts yourself.

To put a more positive spin on it: Look how quickly children learn, simply because they are not afraid of making mistakes and because they do not hold back from trying again and again until they succeed. True, workouts seem hard and problems appear difficult. Yet, as economists know, difficulty, like everything else, is "endogenous". As the Roman philosopher, the younger Seneca (4 BC–65 AD), put it:

> It's not because things are difficult that we dare not venture. It's because we dare not venture that they are difficult.
>
> (Lucius Annaeus Seneca, *Epistulae Morales*, book VII)

2 C The structure of the book

The present volume is structured into four parts. Part I gives an introduction to the basic concepts of *information as an economic good* (Chapter 3) and to its valuation (Chapter 4). It explains how much costly information an individual would optimally acquire if deciding in isolation (Chapter 5) or would produce in a strategic environment (Chapter 6). It very much focuses on the viewpoint of one sole individual (Robinson Crusoe).

Part II concerns the *aggregation of information* held by different individuals. We first show how markets aggregate information (Chapter 7). In this context we also distinguish between information about facts and information about information (Chapter 8). Functioning markets require communication among agents. When communication is imperfect, individuals may fail to coordinate actions (Chapter 9) or may try to learn by observing each others' actions (Chapter 10). We conclude this part with some novel aspects of information acquisition and aggregation in macroeconomic contexts which have arisen in recent years (Chapter 11).

Part III deals with the problems arising under *asymmetric information*. To start with, we introduce the winner's curse (Chapter 12) followed by its sister, adverse selection (Chapter 13). We also show how informational asymmetries can be mitigated by use of optimal contracts (Chapter 14). A helpful device to find optimal contracts is the revelation principle (Chapter 15). All these chapters discuss so-called "hidden information". Another information asymmetry is "hidden action", which leads to moral hazard (Chapter 16).

Part IV leads into terrain only recently discovered by economists: The use of *information within the individual* for identity building and self-management (Chapter 17).

2 D Using the book for teaching

There are several ways of using the book for teaching purposes, some of which we have tried ourselves. The following is a non-exhaustive list of suggestions:

- **Information economics:** Teaching the full book to fourth year or Masters students requires at least three hours per week. Some of the applications can be skipped or left as student assignments.
- **Asymmetric information:** A two-hour standard course would cover Chapters 3, 12, 13, followed by 7, 14, 15, 16.

- **Financial intermediation:** We suggest using the book as background reading and for a discussion of individual sections like 4 E, 6 E, 6 F, 7 E, 9 D, 14 E, 15 D, 16 D, and of complementary topics from Freixas and Rochet (1997).
- **The macroeconomics of information:** The backbone of such a course could consist of Chapters 6, particularly 6 D, 8, 9, 10, 11 and, we would recommend, 17.
- **Information in finance:** To teach finance students the informational background of problems in finance we would use Chapters 3, 12, 13, 7, 8, 10, 9 and 17.
- **Endogenous information:** This challenging subject would most likely be treated in a PhD course, based on Chapters 3, 4, 5, 6, 10, 11, rounded off by reading a few articles which have appeared in journals.
- **Economics and psychology:** The transition from *homo œconomicus* to *homo sapiens* and research into the decision process *within* the individual have been inspired by behavioral economics, psychology and neuro-science. An introductory course, on graduate or PhD level, could be built around Chapter 17 (and some key articles cited therein), with further inputs from Chapters 3, 5, 10, 11 and 16.

2 E Solutions to problem sets and other supporting material

We have set up a website for readers of this book. The address is: http://www.alicebob.info. There we shall post solutions and hints to the problem sets (at least to some) and further material that may be helpful to the reader. The website will also be used to publish any errata, although we hope we shall not have to avail of this space!

Part I

Information as an economic good

3 What is information?

3 A Introduction

The term "information" is not only used very frequently in everyday language, it is also a technical term in a wide range of sciences like physics (see Box 3.1), computer science, statistics, cybernetics, communication theory, linguistics, psychology, and economics. Not surprisingly, information is interpreted in myriad ways.

It has been said that the term "information" is a "semantic chameleon." Indeed, the term seems to adapt itself to its environment automatically like the main character in the Woody Allen movie *Zelig* (1983). The two characteristics of the word "information"—that it is easy to use but hard to define—appear to be two sides of the same coin.

The American philosopher John Searle even concluded that "the notion of information is extremely misleading". This would suggest that we ought to steer clear of any attempt to define it. Yet, in a book like this we cannot get started without endeavoring to explain why we have written such a text. We will introduce the key concept of an "information structure" or "partition". This concept helps us to compare information that different individuals may have. It will also help to describe more accurately the rationale behind the structure of the whole book. In addition, we will introduce a very important concept of updating information using *Bayes' Rule*.

Box 3.1 The information content of the universe

The Australian physicist Paul Davies (b.1946) writes: "Information is playing an increasingly fundamental role in science. It crops up in very different contexts in thermodynamics (in relation to entropy), quantum mechanics (the wave function represents our knowledge of a physical system), in relativity theory (information cannot travel faster than light) and in biology (a gene is a set of instructions)." (http://aca.mq.edu.au/PaulDavies/research/current.htm)

According to quantum mechanics every atom or particle has many attributes (charge, mass, spin, etc.). Each of these can be considered as an elementary unit—one bit—of information. Some scientists therefore consider the universe as being primarily made of information—"it from bit", as the theoretical physicist John Archibald Wheeler (b.1911), the inventor of the term "black hole", put it.

Total information capacity of the universe is estimated to approximately 10^{120} bit (Davies, 2004). This information cannot get lost. When the inputs from digitally dividing one by two, for example, cannot be recovered from the result 1/2, such "lost" information is present in the nature of dissipated energy.

The information production by *homo sapiens* which is estimated to some $5 \cdot 10^{18}$ bytes a year (see Chapter 1), though a large figure, is negligible in comparison.

3 B Main ideas: The strangest good of all

In Charles Dickens' *Great Expectations* (Chapter 7), Pip writes the following letter to Joe:

> MI DEER JO i OPE U R gRWrTE WELL i OpE i SHAL
> soN B HhBELL 42 TEEDGE U JO AN 7HEN wE SHORL
> a sO OLODD hN wEN i M PRENOTD 2 U JO wOT LhRX
> flNBLEvEMErNFxNPrP.

This message is so garbled that it is almost impossible to understand. Yet it perfectly illustrates two basic aspects of information: the engineering and semantic dimensions of information.

The *engineering* aspect of information raises questions, such as how to convey a maximum of information from sender to recipient, given that (i) the transmission channel has limited capacity, and (ii) the transmission may be disturbed by some external noise. Often there is some trade-off between efficiency and robustness against transmission errors (like typos). For example, we could make language more efficient by dropping all the vowels. Ths wld mk txts vr vlnrbl t tps, thgh! In reality, most languages do not drop vowels, and they each use the letters of the alphabet with very unequal frequency. Languages thus have a certain *redundancy*. Redundancy is costly in terms of the length of a text. Its benefit is robustness: Without any understanding of the text, on statistical grounds alone, we can safely bet that when Pip writes "7hen" he means "then" (or perhaps "when"). By contrast, typos in phone numbers (which typically have low redundancy) are hard to discover.

The *semantic* dimension of information shows that messages have a *meaning*, i.e. they refer to concepts known to the sender and—hopefully—to the recipient. Knowing English plus some conventions used in writing letters allows us to deduce that Pip calls Joe "dear", rather than "deer". The reader also benefits from the fact that words can be pronounced; thus "LhRX" may mean "larks". However, the full meaning of Pip's message only becomes apparent to a recipient with sufficient background information.

Economists have borrowed from both the engineering and semantic approaches to information. Early contributions to the economics of information were mostly inspired by engineering issues, analyzed under the label of the "theory of information". The pioneering work in the field, like Shannon (1948), laid the conceptual foundations for cell phones and wireless internet surfing, but also for the economics of information. One task within the engineering problem was measuring information, for example by the number of binary digits or "bits" needed to convey a piece of information. This term was introduced by the statistician J. W. Tukey, who also coined the term "software". Another important concept to measure information is *entropy*, a measure made popular by modern physics (see Shannon, 1948). Entropy, the reciprocal of redundancy, is maximized when all letters in an alphabet (including the blank space) have the same frequency, plus the same conditional frequencies (with respect to all preceding or subsequent letters).

Information economics, as a discipline that is distinct from information theory, took off with authors like Kenneth Arrow (Nobel prize 1972) and George Stigler (Nobel prize 1982), who started to look at the value and at the strategic use of information. Arrow also introduced the concepts of *adverse selection* and *moral hazard*. These have become two key notions to describe human behavior in the presence of asymmetrically informed individuals.

With the literature on asymmetric information, the semantic aspects of information became more important than its engineering aspects. To put it simply: "Actions speak louder than words".

Alice and Bob, walking along the street, pass a Ferrari. "I want one of those," says Bob. "Obviously not," replies Alice with a broad smile. Bob is terribly embarrassed. He knows that if he had really wanted a Ferrari he would have tried to buy or borrow one.[5]

This is the core of revealed preference theory. In the context of asymmetric information, where one individual knows more than others, actions may reveal preferences or other private information. If Fred offers his used car for $7,500, he sends two messages: First, that he is ready to sell for that amount and, second, that the car, at least to the seller, is likely to be worth less than $7,500.

Economic actions thus have a semantic dimension. Buying and selling implies not just an exchange of goods, but an exchange of information. The ability of markets to process information, or their failure to do so, will be an important issue in this book.

While both engineering and semantics try to use precise and clean technology, economics cannot avoid its proximity to everyday language. Information, the semantic chameleon, thus can convey a plethora of different meanings, as Hirshleifer and Riley (1992) point out. They distinguish between several forms of information:

- *Knowledge* is accumulated data or evidence on the world (a stock magnitude), whereas *news* or a message are increments to accumulated evidence (a flow magnitude).
- Information can be *objective* knowledge or *subjective* beliefs.
- A source of information is a *message service*, like a weather forecast service, whereas the *message* is the actual forecast itself.
- *Foreknowledge* is the information about tomorrow's weather, while *discovery* denotes knowledge about the laws of nature or about mathematics (like the millionth digit in the expansion of π), a distinction made by Hirshleifer (1971).
- *Fundamental* information is about the states of nature, while *meta*-information is about other agents' information.

Throughout this book we cannot—and we certainly would not want to—avoid using all these different meanings. In most cases the relevant meaning should be obvious from the context. We hope that this is consistently the case.

As a starting point, a reasonable though vague definition of information might be "a reduction of uncertainty". We will immediately try to be more precise by introducing the concept of an information structure or partition: The finer the partition representing our information structure, the "better" we are informed.

"Definitions give us the mastery of words not of things," the philosopher Francis Bacon (1561–1626) said. To grasp at the mastery of things, economists have found their own strategy: Whatever information is, it can be looked at as an economic good. As a good, it can be produced, stored, consumed, invested (used in production), or sold. And, of course, it has some value. At first glance, these properties, all to be treated in subsequent chapters, are not special to information. However, a closer look reveals a few striking characteristics that distinguish information from most other goods. Information, to take a few examples,

- can be sold without being given away,
- is cheap to reproduce,
- cannot be actively disposed of,
- cannot be detected in a person,
- can often not be prevented from spreading,
- can often not be valued before it is known,
- can be about facts or about other people's information.

We will first try to be more precise when defining information; however, these aspects provide us with the horizon towards which our journey should ultimately lead. A definition is only the first step on this journey.

3 C Theory: Describing, comparing and updating information

3 C a Information structures

An approach towards a precise description and comparison of information has been pioneered by John Harsanyi (Samuelson 2004, p. 375). The starting point is the assumption that the world can take one of several different states. Each state is a complete description of reality; only one of them can hold at any time. Most textbooks call the list or collection or set of all possible states the *state space*. The state space is a set denoted as $\Omega(w)$, containing elements $\{w_1, w_2, \ldots w_n\}$. Rolling a dice (while everything else is held constant) leads to a set of possible states $\Omega = \{1, 2, 3, 4, 5, 6\}$.

Two basic assumptions behind the state-space approach (for details and further literature see Samuelson, 2004) are:

- The set of possible states is finite. This assumption, originally introduced by Leonard Savage, is called the *small-world assumption* (Samuelson 2004, p. 375).
- Agents can list the possible states; there are no possible states of which individuals are not aware. This assumption enables us to describe the information that individuals have. Admittedly, it seems very restrictive to assume that everybody can foresee and list each and every possible future state of the world. The existence of a residual of states of nature that cannot be described in advance may even explain the economic role of ownership (linking residual claims to residual control rights), see Hart (1995).

An *information structure* is the result of splitting the state space into subsets, of a so-called *partition* of the state space. In the above example the state space has six elements associated with the outcomes of throwing a dice. Thus,"even-odd" is a partition; the elements of a partition (like "odd") are *events* or—as often referred to in game theory—*information sets*.

Assume Alice knows if the outcome is odd or even. For her the state space is partitioned into subsets $\{1, 3, 5\}$ and $\{2, 4, 6\}$, as illustrated in Figure 3.1. If the true state of nature is $\{5\}$, e.g., she knows that the true state lies within the *event* "odd", $\{1, 3, 5\}$.

An information *source* telling an individual in which event (subset of the state space) the true state of nature lies—which event has occurred—is a *signal*. The signal can take different values or, equivalently, have different realizations. In the dice example, a signal that can take the values $\{odd, even\}$ indicates which of these events has happened.

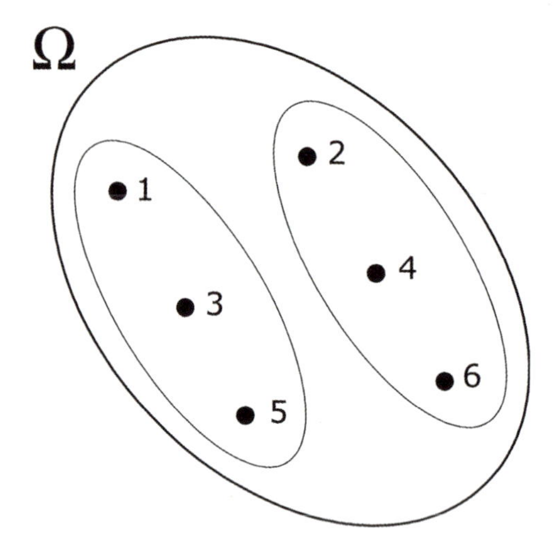

Figure 3.1 An information structure (also known as a "partition") divides a state space Ω into subsets or "events" like "odd" or "even".

Signals can be more or less informative. A *perfect signal* takes different values for each possible state, thus creating an information structure of subsets with only one element in each. A *void signal* returns an information structure consisting of only one set, i.e. the entire state space. In reality, most signals fall between those two extremes. They do contain some information, but are often imperfect. A signal $\{odd, even\}$ is imperfect as both realizations contain more than one possible state of nature. For an individual who wants to know whether a dice shows an odd or an even number, a signal telling whether the outcome is high $\{4, 5, 6\}$ or low $\{1, 2, 3\}$ is not only imperfect, but noisy (see below).

3 C b Heterogeneous and asymmetric information

Information structures help to compare the type and amount of information of different agents. We will use the term *homogeneous information* for information structures whose subsets do not overlap across individuals, while *heterogeneous information* refers to subsets that do overlap. Generally, there is always one homogeneous information structure that is finer than the others (unless they are identical). In this case we speak of *asymmetric information.*

Alice and Bob have the information structures displayed in Figure 3.2. Alice knows whether the outcome of rolling a dice is odd or even, while Bob knows whether it is high or low. They have heterogeneous information. Both have different information structures, but we cannot say which structure is more informative.

By contrast, in Figure 3.3, Bob knows everything Alice knows (i.e. whether the outcome is odd or even), plus he has some private information (whether the outcome is high or low). Information is asymmetric.

Both heterogeneous and asymmetric information raise important economic issues. These will be treated extensively in later chapters.

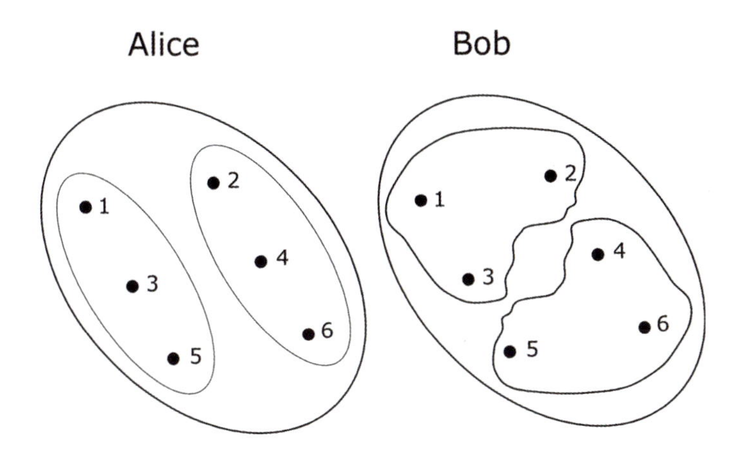

Figure 3.2 Heterogeneous information implies that information structures of different agents overlap. Alice knows whether the outcome of a thrown dice will be "odd" or "even" while Bob knows whether it is "low" or "high".

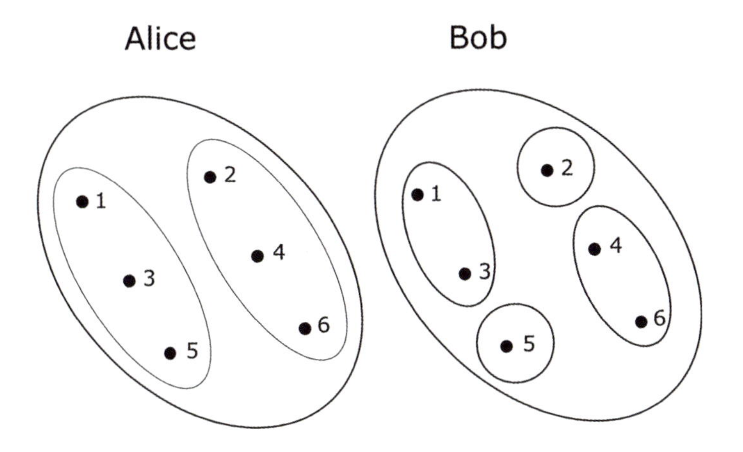

Figure 3.3 Asymmetric information implies that one agent, here Bob, has a finer information structure, i.e. superior information than another, here Alice.

3 C c Noisy signals

The signals described before were not necessarily perfect (one signal occurring in more than one state), but they were not noisy. A noisy information structure is one in which not only can one signal indicate several states, but also several signals can occur in the same state. A signal can only be noisy for the particular kind of information which one would like to obtain. An example of a noisy signal is illustrated in Figure 3.4. Event A involves the prime numbers 2, 3, 5, and 1 (which is not strictly speaking a prime number), i.e. $A = \{1, 2, 3, 5\}$, while $B = \{4, 6\}$ contains the possible faces of a dice that are not prime (and not 1). The signal is a noisy signal with respect to the outcome "even" or "odd", but it is not a noisy signal for the outcome "1 or prime" or "neither 1 nor prime".

A noisy signal is not perfect, but can still be informative. The informative content of a noisy signal derives from the correlation between states of the world and signal realizations. A signal which correctly forecasts the result of rolling a dice (odd or even) in three out of four cases is more informative than the a priori information that, with a fair dice, odd and even are equally likely.

A noisy information structure is best illustrated by the tree of events. Assume we want to know whether rolling a dice has resulted in an odd or an even number being shown. We ask an expert to make a forecast. The expert is not perfect and only observes whether the true outcome is in event A or B, as defined above. In event A, the expert will forecast "odd" and be right three times out of four. In event B, the expert will forecast "even" and will always be right. The expert may see a noisy signal, sometimes forecasting correctly and sometimes not. The degree of accuracy itself depends on the true state of the world (see below).

Conditional probabilities can be defined in two ways. The top part of Figure 3.5, for each of the outcomes "odd" and "even", shows the conditional probabilities of each of the signals (A or B) on the respective branches. For example the probability

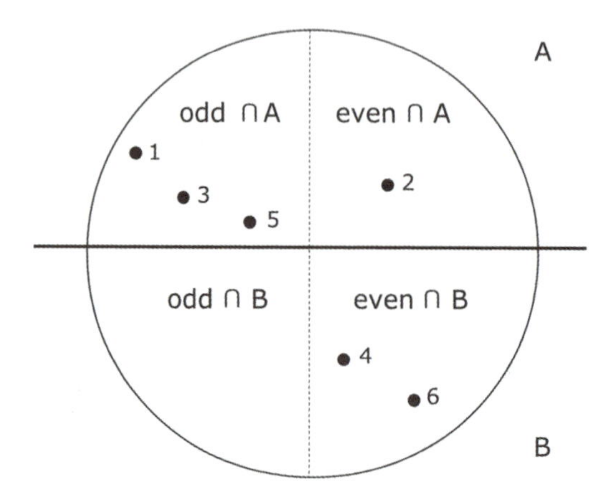

Figure 3.4 A noisy signal is not one to one; predicting "odd" for the event "1 or prime" is only true in three out of four cases.

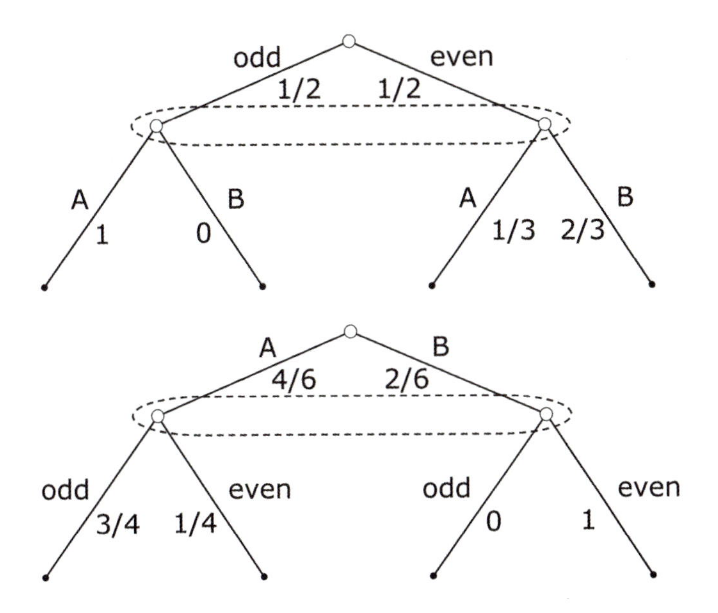

Figure 3.5 The top tree describes the path "even/odd–A/B" and yields conditional signal probabilities as in Table 3.1. Vice versa, the bottom tree follows "A/B–even/odd" and thus yields conditional outcome probabilities as given in Table 3.2.

of getting signal B if the outcome is even is $2/3$. (We will later define these probabilities as the precision of the signal.) In the bottom part of the figure, the same information structure is inverted. For each of the possible signals, the four branches give the conditional probabilities of the two outcomes. For example, when the expert observes signal A, the probability that the outcome is odd is $3/4$. (We will call these probabilities the accuracy of a forecast.)

In Figure 3.5 it is not actually important whether we think of the outcome being realized first and the signal occurring afterwards (top part) or vice versa (bottom part). Both views must lead to consistent results. What is important, however, is to understand that "A, if even" is not the same as "even, if A". In more formal terms, the *conditional outcome probability* of event "even" given signal A ($1/4$) is *not* equal to the *conditional signal probability* of signal A when the true event is "even" ($1/3$).

The conditional probabilities in Figure 3.5 can also be represented as in Tables 3.1 and 3.2. Table 3.1 gives the probabilities of getting the signal A or B, conditional on the outcome being odd or even, respectively. It shows that, for example, you always get a signal A when the event is odd, but in only one out of three cases when the true outcome is even.

Note that the columns in Table 3.1 add up to one, while the rows do not. The probability that the signal is either A or B, when the dice shows any (odd or even)

Table 3.1 Conditional *signal* probabilities are the probabilities of getting numbers 1, 2, 3 or 5 (event A) versus 4 or 6 (event B), given an even or odd outcome. Since either A or B (or strictly speaking both) will certainly happen, the columns add up to one

	odd	*even*
$\Pr(A[=\{1,2,3,5\}])$	1	1/3
$\Pr(B[=\{4,6\}])$	0	2/3
$\Pr(A) \cup \Pr(B)$	1	1

Table 3.2 Conditional *outcome* probabilities are the probabilities of getting an odd or even number in event *A* or *B*, respectively. Evidently, the outcome is either odd or even, and so the columns add up to one

	A	*B*
Pr(odd)	3/4	0
Pr(even)	1/4	1
Pr(odd) ∪ Pr(even)	1	1

number, is one. However, the probabilities that the signal is *A* when the dice shows an odd number and that it is *A* when it shows an even number do not add to one.

The entries to Table 3.2 answer the opposite question: How likely is a particular state of nature if a particular signal has been received? Table 3.2 can be derived from the bottom part of Figure 3.5 and shows that, for example, the signal *A* as a forecast of an odd number is true in three cases out of four. In contrast, the signal "B" as a forecast of the dice showing an even number is always true.

3 C d Updating information: Bayes' Rule

There is a simple relationship between the unconditional and the conditional probabilities of events and signals known as *Bayes' Rule*. The relationship can be read directly from Figure 3.5 or from Tables 3.1 and 3.2. Consider the following example: The (unconditional) probability of an even number (event even) times the conditional probability that the signal *A* occurs when an even number has been drawn is equal to the probability that one observes an even number *and* signal *A*. This must be equal to the (unconditional) probability of signal *A* times the conditional probability that after *A* the outcome will be even. More formally, *Bayes' Rule* states that

$$\Pr(E) \cdot \Pr(F \mid E) = \Pr(F \cap E) = \Pr(F) \cdot \Pr(E \mid F). \tag{3.1}$$

where E is an event, F is a signal related to the event and | means "conditional on". The expression $\Pr(F \cap E)$ denotes the probability that we observe event E

and signal F. Although the rule is actually no more than an expression of the consistency of probabilities, it is still very useful.

Bayes' Rule tells an agent how to update his information after receiving a signal. Assume that an agent originally thinks event E will occur with probability $\Pr(E)$, the so-called *prior* probability. Now a signal F occurs, and the prior information has to be corrected for the additional knowledge contained in signal F. From the prior probability $\Pr(E)$ and the likelihood of signal $\Pr(F)$, the agent can infer the *posterior* probability of event E, $\Pr(E \mid F)$, by using Bayes' Rule (3.1).

$$\Pr(E \mid F) = \frac{\Pr(E) \cdot \Pr(F \mid E)}{\Pr(F)},$$

provided the agent knows the relevant probabilities for the signal F on the right hand side of the equation. The fraction $\Pr(F \mid E)/\Pr(F)$ can be interpreted as a correction factor affecting the prior probability $\Pr(E)$ after observing the signal F: It exceeds 1 if $\Pr(F \mid E)$ is greater than $\Pr(F)$, that is, if the signal is closely related to the event E. The correction factor falls below 1 if the occurrence of the signal hints at a low probability of the event E.

Bayes' Rule will be particularly helpful when we try to find the *value* of a piece of information in Chapter 4.

3 C e Information quality

Fineness . . .

The concept of an information structure allows us to describe and compare the information different agents have. One criterion is *fineness*. An information structure is said to be more informative than another if it leads to a finer segmentation of the state space. The information structure on the left-hand side of Figure 3.6, for example, is less informative than the slightly finer information structure represented on the right-hand side.[6] Like information, questions that are targeted at receiving information can be more or less informative. Thus question Q is said to be better than question Q' if it is represented by a finer information structure. In that respect, linguists and economists here use very similar concepts (Van Rooy, 2003).

. . . and its limits

Unfortunately, the fineness criterion does not make it possible to rank all possible information structures. Among the two information structures represented in Figure 3.7, neither is necessarily better. Although the right-hand segmentation looks finer (it has more elements), not all of its individual elements are finer than those on the left-hand side. The fineness criterion fails, for example, when across two information structures some of the subsets overlap. It also cannot be used when one information structure is finer than the other in one half of the state space; the

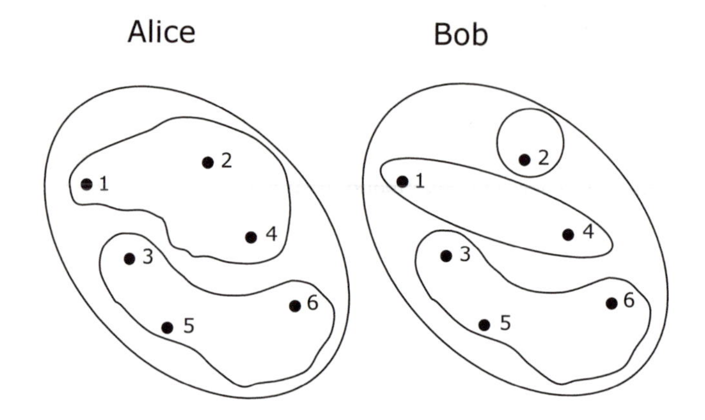

Figure 3.6 The right structure is finer than the left: The higher degree of informative (useful) segmentation makes the information finer.

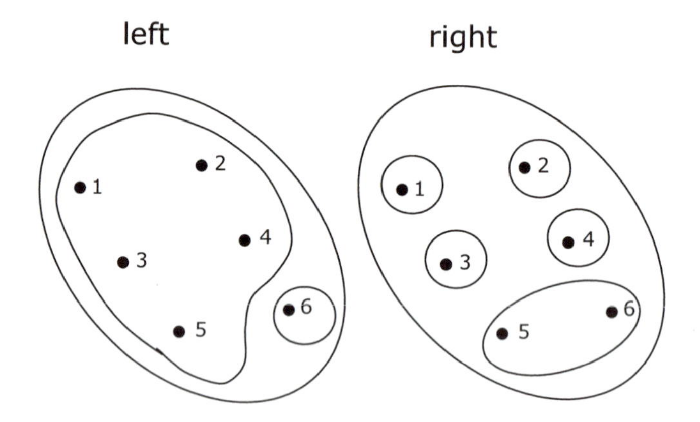

Figure 3.7 The right structure is not finer than the left, because the partitions overlap.

opposite is true for the other half. In other words, the fineness criterion applies to asymmetric information, but not to heterogenous information.

Another limitation of the fineness criterion becomes apparent when we try to rank noisy signals. Take the example illustrated in Figure 3.8 of two experts forecasting the weather as either "dry" or "rain." Both experts can only distinguish two states of the world; their forecasts have the same fineness. Yet, Expert 2 is who is right with probability q_2 and wrong with probability $1 - q_2$ is better than Expert 1 if $q_2 > q_1$.

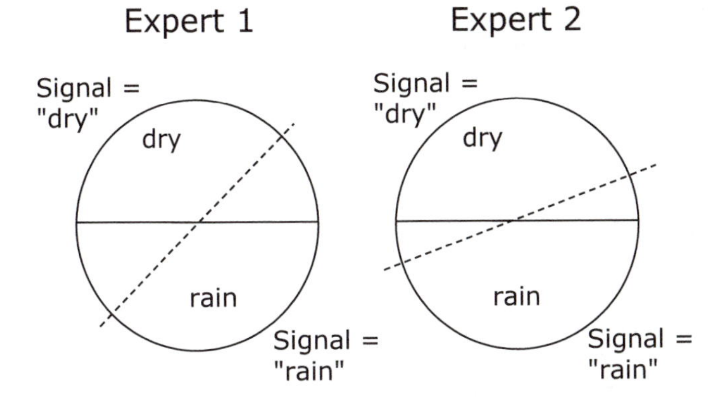

Figure 3.8 Expert 1 is less precise, because her signal is more noisy than that of Expert 2. The precision of signal "rain", for example, is equal to the fraction between the obtuse slice (in which the outcome "rain" is announced by the corresponding signal) and the bottom half (outcome "rain") of the cake.

The precision of a signal

The two experts can be ranked by an additional criterion, the so-called *precision* of a signal. The precision, q, is equal to the probability that a state of nature (rain) leads to the respective signal ("rain"). In the dice example, a signal with the structure ("odd"/"even") as an information about "high" or "low" has a precision of $q = 2/3$. In terms of Bayes' Rule the precision of signal F in state E is $\Pr(F = E \mid E)$. As we will discuss later (Chapter 4), the precision of a signal is related but not in general identical to the accuracy of a forecast, the probability that the state of nature is equal to the signal, $\Pr(E = F \mid F)$.

A utility-maximizing individual is better off, the higher (all other things equal) the precision of a signal. The precision thus is a useful measure for the quality of a signal when fineness alone fails to discriminate. Unfortunately it only makes sense in special cases like binary signals.[7] Moreover, not all binary signals have a defined precision. One example is our expert who cannot say whether a dice shows an even or an odd number, but only knows whether it is in group $A = \{1, 2, 3, 5\}$ or $B = \{4, 6\}$. The expert has precision $2/3$ in state "even" and of one in state "odd". We can say that this expert is better than an expert with a non-state-contingent precision of $2/3$ or worse than an expert who is always right. However, we cannot rank the expert in comparison to a fellow expert with a non-state-contingent precision of $5/6$, say.

Further criteria

In some cases, information structures can be ranked by comparing the distributions of payoffs over the state space. Two standard criteria are the so-called *first-order stochastic dominance*, or the *monotone likelihood ratio property*. We will not look at these concepts here.[8] Instead we will compare information structures by their most important economic attribute, their *value* which will be treated in Chapter 4.

3 D Conclusions and further reading

In this chapter we have treated information as a cluster of states of nature that we called an information structure. An information structure (or partition) is a collection of non-overlapping subsets of the state space. All possible states of nature fall into one of these subsets. The finer the information structure of an agent, the more precisely the agent knows the true state of nature.

Most signals in real life are noisy; they only tell the true state of nature with some probability. States of nature usually have an important probability dimension. Learning from a noisy signal can be seen as updating the probabilities on the underlying states. Bayes' Rule is a powerful instrument to update (prior) probabilities with the help of a signal into posterior probabilities. Bayesian learning is used in many economic contexts such as finance, insurance, and many game theoretic problems. It is particularly important in assessing the value of additional information, as we will see in Chapters 4 and 5. It is also well known that human beings are far from perfect Bayesians in real life. There are several empirically well documented cognitive biases preventing sober Baysian updating of probabilities like, for example, wishful thinking. Bayes' Rule thus provides us with a benchmark for what an individual would believe in the absence of any such biases, rather than a strong prediction of what the individual will in fact believe.

We also introduced criteria for ranking information sets. Under simplifying assumptions, the fineness of an information structure is a useful criterion, and so is the precision of a signal. Usefulness means that the "better" information structure is always preferred by a utility-maximizing agent. Unfortunately, for many information structures there is no simple criterion short of computing the value of a piece of information. This, however, requires more input than the *objective* properties of information sets. The value of a piece of information depends on *subjective* properties of the potential recipient, in particular on preferences and on available actions.

In the present chapter we have not gone into much technical detail. There are excellent texts offering more stringent treatment, such as Hirshleifer and Riley (1992, Ch. 5), Laffont (1993, Chs. 1, 4) and Brunnermeier (2001, Ch. 1).

Checklist: Concepts introduced in this chapter

- The engineering aspect of information
- The semantic aspect of information
- The characteristics of information as an economic good
- The state-space approach
- Information structures, partitions

- Events, information sets
- Signals, perfect and noisy
- Homogeneous, heterogeneous and asymmetric information
- Decision tree
- Bayes' Rule
- Precision of a signal

3 E Problem sets: Medical and financial testing

For solutions see: www.alicebob.info.

Problem 3.1 Bits and bites

Compare an apple and a piece of information in economic terms. Find three common and three distinctive properties.

Problem 3.2 What was the signal?

Alice has the prior information structure on the outcome of rolling a dice $\{(1,3,5),(2,4,6)\}$. After receiving a signal, her information structure is $\{(1),(2),(3,5),(4,6)\}$. What could the structure of the signal have been if the signal only distinguishes between two events?

Problem 3.3 Sunshine or rain?

The probability of rain is $3/4$ and of sunshine $1/4$. A weather service has the following forecast probabilities: if it is to be sunny, the service announces "fair" with probability $3/4$ and "bad" with probability $1/4$; if it will rain, the service announces "fair" and "bad" with probability $1/2$ each. What is the probability of a sunny day after a forecast "fair"?

Problem 3.4 Debtor solvency

A bank would like to know the probability with which a debtor fails. From the debtor's profit and loss statement the bank can derive an estimate of these probabilities. We denote the probability of solvency $\Pr(S)$ by $p > 1/2$ and the probability of default $\Pr(F)$ by $1-p$. In addition, the bank can conduct a special audit leading to a forecast f or s. The forecast reflects the true state with precision $\Pr(s \mid S) = (f \mid F) = q$.

(a) What is the overall probability that the auditor will forecast a failure?
(b) What is the probability that a bank is solvent, but the auditor forecasts a failure?
(c) What is the probability that after an auditor's forecasts of a failure, the bank turns out to be solvent?
(d) Now, assume that the bank cannot conduct audits, but it can distinguish whether a due interest payment has been paid by the debtor or not. Which signal would you expect to be better: the audit or the payment/non-payment of interest?

Problem 3.5 Medical testing

Tests for HIV have a very high accuracy. An infected (non-infected) individual has a 99.9 per cent chance of getting a positive (negative) test result. Some people have proposed that everyone should be tested for HIV. Others worry that too many healthy individuals would get a positive result.

In Switzerland, 20,000 of the 7 million resident population are HIV positive. (The infection rate is higher compared to Germany, for example, where there are 40,000 cases in a resident population of 82 million.)

(a) Draw a square the sides of which represent the probabilities (i) of carrying or not carrying the virus and (ii) of testing positive or negative. (Hint: If you do not succeed read Chapter 4 first.)
(b) Would the probability that a positive test result is actually wrong (i.e. the individual does *not* carry the virus) be (i) 1% or below, (ii) 1–10%, (iii) 10–20%, or (iv) above 20%?
(c) What would be the corresponding probability that a person with a negative test result is in fact HIV positive?
(d) What is the probability of being HIV negative despite a positive test result in Germany?

Problem 3.6 The minimum redundancy alphabet

How would a game of SCRABBLE work in a redundancy-free (i.e. maximum entropy) language?

Problem 3.7 The exam fairy

Assume you just wrote an important exam. Now a fairy appears and asks: "How much would you pay for the information that you did indeed pass the exam?" What should you answer?

Problem 3.8 Do all Cretans lie?

The philosopher-poet Epimenides from the sixth century BC reportedly wrote: "The Cretans are always liars." (Titus 1:12). As Epimenides was a Cretan himself, this sentence became a famous logical paradox.[9] Assume that Cretans always lie. You are hiking in Greece and ask a person for directions on how to get to Athens. The person may reply "left" or "right" and may be from Crete or not.

(a) What are the possible states of the world before the answer? What is your information structure?
(b) How do you optimally phrase your question?
(c) What is your information structure after the answer?

Problem 3.9 Expected surprises

According to data collected by Thomson First Call for the period 1994 to 2003 (1*st* quarter), quarterly profits of S&P-500 companies exceed analysts' forecasts by an average 2.8 per cent. In the press this difference was explained by companies' (successful) attempts to push up their share prices. Companies, it was said, spread some slightly too pessimistic information during the quarter in order to surprise markets positively when profit figures are released at the end of a quarter. A positive surprise regarding profit figures normally leads to an increase in the share price. However, in one case, the stock of a company fell after it released profits, even though these slightly exceeded analysts' forecasts. One newspaper explained that the company had a reputation for surprising markets more than just a little.

(a) Can companies in the long run keep their share price on a higher level by optimally timing the release of information?
(b) Discuss the notion of an expected surprise.

4 The value of information

4 A Introduction

"Knowledge itself is power," the philosopher Francis Bacon (1561–1626) said. In economic terms, knowledge or information has a value. By reducing uncertainty it helps us to take the right action. For example, knowing tomorrow's closing prices would help to pick the right stocks today. In this chapter we show where exactly the value of a piece of information comes from and, in particular, how utility-maximizing individuals *quantify* its value. Let us take the simple world of Robinson Crusoe, coping with "nature" (see Box 4.1) being his only source of uncertainty. Of course, this situation would change when he met Man Friday; but the issues and problems arising when several agents are present will be dealt with in later chapters.

Box 4.1 Man against nature

Economists use the term "nature" as a synonym for "the non-strategic player" in a game. A *strategic player* maximizes something—pleasure, prestige, income, the well-being of others, etc.—, commonly referred to as "utility", a joker card that can represent any given goal.

By contrast, *non-strategic play* is blind. It has no objective, no preferences, and no foresight. Or, if there is a plan or a strategy, we do not know it, and economic agents cannot use it to their own ends. In game theory all non-strategic influences are attributed to one single "player" called "Nature" (game theorists try to build parsimonious models). Nature stands for the source of all forces and uncertainties that do not have their origin in some conscious strategic behavior. Its decisions are partly predictable (thanks to natural sciences), partly random. But they are always blind, impartial and unconscious.

In real life, though, humans often treat nature nature as a conscious being. Swiss mountain farmers in some places still practise the "Alpsegen" (traditional chants sung at dusk to bless pasture and cattle, and to fend off demons).

Sports stars are notorious for their habit of wearing the right cap, shoes or even underwear during important games. And from the Trevi Fountain in Rome to the penguin enclosure at the local zoo, we find the water full of coins carrying their visitors' wishes.

All this would not work with a truly non-strategic partner. The human habit to persuade, conjure, even bribe nature, is a bit of a puzzle. We will revert to it in Chapter 8.

4 B Main ideas: The source(s) of information value

An investor, jumping to the stock market page of the morning newspaper is not just eager to find out how shares closed; first of all, he would like to learn that they *rose* during yesterday's trading. Most investors would be prepared to pay quite large sums of money for good news. Unfortunately, there is no market for good news. Nor is there a market for bad news. There is only a market for news. "How much do you pay to know that share prices went up 2.8 per cent?" is a paradoxical question. It gives away the very information it had meant to sell. News can only be sold before it is known. To be precise, what is sold is the source of news—a newspaper, internet access, an investor's letter—but never the actual news.

The expression "value of information" therefore always refers to the value of a piece of information *source*, not the value of the actual *news*. In other words, the value of information is an *ex ante* concept; *ex post* it does not make any sense.

4 B a Information as a guide to action

The main source of the value of information is action. The possibility of *reacting* to information increases an individual's expected utility. A simple example is a weather forecast service. The service is valuable if it reduces the probability that we are caught in the rain without an umbrella and/or the probability that we carry an umbrella on a sunny day.

Here is the principle of information valuation:

> **The value of information as a guide to action:** *The value of information (= of an information source) is the increase in utility an individual expects from receiving the information (= the actual news) and from optimally reacting to it.*

We will explain this in more detail in the theory section. A simple example may clarify the principle.

Alice has invited Bob to dinner. He would like to bring a bottle of wine, but he does not know whether Alice will cook fish (F) or meat (M). In the first case, Bob would prefer to bring white wine (w), in the second, he would bring red (r). Bob derives utility of 1 from bringing the right wine and utility -1 from bringing the wrong wine. In the absence of any further information, the odds of F and M are fifty-fifty respectively. The same is true for the chances of Bob's choice being right (Fw, Mr) or being wrong (Fr, Mw). Whatever Bob brings, his expected utility is $U_0 = 0\,[= (1/2)(1) + (1/2)(-1)]$.

Bob could call a friend of Alice's who knows whether she is rather a fish or a meat person. Bob thinks that the friend's advice would signal Alice's choice with a three out of four probability. This means that it would increase Bob's expected utility to $U_i = (3/4)(1) + (1/4)(-1) = 0.5$. Then, the difference between the two utility levels — $U_i - U_0 = 0.5\,[= 0.5 - 0]$— is the value (in expected utility) of the friend's advice.

The example illustrates three things:

- A piece of information is meaningful because it allows probabilities to be revised (using Bayes' Rule, as introduced in Chapter 3).
- The information is valuable because the revised probabilities have an influence on the optimal action.
- As a basis for action, a piece of information cannot have a negative value; the recipient can always ignore it.

Box 4.2 Information in payment systems

In payment and settlement systems two main architectures are used. One is *real-time gross settlement* (RTGS), the other is *clearing*. RTGS systems have the advantage that payments are legally final, while clearing may have to be unwound in the event of a participant failing during the day. The drawback of the RTGS system is the higher cash levels participating banks need in order to settle payments; in a clearing system, most payments are directly cleared against each other without the need for cash balances. As cash holdings (in this case: accounts with the central bank) do not bear interest in most countries, RTGS systems tend to be more expensive for banks than clearing systems, *ceteris paribus*. There are, however, two strategies to lubricate RTGS systems: (i) intra-day credit offered to banks by the central bank (which would be more risky in a clearing system), (ii) information about the queue of incoming payments.

One example is the Swiss Interbank Clearing System (SIC), which, notwithstanding its name, is an RTGS, rather than a clearing system. Banks participating in SIC can see online the aggregate value of payments in their favor that have already been entered into the system but not yet executed. This does not remove uncertainty altogether: some further payments to a bank may be made in the course of the day and some payments to be made by the bank may not be known much in advance (like the use of credit lines or spontaneous withdrawals of demand deposits). Still, the bank considers information on payments that are already on their way as valuable. (The value of this information is the subject of Problem 4.7.)

The value of better information on queued payments under SIC is illustrated by the reduction in (non-interest-bearing) cash holdings by banks. For the purpose of interbank payments in SIC, banks hold cash accounts with the Swiss National Bank (SNB), the country's central bank. The level of these so-called giro accounts fell from a monthly average of CHF 8 billion in 1987 (prior to the introduction of SIC) to CHF 3 billion by the end of 1989 (the second year of SIC). The SNB's difficulty in forecasting and compensating this reduction contributed to the inflationary pressures of the following years.

4 B b Can information have negative value?

Would an individual pay for not receiving a piece of information? Everyday experience suggests that, contrary to our claim made above, there are indeed instances where an individual would prefer not to know. Within the "information as a basis for action" paradigm there are two potential cases, where an individual would prefer not to have an information.

Known knowledge

The first case concerns what we call *known knowledge*:

In fact, Bob would not mind drinking red wine with fish or white wine with meat. But he would like to surprise Alice with the right choice. If she was to learn that Bob has consulted a friend, the surprise effect would be ruined. Bob concludes that he would be better trying his 50 per cent chance of blindly picking the right wine.

This seems to show that information, seen as a guide for action, may have a negative value. Yet, this is an illusion created by confusing two issues:

- Having a piece of information;
- Being known to have a piece of information.

While having a piece of information *is* an advantage *ceteris paribus* (just imagine Bob blindly picks the wrong wine, while Alice thinks he asked her friend), *being known* to have a piece of information can be a handicap (see Chapter 13). The information *per se* still has a positive value.

Public knowledge

A piece of information can have negative value to an individual if it is only available in the form of *public knowledge*. A farmer would certainly want to know whether his farm is going to be hit by a hailstorm. But all farmers together might prefer not to know in advance who will be hit, as such knowledge would make insurance impossible. This is the insurance destruction effect to be discussed in Chapter 13.

Strategic ignorance

The second case regards a phenomenon known as *strategic ignorance*.

When Alice decided to write a PhD thesis, she greatly overestimated her abilities. At least, she would realize this later. However, this also led her to conclude that with a realistic picture of her abilities she would never have started in the first place. In fact, she even congratulated herself for her ability to underestimate future tasks in relation to her abilities. That trait might help her to get rid of bad habits later, maybe even to quit smoking.

Strategic ignorance assumes a certain degree of time inconsistency. All things considered, Bob would prefer to have a nice time now; but later he would prefer to have finished his thesis. Present Bob knows perfectly well how he will feel in the future; only, his unfettered will may not be strong enough to behave accordingly. Ignoring the difficulties of the task or his personal limits helps Present Bob comply with the preferences of Future Bob. Full information would indeed have a negative value. In terms of standard economic theory, Bob is not perfectly rational. Only recently, economists have started to model seemingly irrational phenomena such as strategic ignorance (see Chapter 17).

4 B c Information love and information aversion

Information can have a negative value when it enters the utility function directly, rather than as a basis for action. The proverb "Ignorance is bliss" (Thomas Gray, 1716–1771) describes many real-life situations. Watching the World Cup final on TV loses much of its fun if the final result is already known, and few people would read the last page of a detective story first. Information may destroy suspense, but it may also destroy the quality of life. An individual may prefer not to take a medical test, particularly if there is no cure for the disease (we will meet an interesting example in Box 13.1). Therefore, information may be damaging. But information may also be a pleasure quite of its own, even though there is nothing in it to react to. After all, who can honestly say that they do not enjoy a bit of gossip from time to time?

Information can have a positive or negative value depending on whether it is a source of pain or pleasure. This is a source of value quite different from possible action. It can be modeled in two ways. Information may enter a person's *utility function* directly—with a positive sign for good gossip or a negative sign for information which destroys suspense. It would thus seem that Gray's proverb refers to a somewhat different case.

Bob recently went through an exam and has just found the envelope containing his exam results in his mailbox. Should he open the letter immediately, or should he first go to dinner with Alice? Bob is torn between the two choices. On the one hand he is impatient to know whether he has passed; on the other, he is afraid of spoiling his dinner date with Alice. While Alice, on the phone, suggests having dinner first, Bob rips open the envelope . . .

There are two opposite forces at work in the above example. Fear, suggesting that it is better to wait, and impatience, driving the person on to find out. The respective technical terms are the degree of risk aversion and the elasticity of intertemporal substitution. If both coincide (in the case of a so-called "Von Neumann–Morgenstern utility function"), timing is no issue. If they are different, an individual is information-loving or information-averse. Information then has a positive or negative value. Here, information does not affect utility as an argument in the utility function, but rather through the very shape of the utility function. This assumes some special (or rather, a generalized) class of preferences. We will discuss preferences regarding the early or late resolution of uncertainty below as an Application (4 D).

4 C Theory: Knowledge is power

4 C a "Elementary, my dear Watson"

To show the principles of information valuation we start with a simple example.

The example is a slightly generalized version of Bob's "red or white wine" problem. We will call it the "Elementary Game" (Box 4.3), as it is a simple molecule from which more complicated structures can be built. The Elementary Game combines two elements: a simple lottery of winning X dollars with a probability p and of losing one dollar with a probability $(1 - p)$, plus a signal forecasting the outcome before a player has to decide whether to play or not.

Box 4.3 The Elementary Game

Assume Bob is invited to play a game with two possible outcomes. With probability p Bob wins X dollars, with probability $1 - p$ he loses one dollar. Before Bob decides whether he will play the game or not, he can obtain a signal on the outcome, i.e. a signal that predicts the true outcome with a certain degree of precision. For simplicity's sake, we assume that the signal has a probability q to indicate "good" if the true outcome is X, as well as a probability q to signal "bad" if the true outcome is -1. More formally, $\Pr[\text{good}|X] = \Pr[\text{bad}|-1] = q$. (A perfect signal would predict the true outcome with probability $q = 1$.)

The game as a decision tree

The Elementary Game can be represented in two ways. The first is a decision tree taken from game theory and depicted in Figure 4.1. This game tree is drawn such that nature decides first whether Bob will win (with probability p) or lose (with probability $1 - p$) if he plays the game. As long as Bob does not have any signal, he may be in either the left or the right node (arrived at after nature draws "win" or "lose", respectively); for Bob the two nodes lie within the same "information set", as game theorists would say. Before Bob has to decide about accepting or rejecting the game, he now can obtain a signal. His optimal reaction to the signal will depend on the possible payoffs from the game. Assume that Bob, prior to the signal, was indifferent. Then he will play the game if the signal says "good", and rejects it if it says "bad".

The probability square

The second representation of the Elementary Game is a probability square as in Figure 4.2. The square indicates the probabilities p and $1-p$ of the two outcomes as well as the precision of the signal q. It is immediately clear from this representation

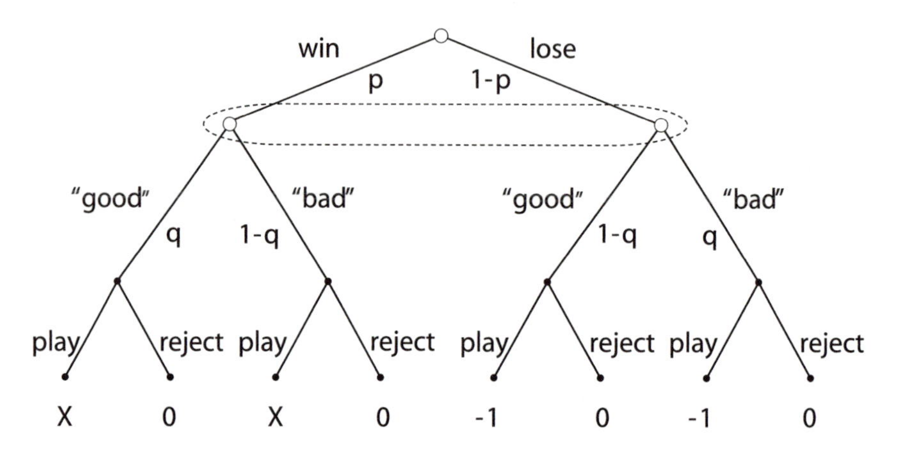

Figure 4.1 The Elementary Game: In a take-it-or-leave-it game the player wins with probability p or loses with probability $(1-p)$. In each case the player may receive a signal ("good" or "bad") with precision q. A player who is indifferent in the absence of a signal will play the game if the signal is "good" but not if it is "bad".

that the probability of a favorable outcome after a good signal $\Pr[X|\text{good}]$ is not equal to the probability of a loss -1 after a bad signal $\Pr[-1|\text{bad}]$, unless p equals $1/2$.

When the two outcomes are equally likely *ex ante* (like tossing a coin with $p = 1/2$), the game becomes much simpler. The precision q (the probability of a good signal in the case of a favorable outcome) is then also equal to the probability of a favorable outcome after a good signal. At the same time, q is also the probability of a loss after a bad signal. Both signal realizations ("good" and "bad") are equally likely, since the realizations of a perfect signal must have the same probabilities as the respective outcomes, in this case $1/2$. With $p = 1/2$, it is also easy to illustrate the concept of risk aversion: X must be equal to 1 to make risk-neutral Bob indifferent to playing the game or not. If Bob is risk-averse (-loving), X must be larger (smaller) than unity.

Signal value: a cookbook approach

What is the value of the signal in the Elementary Game? There are two possible ways to formulate the answer: (i) What is the increase in utility that Bob expects from the signal? and (ii) How much would Bob be ready to pay for the signal? The first answer gives the value of information in units of *utility*, the second in units of *money* or of a commodity. The principle of information valuation is the same in both cases, but the expressions differ somewhat. The utility value of a piece of information is the difference between the expected utility a decision-maker can achieve by choosing the best action *conditional* on the information received, and

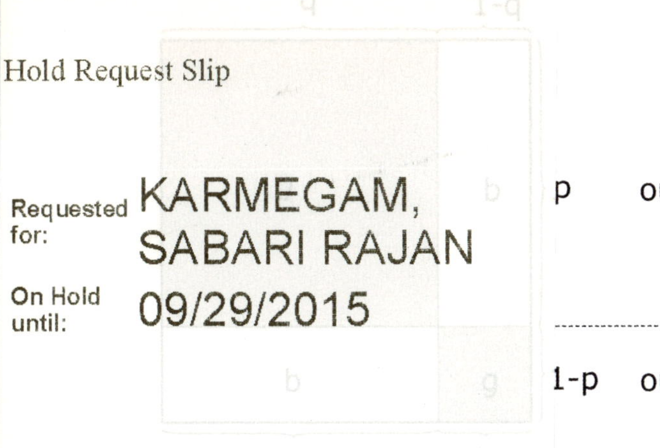

Figure 4.2 A probability nature shows how signals are related to prior probabilities and to actual outcomes. The signal takes the "good" realization (g) in the two shaded areas: either as a true signal of a positive outcome (top left) or as a wrong signal of a negative outcome (bottom right). The "bad" signal realization (b) occurs in the white areas: either as a true signal of a negative outcome or as a wrong signal of a positive outcome.

the expected utility from the action that is best in the *absence* of information. The money value is the amount of money that renders the decision-maker indifferent to both [illegible] information value in utility ("utils") or in money, the calculation involves four steps:

1. Determine the best actions for each state of information;
2. Calculate the expected utility from the best action(s) conditional on the information received (or its money equivalent);
3. Calculate the expected utility from the action that is best in the absence of information (or its money equivalent);
4. Take the difference between (2.) and (3.) (or between the respective money equivalents).

The value of a perfect signal

To illustrate the principle, we look at the simplest possible representation of the Elementary Game. Risk-neutral Bob is invited to play the Elementary Game with equal probabilities ($p = 1/2$) to win $X or to lose $1. He can obtain a perfect signal ($q = 1$) about the outcome of the game. Bob rationally chooses the action that maximizes his expected utility based on the information available. As we want him to be indifferent a priori between playing the game and not playing, we set $X = 1$. Moreover, as Bob is risk-neutral, the utility and the money value of information coincide. Following the four steps outlined above, we obtain:

1. After signal "win": play; after signal "lose": reject.
2. The expected utility with information is the signal-probability weighted sum of utilities after each possible signal:

$$u_I = \frac{1}{2}1 + \frac{1}{2}0 = \frac{1}{2}.$$

3. The expected utility without information is:

$$u_0 = \frac{1}{2}1 + \frac{1}{2}(-1) = 0.$$

4. The difference between the expected utility with and without information is $1/2$. The money value of information (for a risk neutral player) would be \$0.5. This is how much a rational player would be ready to pay for the information.

The result has an intuitive interpretation. If Bob decides blindly, i.e. without consulting the signal, he has a fifty per cent chance of taking the wrong decision (playing when he should not, not playing when he should). In each case he loses (or fails to win) one unit (dollar or utility). The value of the signal (or the opportunity cost of not consulting it) is 0.5 units.

4 C b The value of an imperfect signal: A numerical example

An imperfect signal is right most of the time, but it may also be wrong. In order to determine the value of an imperfect signal we will again use the Elementary Game.

It may be helpful to illustrate the value of an imperfect signal with an example, before we carry out a more general analysis. The example is the game "Alice's dice".

Box 4.4 Alice's dice

Alice (who is known to be honest) rolls a dice behind a curtain. She offers Bob a signal about the outcome. Bob, after receiving the signal (or after saying he does not want it), can choose between accepting or rejecting the game. If he rejects, he neither wins nor loses. If he accepts, he wins one cookie if the outcome is $H =$ "high" ($H \in \{4, 5, 6\}$) and loses one cookie if the outcome is $L =$ "low" ($L \in \{1, 2, 3\}$).

A signal on the outcome of the dice can take different forms. Alice might offer to say whether the outcome is odd or even or whether it is a 6 or not. Unfortunately for Bob, Alice does not give away her information for free.

Alice invites Bob to play one of her favorite games of dice (see Box 4.4). Alice offers to disclose whether the outcome of rolling the dice is even or odd before Bob hast to accept or reject the game. "How many cookies would you be ready to pay for the information?" Alice asks.

Bob faces a fair game with an expected value of zero and an equal probability for both events. In the absence of a signal, risk-neutral Bob would be indifferent to accepting or rejecting the game. The signal is symmetric with precision 2/3, two out of three high (low) numbers being even (odd). What is the signal worth?

It is a good idea for students to try to find the answer before reading on. We follow the routine developed above. The starting point is: The value of a piece of information is the increase in utility (or its monetary equivalent) an agent can expect from the information if he makes the best use of it. This difference of returns from playing with, rather than without the information is easy to compute.

- If Bob is *uninformed*, the game has a value of zero (as it is fair). In fact, Bob is indifferent to playing or not playing.
- If Bob is *informed*, he faces the game tree represented in Figure 4.3. There are two possible states of nature, H and L. Depending on the outcome, the signal is either "even" or "odd". In the former case, Bob would accept the game. Consequently, his odds of winning rather than losing a cookie are 2/3 which translates into an expected profit of 1/3 of a cookie. In the latter case

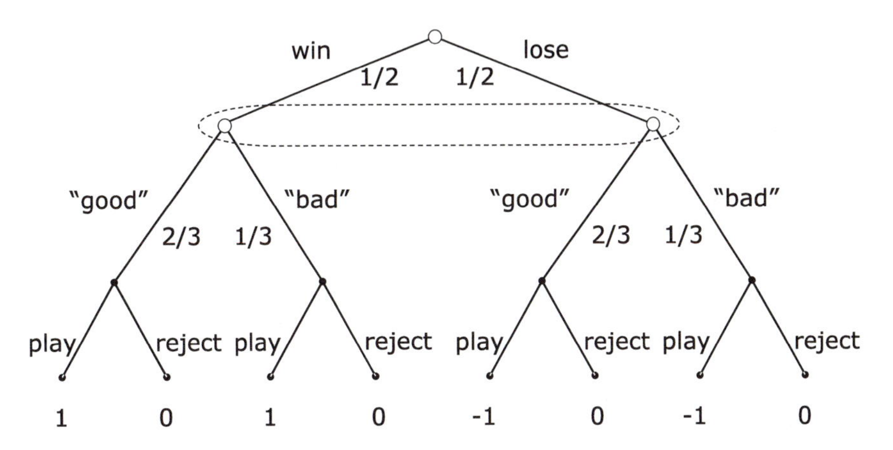

Figure 4.3 In the game "Alice's dice" the player wins if a dice shows a high number (H) and loses with a low number (L). A signal discloses whether the outcome is even or odd.

the odds are reversed. Rather than facing an expected loss of 1/3 of a cookie, Bob would reject the game.

- In this particular case, both signals are equally likely *ex ante*. The expected payoffs (1/3 and 0) have to be weighted by 1/2 each. The expected payoff for an informed player is thus 1/6.
- The value of the information is equal to the *difference* between the value of the game with and without the information. In the present example, this is equal to the value of the game with information, as the game without information has zero value. A risk-neutral player would be ready to pay 1/6 of a cookie.

The example does not reveal all potential complexities. One simplifying assumption was the *symmetry on payoffs*: the potential gain being equal to potential loss. A consequence of this assumption was the (risk-neutral) *uninformed player's indifference* between playing and not playing. Another restrictive assumption was the player's *risk neutrality*. In the following sections we will relax these assumptions. We will retain a third assumption, though: that of a symmetrical signal with equal probability of being right or wrong in either state of nature.

4 C c *The value of an imperfect signal: The Elementary Game generalized*

Next, we will look at a game with asymmetrical payoffs. We will use the Elementary Game with a slight modification: The potential gain is X units instead of one unit of a good.

Next time, Bob wants to be prepared. So he thinks in general terms: A risk-neutral player is offered a game. With probability $p = \Pr[X]$, he wins X dollars, with probability $1 - p = \Pr[-1]$ he loses one dollar. Before deciding (accepting or rejecting the game), he can receive a binary signal predicting the outcome. The signal is either g ("good") or b ("bad"), suggesting that the player will win or lose. In both cases the signal has precision q. "What should I pay for such a signal?"

The valuation principle

At the risk of repeating ourselves, let us start from the basic principle: The value of a piece of information is the increase in utility from receiving the information and from optimally reacting to it. In other words: It is equal to the expected utility *with* the information (under the assumption of optimal reaction) minus expected utility *without*.

The expected utility of a player depends on three things:

- the stakes (the payoffs in the possible outcomes);
- the probabilities of the possible outcomes;
- optimal actions.

Let us examine these three ingredients of information value in turn.

The stakes

The payoffs from the two possible outcomes (win, lose) are known to be X and -1, respectively. The payoff from not playing is assumed to be zero.

Probabilities

An uninformed player expects to win or lose with the prior non-contingent probabilities p and $(1-p)$, respectively. For an informed player, the posterior probabilities are contingent on the signal realization. These can be derived from the known contingent probabilities of the signal realizations. We know that $q = \Pr[g|X] = \Pr[b|-1]$, i.e. the signal has equal precision q under both possible outcomes.

What we want to know, though, is not the probability of, say, a good signal in the event of a "win", $\Pr[g|X]$, but the opposite probability of winning after observing the signal "good", $\Pr[X|g]$). Yet, using Bayes' Theorem, we can convert the former into the latter. The resulting contingent or posterior probabilities of winning or losing (or of the respective payoffs) are:

$$\Pr[X|g] = \frac{\Pr[X] \cdot \Pr[g|X]}{\Pr[g]} \tag{4.1}$$

$$\Pr[-1|b] = \frac{\Pr[-1] \cdot \Pr[b|-1]}{\Pr[b]} \tag{4.2}$$

We know the two terms in each nominator, the unconditional probabilities of the two events, $\Pr[X] = p$ and $\Pr[-1] = (1-p)$, as well as the signal's precision, $\Pr[g|X] = \Pr[b|-1] = q$. Yet, we still need to compute the denominators, that is the probabilities of observing a good signal ($\Pr[g]$) and a bad signal ($\Pr[b]$). Fortunately, this is simple. A good signal occurs in two cases: When things go well and the signal is right, or when things go badly and the signal is wrong (remember the grey areas in Figure 4.2). The same logic leads to the probability of a bad signal (the white areas in Figure 4.2).

The overall (unconditional) signal probabilities thus are:

$$\Pr[g] = pq + (1-p)(1-q),$$

$$\Pr[b] = p(1-q) + (1-p)q.$$

Putting everything into (4.1) and (4.2) yields:

$$\Pr[X|g] = \frac{pq}{pq + (1-p)(1-q)},$$

$$\Pr[-1|b] = \frac{(1-p)q}{p(1-q) + (1-p)q}.$$

These are the probabilities that *the signal is right*, (i.e. that a correct forecast is made). It is straightforward to compute the complementary probabilities that the signal is wrong. However, it is important to stress once more the difference between (i) the probability that a signal is right, which we may call the *accuracy of the forecast* and (ii) its mirror image, the probability of a right signal, which we called the *precision of the signal*. The precision here is $\Pr[g|X] = \Pr[b| - 1] = q$. The two probabilities coincide in a special case where $p = 1/2 = (1-p)$. If $p > 1/2 > (1-p)$, and the relation between the probabilities is:

$$\Pr[X|g] = \frac{pq}{pq + (1-p)(1-q)} > \frac{pq}{pq + p(1-q)} = q$$

$$\Pr[-1|b] = \frac{(1-p)q}{p(1-q) + (1-p)q} < \frac{(1-p)q}{(1-p)(1-q) + (1-p)q} = q$$

In other words, if the probability of an event is greater (smaller) than $1/2$, the probability of the signal being right is greater (smaller) than the probability of receiving the right signal (the precision).

Optimal actions

Optimal actions are the third and final input we need to compute the value of information. "Optimal" means the best action, given the agent's state of information. In our example, there are three possible states of information: good news (g), bad news (b) or no information (n). Which action is optimal in each of these three states of information depends on payoffs and on the relevant probabilities.

- With good news: play if $X \geq \dfrac{(1-p)}{p}\dfrac{(1-q)}{q}$.
- Without signal: play if $X \geq \dfrac{(1-p)}{p}$.
- With bad news: play if $X \geq \dfrac{(1-p)}{p}\dfrac{q}{(1-q)}$.

The same story can be told in a different way. Instead of expressing optimal actions as a function of signals, we can express them as a function of X, that is the gain from winning. If X is large enough, a risk-neutral player would accept the game regardless of the signal, as the game would yield positive expected utility even after bad news. Conversely, if X is low enough (for example, zero), a player

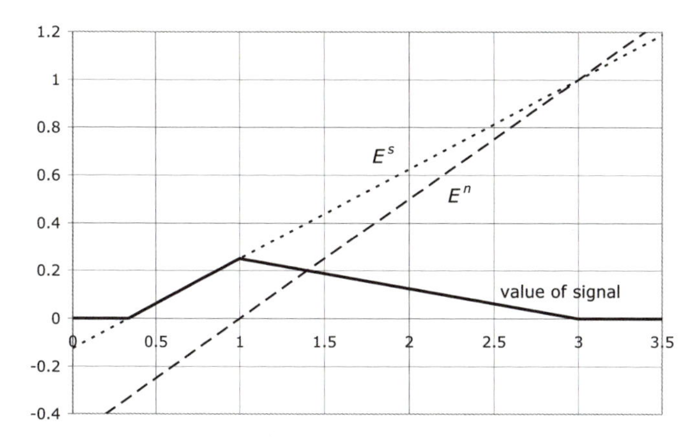

Figure 4.4 Value of a signal and the corresponding expected payoffs with (E^g) and without (E^n) a signal as a function of X in the Elementary Game under risk neutrality. The prior probability of X is $p = 1/2$, the signal's precision is $q = 3/4$.

would not care about the signal either, but would reject it straight away. In both cases the signal is worthless as it has no impact on a rational player's actions.

There are thus four regions for X, each characterized by some optimal actions. We will describe them in formal terms, but refer to Figure 4.4, which is drawn for the example of $p = 1/2$ and $q = 3/4$ and which we will explain later:

- If $X < \dfrac{(1-p)(1-q)}{pq}$:
  Never play, as the expected payoff is negative, even after good news (in Figure 4.4, region $X \leq 1/3$).
- If $\dfrac{(1-p)(1-q)}{pq} \leq X < \dfrac{1-p}{p}$:
  Play in the event of good news; without a signal, or with bad news, reject (in Figure 4.4, region $1/3 < X \leq 1$).
- If $\dfrac{1-p}{p} \leq X < \dfrac{(1-p)q}{p(1-q)}$:
  Play with good or no news; reject after bad news (in Figure 4.4, region $1 < X \leq 3$).
- If $\dfrac{(1-p)q}{p(1-q)} \leq X$:
  Always play regardless of the signal (in Figure 4.4, region $3 \leq X$).

The first and last cases are the trivial cases already mentioned in the text. The game is either so attractive or unattractive that even a bad or a good signal could not change the optimal action. The signal in these cases is obviously worthless. The two interesting cases are the intermediate ones, where a player accepts only after good news, or where a player accepts except after bad news.

The value of the signal

Before we move on, we will take a slightly different angle. Instead of asking: "What is the optimal action, given X?" we ask: "How does the value of the signal vary with X under different strategies (action rules)?" For the sake of simplicity we will take the case where there is equal odds of winning or losing, i.e. $p = (1-p) = 1/2$, and where the signal precision, q, is 0.75. These are the assumptions behind Figure 4.4.

In the figure we compare expected payoffs from two mechanical strategies:

1 Playing without information
2 Playing with a good signal

We ignore the two other (but trivial) cases where a player would play even with a bad signal or reject even with a good signal.

The payoff from the first strategy, uninformed play,

$$E^n(X) = pX + (1-p)(-1) = pX - (1-p), \tag{4.3}$$

is illustrated by the dashed line in Figure 4.4. The dashed line reflects the expected payoff from unconditional play by an uninformed player. The expected payoff is positive when $X > \frac{1-p}{p} = 1$, and the slope of the line is p. For each additional dollar in the outcome X, the expected payoff from playing the game increases by p dollars. To the left of $X = 1$, expected payoff would be negative; an uninformed player rejects the game. The expected payoff to an uninformed player who behaves rationally can thus be written as $\max[0, E^n(X)]$.

Let us now look at the expected payoff from the second strategy, playing only with a good signal, $E^g(X)$. $E^g(X)$ is the weighted sum of gain and loss, the weights being the probabilities that the "good" signal is right or wrong, respectively. Using Bayes' Rule we get:

$$\begin{aligned}
E^g(X) &= \Pr[g]\,\{\Pr[X|g]X + (1 - \Pr[X|g])(-1)\} \\
&= \Pr[g]\Pr[X|g](X+1) - \Pr[g] \\
&= \Pr[X]\Pr[g|X](X+1) - \Pr[g] \\
&= pq(X+1) - \{pq + (1-p)(1-q)\} \\
&= pqX - (1-p)(1-q)
\end{aligned}$$

The expression $E^g(X)$ defines the dotted line in Figure 4.4. The expected payoff is positive for $X > \frac{(1-p)(1-q)}{pq} = 1/3$, that is at a lower value of X than the payoff without a signal E^n. $E^g(X)$ also has a lower slope (pq) than $E^n(X)$ (p)

reflecting the imperfect precision of the signal ($q < 1$). The two lines intersect at $X = \frac{(1-p)q}{p(1-q)} = 3$; if X exceeds that value, it is obviously advantageous to play the game even with a bad signal.

The expected payoff to a player who has received a good signal and behaves rationally (rejects if X is too small) can be written as $\max[0, E^g(X)]$. The value of the signal then is equal to the difference $E^g(X) - E^n(X)$, provided this difference is *positive*. If the difference is negative (here if $X > 3$), the player should play blindly, as even a bad signal would indicate a positive expected return.

Bringing everything together, we can write the value of the signal as a function of X using the expected payoffs $E^g(X)$ and $E^n(X)$ as

$$V_I(X) = \max[0, \underbrace{\max[0, E^g(X)]}_{\text{informed}} - \underbrace{\max[0, E^n(X)]}_{\text{uninformed}}].$$

This is easily translated into words, keeping an eye on the solid line "value of signal" in Figure 4.4. To begin with, the value of the signal or of information, V_I, is either zero or equal to some difference (if non-negative).

- V_I is zero if the game, even after a bad signal, has a positive expected return. In Figure 4.4 this trivial case is reflected in the right horizontal section of the solid line (if $X > 3$), where a player prefers to play uninformed anyway.
- Otherwise, V_I is equal to the difference between two maximands. The first is the value of *informed* play. Expected return after informed play (playing after a bad signal is already ruled out) is:

  - 0 if even after a good signal, the game has a negative expected return. This is the other trivial case, illustrated by the left horizontal leg of the solid line in Figure 4.4 (for $X < 1/3$).
  - $E^g(X) > 0$ if it pays to play after a good signal (for $X \geq 1/3$).

  The second is the value of *uninformed* play. This is:

  - 0 if the game has a negative expected return for an uninformed player. This is illustrated by the rising leg of the solid line in Figure 4.4, relevant in the region $1/3 \leq X < 1$.
  - $E^n(X) > 0$, if the game has a positive expected return for an uninformed player. This is shown by the falling leg of the solid line in Figure 4.4, relevant in the region $1 \leq X < 3$.

Solutions

Finally, we can solve Bob's problem of computing the value of a piece of information in the game before him. In the two extreme cases where he would even accept to play after a bad signal or not even after a good signal, the value of information is obviously zero: the signal has no impact on Bob's optimal decision. We can thus focus on the two intermediate cases where Bob only plays after good news (in Figure 4.4, region $1/3 < X \leq 1$) or where he plays except after bad news (in Figure 4.4, region $1 < X \leq 3$). In these regions of X the difference between expected utility with and without information is:

$$
\begin{aligned}
V_I^A &= \Pr[g]\{\Pr[X|g]X + (-1)(1 - \Pr[X|g])\} \\
&= pqX - (1-p)(1-q) \\
V_I^B &= \Pr[g]\{\Pr[X|g]X + (-1)(1 - \Pr[X|g])\} - \{pX - (1-p)\} \\
&= \Pr[b]\{-\Pr[X|b]X + (1 - \Pr[X|b])\} \\
&= (1-p)q - p(1-q)X
\end{aligned}
$$

V_I^A or V_I^B are the maximum amounts Bob should pay Alice for the information she offers (abstracting from the possibility that Alice might settle for less for some reason or other).

The intuition of these results is straightforward: In the "play-only-after-good-news" region, the signal is relevant for action if it is good, a good signal having probability $\Pr[g]$. Bob plays and wins X dollars with probability $\Pr[X|g]$ or loses 1 with probability $(1 - \Pr[X|g])$. In the "play-except-after-bad-news" region, the signal is relevant for action if it is bad, a bad signal having probability $\Pr[b] = 1 - \Pr[g]$. Bob rejects and improves his chance to get one dollar from $(1 - \Pr[X|b])$ to 1, thereby sacrificing a chance of $(1 - \Pr[X|b])$ to get X.

A non-monotonicity in information value

These results are also consistent with the peak in information value as a function of X, characterizing the solid line in Figure 4.4. If Bob only accepts the game after good news, the value of the information *increases* in X. If he plays the game in the absence of a signal but rejects it after bad news, the value of the information *decreases* in X. The information is most valuable where Bob is undecided between playing or not, i.e. at $X = (1-p)/p$.

Indifference as the source of information value

Information about the consequences of two actions is most valuable if, without it, the recipient would be indifferent to both actions. The value of information therefore increases with the degree of indifference. As a consequence, we should spend most money or effort on information in situations where we are perfectly indifferent to the two alternatives. The intuition is quite simple: A teacher who has to grade students' work normally spends more time on those that are on the verge of passing and failing than on the best and worst. We will elaborate on this example in Application 4 E.

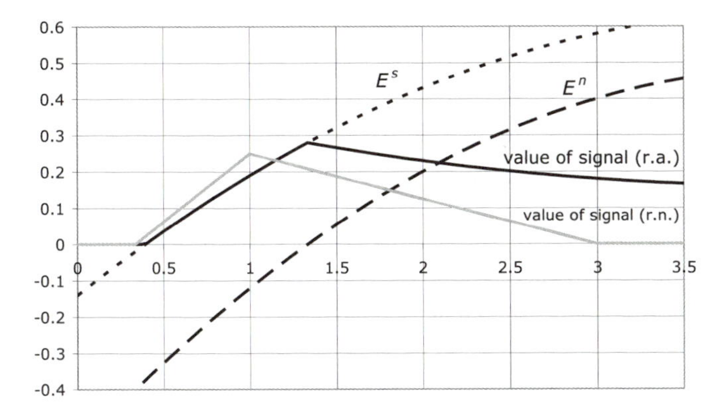

Figure 4.5 The case of risk aversion (r.a.): Value of a signal and the corresponding expected payoffs of informed play (E^g) and uninformed play (E^n) as a function of X in the Elementary Game with $u(c) = c - 0.12c^2, p = 1/2, q = 3/4$. The grey line draws the signal value under risk neutrality (r.n.).

4 C d Risk aversion and the value of information

How would the value of information change if an individual like Bob became risk-averse or risk-loving?

Spontaneously, most people think that risk aversion increases the value of information. However, this is not generally the case. Unfortunately, the impact of risk aversion on the value of information is quite hard to formalize. It is easy to show why. As before, the (utility) value of information can be computed as the difference between the expected utility of the game with a good signal ($E^g(X)$), and the expected utility of the game without information ($E^n(X)$), provided the two are positive and the former exceeds the latter.

Figure 4.5 represents the utility values of the different strategies (always play, $E^n(X)$, and play only with a good signal, $E^g(X)$), as well as the value of information for a risk-averse individual with quadratic utility $u(c) = c - 0.12c^2$. The grey line (borrowed from Figure 4.4) represents the information value for a risk-neutral individual ($u(c) = c$) playing the Elementary Game. It is easy to see why the value of the signal under risk aversion lies to the right of the risk-neutral value.

Imagine Bob who used to be risk-neutral has an off day, when he feels more cautious. He suddenly becomes slightly risk-averse. Even a marginal degree of risk-aversion would let him reject the game at the critical value for a risk-neutral player, $X = (1 - p)/p$. To make Bob indifferent again, X has to be slightly higher than this value, and the point of indifference (where the value of information peaks) moves to the right.

On the right-hand side of Figure 4.5, in the region where a bad signal prevents Bob from playing, the value of information increases in risk aversion as the utility loss in the event of a bad outcome -1 is greater than the utility gain with a good outcome X. On the left-hand side, in contrast, the value of information falls with increasing risk aversion. In that region, an increase in the degree of risk aversion makes Bob refuse to play the game up to a larger X, reducing the value of the signal. In the middle, the impact of risk aversion is ambiguous, as one impact of risk aversion on the value of an informed action is compounded with the effect that risk aversion may change how Bob would act in the absence of information.

4 D Application: The resolution of uncertainty

The letter

We have stated above that information may have an intrinsic (positive or negative) value even if it neither enters an individual's utility function directly nor permits the individual to take some action. Remember Bob's letter detailing the results of an important exam.

Bob has just received a letter with the results of an important exam that will affect his future income. He is invited to have dinner with Alice the same evening. Having the letter in front of him, he is faced with the dilemma: Would I like to know the results of the exam before or after the dinner with Alice?

Bob's choice

Bob's choice is illustrated in Figure 4.6: Either he waits until *after* dinner (left part) or he opens the letter *before* dinner (right part). The exam result does not affect utility from the dinner directly (payoff $D = 1$), nor can the information (to have passed or not) *per se* trigger any action that might affect Bob's payoffs in each of the two states (payoffs $S = 2$ and $F = 0$ in case of success and failure, respectively). From the game tree in Figure 4.6 one thing is clear: Bob's expected utility in period 1 is not affected by Bob knowing his period 3 payoffs before or after period 2 (i.e. dinner). *Expected utility functions* (which are also labeled Von Neumann–Morgenstern preferences) are silent on whether an early or late resolution of the uncertainty is better, as long as it lies in the future.

One of the two strategies might be more appealing to Bob than the other. If this is the case, we say that the timing of the resolution of uncertainty matters to the

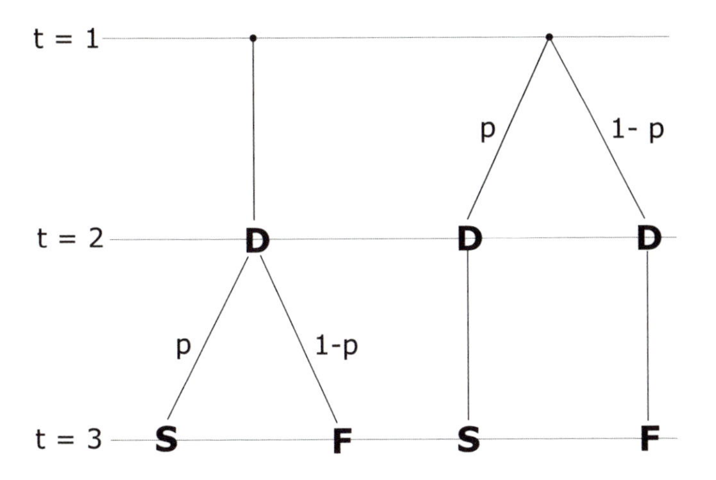

Figure 4.6 Dinner (D) takes place at $t = 2$. Uncertainty regarding having passed the exam successfully (S) or having failed (F) is resolved either after dinner (left side) or before dinner (right side). Expected utility as of $t = 1$ may depend on the timing of the resolution of uncertainty.

individual. Assume Bob has the following utility function:

$$U = [u_2 + E(u_3)]^x,$$

where u_2 is the utility from the dinner (in $t = 2$) and $E(u_3)$ is the expected utility from future income (depending on having passed or failed). The utility from having dinner is $u_2 = 1$, while the utility from the exam is either $u_3(S) = 2$ (successfully passed) or $u_3(F) = 0$ (failed). The odds that Bob has passed or failed are fifty-fifty (i.e. in Figure 4.6 $p = 1/2$), say.

Bob now compares the utility of opening the letter before and after dinner:

$$EU(open) = \frac{1}{2}[1 + 2]^x + \frac{1}{2}[1 + 0]^x$$

$$EU(wait) = \left[1 + \left(\frac{1}{2}(2 + 0)\right)\right]^x$$

The option with the greater utility value depends on the parameter x. It is easy to verify that with $x = 1$ (Von Neumann–Morgenstern preferences), one is indifferent, obtaining an expected utility of 2 in either case. However, we assume that Bob has $x = 2$, while Alice has $x = 0.5$. Bob's utilities thus are: $EU(open) = 5$ and $EU(wait) = 4$. Alice, by contrast, would find that $EU(open) = (\sqrt{3} + 1)/2 \approx 1.37$ and $EU(wait) = \sqrt{2} \approx 1.41$. While Alice suggests they open the letter after dinner, Bob tears it open before they even sit down.

Information love and information aversion

The issue is especially important in macroeconomics. Von Neumann-Morgenstern preferences (or expected utility) automatically lead to a setting in which risk aversion is the inverse of the elasticity of intertemporal substitution, although the two concepts are very different. The elasticity of intertemporal substitution measures how much the individual cares about spreading out consumption over different points in time. By contrast, the attitude towards risk measures how much the individual cares about spreading consumption across states of nature.

The formalization of *non-expected* utility is not trivial and is usually done in a recursive fashion. A relatively "simple" example (but still fairly sophisticated) are the RINCE preferences formalized by Farmer (1990). RINCE stands for RIsk Neutrality and Constant Elasticity of Substitution. Formally

$$U_t = \omega(c_t, E_t U_{t+1})$$

where $\omega(\cdot)$ is an aggregator function linking current payoffs c_t to the future utility level U_{t+1}. The RINCE aggregator takes the following form for an intertemporal problem in finite time T:

$$\omega_T = \omega(x, \cdot) = x$$

$$\omega(x, y) = \left(x^\rho + \beta y^\rho\right)^{\frac{1}{\rho}}$$

If $\rho = 1$, we are back to the standard sum of period payoffs with a discount factor β, $U_0 = \sum_{t=0}^{T} \beta^t c_t$. With $\rho \neq 1$ there are two cases:

- an individual with $\rho > 1$ (high elasticity of intertemporal substitution) has a preference for an early resolution of uncertainty; such a person is *information-loving*.
- an individual with $\rho < 1$ (low elasticity of intertemporal substitution) has a preference for a late resolution of uncertainty; such a person is *information-averse*.

Early versus late resolution

Figure 4.7 depicts the utility difference between an early and a late resolution of uncertainty as a function of ρ. In the setting of our example, a student with a low elasticity of intertemporal substitution will want to open the letter with the exam results after going for dinner (left side), while a student with a high elasticity of intertemporal substitution will want to open the letter before dinner (right side).

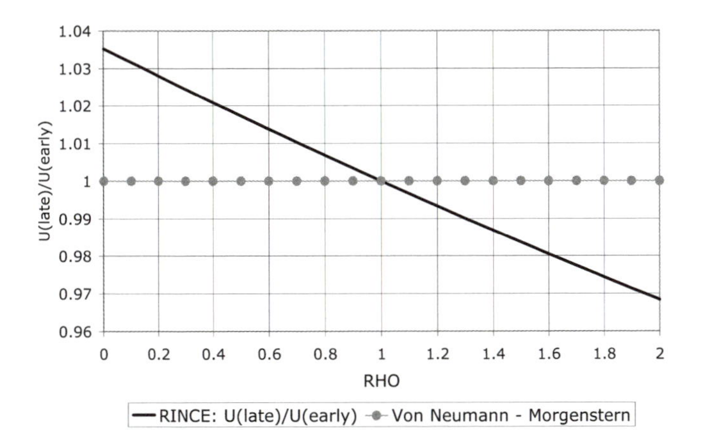

Figure 4.7 With a Von Neumann–Morgenstern utility, the timing of information arrival does not matter. However, with RINCE preferences resolution of uncertainty does affect utility. An individual with high (low) ρ is information-loving (-averse).

4 E Application: The informational cost of mediocrity

A common experience

Both authors of this book regularly have to grade exams and term papers. Like many of their colleagues, they have observed first-hand that very good and very poor answers are relatively easy to grade. The time-consuming answers fall into the intermediate category. Most time often goes into grading exams straddling "pass" and "fail".

From the above results on the value of information in the Elementary Game, we suddenly understood why mediocrity is so costly in terms of information acquisition. The peak of information value (suggesting careful grading) in Figure 4.4 echoes exactly our experience.

A model

Assume a teacher has to grade exams in order to decide who passed and who failed. The teacher uses some benchmark for what should be considered the minimum quality for passing (rather than allowing a particular percentage of candidates to pass or fail). The task is roughly the same as that of an investor examining investment projects in order to decide whether to invest or not.

The teacher's incentive to look closer at one particular project is measured by the value of information. The teacher is in fact playing the Elementary Game (against nature). In the present context this means that the teacher grades one answer after the other. With each answer she acquires an overall first impression.

We will denote the initial impression for student i, which we in turn will denote by X_i. Based on this first impression X, she decides two things: (i) whether, in the absence of any further information, she would let the student pass or fail, and (ii) how much time she is going to invest in grading the answer (of course, in reality this decision is taken on a somewhat *ad hoc* basis, but we are trying to keep things simple).

Of course, the teacher would like to let all students pass. From a rejected answer she derives zero utility. From an accepted answer she derives the following payoff: She wins X "utils" if she objectively judges the student as good, but she loses 1 util if the student should in fact have failed. Although she would prefer to let everyone pass, she has an incentive to be careful. She also knows from long experience that among a very large number of students a fraction $p < 1$ should actually pass.

The teacher's strategy

The teacher would pass all exams, which on first impression (X_i) satisfy the conditions (note that whenever she is indifferent she chooses "pass"):

$$pX_i - (1-p)1 \geq 0.$$

However, she can invest some time on an exam paper in order to revise her opinion. As a result, she revises the probability that the student should in fact pass, up to q or down to $1-q$. (We could also assume she revises her initial assessment X, but the chosen presentation is easier to represent graphically.) We take the value of such information as a proxy for the incentive to spend time on grading the respective answer.

Information value

The value of information, $V(X)$, is actually identical to the value of information in the Elementary Game, indicated by the solid line in Figure 4.4.

The striking feature of the value of information function is its peak around $pX/(1-p)$. Another feature is that beyond some upper and lower bounds, the value of information falls to zero. Both are easy to understand. Very poor and very good answers are easy to grade, the former probably slightly easier than the latter (the extreme case being the blank sheet). A student with an initial assessment between the lower bound and the peak would be rejected in the absence of any further information. This information becomes more valuable, the more expensive the mistake is (fail instead of pass). After the peak, students would pass with their initial assessment. In this area, the value of information falls, as a mistake (pass instead of fail) becomes less and less expensive (in expected terms). The difference in the slope of $V(X)$ to the left and right of the peak thus reflects the difference in mistakes an uninformed teacher would make: on the left side it is a type II error (reject good student), to the right it is a type I error (accept bad student).

A field experiment

So much about the theory. Is the effect discernible in reality? To get an idea, the authors conducted an experiment at the University of St. Gallen during summer term 2005, in which many graders on a variety of exams participated. For each answer the number of points and the time spent for grading one problem (in minutes) was recorded. The results of one representative teacher are reported in Figure 4.8.

The results basically confirm the forecast from our simplified model. Mediocre exams consume more of the teacher's time than very good or very bad ones. Yet, the peak is not quite as acute as in the model. Furthermore, good exams are somewhat harder to grade than the model would predict. The plausible explanation is that better answers are on average longer; a serious empirical test should thus control for length. Also, grading was more than a pass/fail decision; grading according to a points' system is more difficult with a longer than with a shorter text.

Of course, the issue would need a more thorough empirical test. However, there are many natural experiments which could confirm the form of the value of information function as an inverse U. We would speculate that several of our readers have lost some patience with mediocre candidates in exams, in the job market or in the market for corporate takeovers. They all know what we mean by the informational cost of mediocrity.

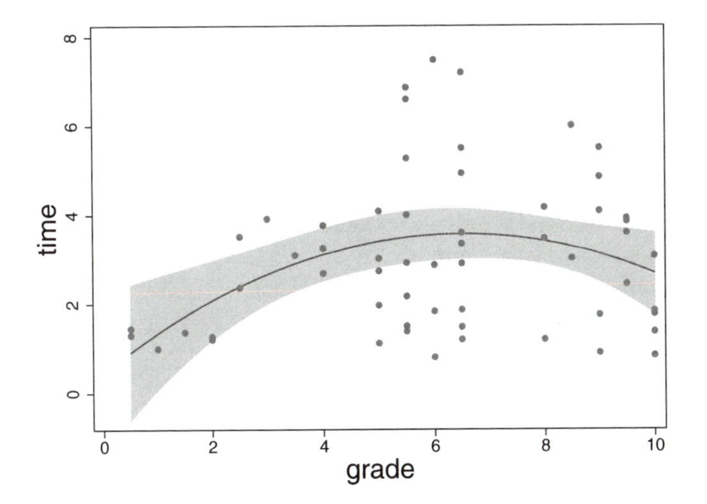

Figure 4.8 Time used (vertical axis) and points granted (horizontal axis) by a teacher grading 60 exams. The quadratic fit has an inverse U-shape (the shaded area indicating the 95 per cent confidence interval).

4 F Conclusions and further reading

The valuation of information is relevant for a wide range of problems—from subscribing to a newspaper to drilling for oil or waiting for the next labor market statistics. We have tried to show the general principles of information valuation. We have demonstrated that the value of information, represented by a signal to the player, depends on:

- the *stakes*, i.e. the size of the payoffs (here X and -1);
- the prior *uncertainty* p;
- the precision of the information q;
- the degree of risk aversion.

While these are the factors that jointly decide on the value of a piece of information, their individual impact is not always unambiguous. It is not generally true that "when stakes are high, information is more valuable". The intuition that a risk-averse individual always pays more for information is also wrong.

For further reading we recommend the classic works by Laffont (1993, Ch. 4) and by Hirshleifer and Riley (1992, Ch. 5). Readers looking for a game theoretic presentation may find Bierman and Fernandez (1995, Ch. 3) most helpful.

In most economic applications, valuing information is only part of the problem. The other part is deciding on the optimal *quantity* of (costly) information an individual would like to receive. This will be the subject of the following chapter.

Checklist: Concepts introduced in this chapter

- Information as a guide to action
- Probability square
- Signal value
- The valuation principle
- Precision of a signal versus accuracy of a forecast
- Optimal action
- Indifference and information value
- Resolution of uncertainty
- Information love and information aversion

4 G Problem sets: Precious advice

For solutions see: www.alicebob.info.

Problem 4.1 Signal value

Alice throws a dice. If the outcome is either 1, 2, or 3, Bob wins one cookie. If the outcome is either 4, 5, or 6, Bob loses one cookie. Before risk-neutral Bob has to accept or reject the game, he can choose among three signals. Signal 1 shows whether the outcome is even or odd; Signal 2 shows whether the outcome is 5 or 6; Signal 3 shows whether the outcome is a 6. Which is the most valuable of the three signals?

Problem 4.2 A perfect forecast
Risk-neutral Bob considers investing \$1 in a project that yields \$$X$ if he has good luck, and nothing if he has bad luck. The probabilities of good and bad luck, respectively, are (i) 75 per cent and 25 per cent in a boom and (ii) 50 per cent each in a recession.

(a) What would he pay for a perfect forecast of good and bad luck in a boom and in a recession, respectively?
(b) What would Bob pay for a perfect forecast of a boom or recession?

Problem 4.3 Knowing one's bank
A depositor holds a deposit of \$1,000 in a bank. The deposit could be withdrawn at a cost of 0.5 per cent at any time (the depositor considers the deposit and cash as perfect substitutes, but she must walk to the bank in order to withdraw the deposit). If the bank becomes insolvent before the deposit is withdrawn, the depositor gets nothing. The depositor thinks that a year from now the bank will be solvent with a probability of 0.99.

(a) What is the value to the depositor of a perfect signal revealing whether the bank will be solvent in a year?
(b) What would be the value of the signal if the depositor were insured for x per cent of the deposit? (Draw a graph showing the value of the signal as a function of x.)

Problem 4.4 The value of precision in the Elementary Game
Reconsider the Elementary Game (from Box 4.3). You are offered a lottery with two possible outcomes. With probability p, you win \$$X$, with probability $1 - p$ you lose \$1. Before deciding (accepting or rejecting the lottery), you can obtain a binary signal that predicts the outcome as either g ("good") or b ("bad"), the former suggesting you will win, the latter you will lose. The signal has precision q in both cases ($q = Pr(g|win) = Pr(b|lose)$).

(a) Assume $p = 0.5$. Which value of X makes the signal most valuable? How much is that value?
(b) How does the signal value vary with q?
(c) How does risk aversion affect the value of the signal? (Give your answer in words, perhaps using a graph).

Problem 4.5 Information and insurance
An individual with utility $U = \sqrt{w}$ faces an investment opportunity. An investment of \$100 yields \$64 or \$144 with equal probability.

(a) Would the individual find the investment attractive?
(b) How much would the individual pay for perfect insurance against the outcome?
(c) How much would the individual (approximately) pay for a perfect forecast of the outcome (in the absence of insurance)?

Problem 4.6 Who is the better expert?

A risk-neutral investor faces two alternative investment opportunities: Investment 1 pays \$2 in a boom and nothing in a recession. Investment 2 yields \$3 in a boom and −\$2 in a recession. The probabilities of a boom and a recession are $3/4$ and $1/4$, respectively. Before the investment takes place, the investor is contacted by two experts: Expert A forecasts all booms and one in two recessions. Expert B forecasts all recessions and one in two booms.

(a) Who is the better expert?
(b) How should the investor invest?

Problem 4.7 Information as a substitute for liquidity

Assume that the Swiss Interbank Clearing System (SIC) works along the following stylized rules (remembering that reality is more sophisticated):

- At the beginning of each day, a bank can put money into its SIC account with the Swiss National Bank. We call the respective amount chosen by a bank L. Once put into the SIC account, L cannot be used for any other purpose than for SIC transactions during the day. The opportunity cost of funds in the SIC account (the interest that could be earned when put to another use) is denoted by i.
- Banks face random inflows and outflows, which are executed during the day. Inflows and outflows are credited and debited against banks' SIC accounts. During the day banks may freely overdraw their accounts. At the end of the day, however, balances must not be negative. If a bank has a shortfall at the end of the day it has to borrow the respective amount at rate r. For simplicity's sake, we assume that $r = 2i$.
- The aggregate values of incoming and outgoing payments, $\tilde{Y}$ and $\tilde{Z}$, respectively, are random variables drawn from two identical and independent Uniform distributions, $Y \sim U(0, \overline{X})$ and $Z \sim U(0, \overline{X})$.
- Banks are informed about the volume of incoming payments, Y, at the beginning of the day, before they have to decide about L.

(a) What is the optimal value L (at the beginning of a day) if a bank knows the realization Y (but not Z) for the coming day?
(b) What is the optimal value of L if a bank knows the realization of neither Y or Z?
(c) What are expected liquidity costs in both cases?
(d) What is the value of knowing Y at the beginning of a day when L has to be set?

Problem 4.8 Bad news

Over the last few years, the so-called baby-boomers (individuals born in the 1950s and 1960s) had to learn that, due to demographic reasons, their present pension systems are not viable in the long term. Their expected utility thus fell with the news. Does this mean that the news had a negative value?

Problem 4.9 "*Der Ritt über den Bodensee*"

(Riding across Lake Constance)

There is a famous fable of the man who rides home on a stormy winter night. He whips his horse across what he thinks is a large plain. After having crossed the plain he becomes aware that, in fact, without noticing he just rode across a (thinly) frozen lake and falls dead from his horse. Can good news such as learning that you have safely crossed a frozen lake have a negative value?

Problem 4.10 Should doctors tell the truth?

A patient tells the doctor that he wants to know his diagnosis, whether good or bad. Assume the doctor knows he has bad news. Assume also that the doctor knows from experience and perhaps from knowing the patient that the desire to know the truth is very much driven by (perhaps excessive) hope for good news. Should the doctor tell? (Check your answer with Caplin and Leahy (2004).)

5 The optimal amount of information

5 A Introduction

> And it came to pass at the end of forty days, that Noah opened the window of the ark which he had made: And he sent forth a raven, which went forth to and fro, until the waters were dried up from off the earth. Also he sent forth a dove from him, to see if the waters were abated from off the face of the ground; But the dove found no rest for the sole of her foot, and she returned unto him into the ark, for the waters were on the face of the whole earth: then he put forth his hand, and took her, and pulled her in unto him into the ark. And he stayed yet other seven days; and again he sent forth the dove out of the ark; And the dove came in to him in the evening; and, lo, in her mouth was an olive leaf pluckt off: so Noah knew that the waters were abated from off the earth. And he stayed yet other seven days; and sent forth the dove; which returned not again unto him any more.
>
> (Genesis 8:6–12; King James version)

In most practical situations an individual has more than the choice between no information and information. Often, the quality of the information varies, with better information normally being more costly.

How long should an oral exam be? What is the optimum size of a sample used in drug testing? How much exploratory drilling in an oilfield is optimal? When is the right time for an irreversible investment? How many partners should one look at before deciding with whom to raise children? These are all variations of the basic question: How much costly information should I receive? This is the issue we will analyze in the present chapter.

This chapter is a logical sequel to Chapter 4 on the value of a piece of information. Once we know how to value a piece of information, we can also calculate the difference in the values of two different sources of information. If this difference is worth the additional cost of the more expensive information, a rational individual would want to receive it.

This may sound simple, and the theory in this chapter indeed looks simple, particularly to a trained economist. Yet, the theory has some clearly non-trivial applications. In one type of model, an individual has to decide *in advance* what the optimal amount of costly information might be. The number of subjects to be interviewed in an opinion poll has to be fixed before interviews start. Once a satellite to explore Mars is under way, scientists cannot load it with additional measurement equipment. We will look at a specific example in an Application section (5 D) on the optimal precision of a central bank's inflation forecast. In another type of model, the amount of information to be acquired can be decided *on the fly*, so to say. A shopper, at any time, can choose between continuing his search or stopping (buying or going home). We will present such models of sequential search in an Application (5 E), which is devoted to the search for the lowest price and to the value of waiting.

The optimal amount of information is reached when further efforts are not worth their cost. This logic of optimization is the same in both the "in advance" and the "on the fly" types of models. But, in the latter case it seems easier to apply: Noah could hardly have said in advance how many times he should send off the dove (in fact, he started with a raven). Yet, when the dove did not return from the third flight, Noah knew that she had found solid ground—there was no need to send off any further birds.

5 B Main ideas: Is it worth the cost?

Why not stop here? Is it really worthwhile for the reader to continue, to turn the page, and the one after that, and the one after that—until she eventually finishes this book? Though we hope that some readers may finish, we are realistic enough to know that some will not. Like any other activity, reading has a cost. The text may be dull, sitting with your nose in a book for a long time is painful—but most importantly, life offers quite a few alternatives to reading a textbook (of which only one can be chosen at a time, though).

Those readers who decide to stop here and those who go on have arrived at different conclusions. But the logic they used is probably the same: All have weighed, more or less consciously, the benefit from reading one more page against the cost of not doing so.

That is, at least, the prescription found in microeconomic textbooks: (i) think in marginal changes, and (ii) push every activity to the point where the last small increment yields a benefit that just covers the additional cost.

Alice wants to find out more about this guy Bob she is supposed to meet again in half an hour. With the few bits of information she has, she tries an internet search. The first page of returns does not yield much. Nor does the second. Should she try a third? The problem is: If she continues searching, she may lose track of time. Would the chance of learning more warrant the cost of being late? No, Alice decides, and runs for the bus.

The problem of deciding the optimal amount of costly information comes in two different versions: as a one-shot *ex ante* decision and as a sequential decision. In a one-shot *ex ante* decision, a pharmaceutical company, for example, decides the number of individuals on whom a new medication shall be tested. Or a university department decides that oral exams of half an hour are sufficiently informative.

In a *sequential* decision, a job market candidate may, for instance, reject the first two offers and accept the third. In some situations involving sequential decisions, the individual undertaking the search may be able to return to a shop already visited. In others, typically in the search of marriage partners, rejecting an option tends to be final.

While the present chapter focuses on the optimal information *acquisition*, there exists a similar problem of optimal *processing* of available information. An individual could take a decision like a consumption-savings-plan in daily, monthly, annual, or variable intervals. Short planning intervals involve a lot of (probably costly) information processing, long intervals may lead to planning errors in the form of sluggish adaptation of plans to new circumstances. This issue will be taken up in Chapter 11 under the heading of "rational inattention".

5 C Theory: Deciding at the margin

5 C a A starter: Optimal browsing

A typical problem of optimizing the quantity of information is an internet search, as in the example of Alice used above. A search engine often responds to a submission with a list of "hits" presented on several pages in an order of decreasing relevance. Assume that the chances of finding the required source on the first page of hits is 1/2; on the second page it is 1/4, on the third 1/8, etc. The sequence of stacked bars in Figure 5.1 illustrates these decreasing marginal probabilities. The figure

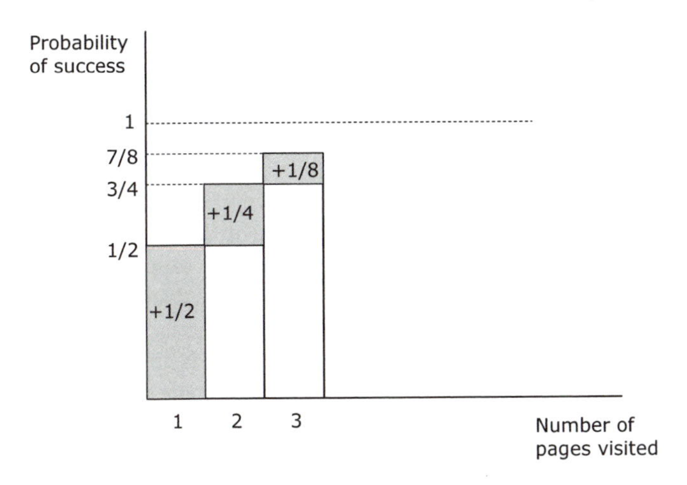

Figure 5.1 Internet search provides several pages of results. The utility gain (measured by the probability of a successful search) from looking at one more page is positive but decreasing (by half, in the example).

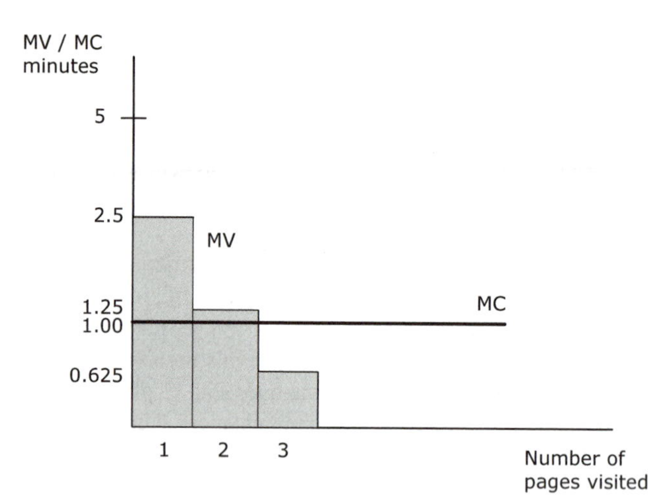

Figure 5.2 The optimal amount of information is given at the intersection of the marginal cost of searching and the marginal value of information both measured in minutes of searching. In discrete terms, this means that it is worth visiting sites found on the first and second pages of hits.

also makes clear that the marginal probabilities add up to one. If the number of pages goes to infinity, the probability of finding the right source goes to one.

We assume that the value of finding the right source is the equivalent of five minutes. The expected value of viewing a page (the marginal value of information), expressed in minutes, thus is five minutes times the probability of a "hit" (assumed to be the inverse of the square of pages visited). In numbers, the expected value of viewing the first three pages is $5/2 = 2.5$, $5/4 = 1.25$, and $5/8 = 0.625$, respectively. Browsing through all (which means a large number of) potential results would be worth five minutes (the probability of a hit going to one).

This makes clear that perfection is too expensive. Browsing through a large number of websites would be too costly in terms of time and effort. Assume that examining one page of results (including following the links listed) takes one minute. Figure 5.2 illustrates that it is optimal here to view both the first and second pages of results, but not the third. The marginal value of the second page is more than a minute, while the third page is worth less than a minute.

The parameters assumed for this illustration are arbitrary, of course. They do not seem to be completely off the mark, though. It is often said that people only care about the first page of results, and sometimes go to the second.

5 C b The optimization rule

One of the first things economists are taught is thinking in increments, rather than in levels. When increments can be made arbitrarily small, we speak of marginal changes. The comparison of marginal changes is relevant for optimal decisions. The optimal level of any economic activity can be described by the criterion:

marginal value = marginal cost

Let $v = v(q)$ be the value of a signal with precision q. The cost of receiving the information is $c = c(q)$. The net utility of the signal is $u = v(q) - c(q)$. For a maximum, the first derivatives of the value and the cost functions have to be equal:

$$\frac{\partial v}{\partial q} = \frac{\partial c}{\partial q}.$$

This is called a *first-order condition.* It is *necessary* for a maximum, but not *sufficient.* The equality of marginal value and cost describes maxima as well as minima. For a maximum, a further condition is required: With a further increase in precision, the marginal value generally must fall more quickly than the marginal cost. The second derivative of the value function must be smaller than the second derivative of the cost function:

$$\frac{\partial^2 v}{\partial q^2} < \frac{\partial^2 c}{\partial q^2}.$$

This is called a *second-order condition.* In the above example, the second-order condition is met. In Figure 5.2 the bars representing marginal value decrease from the left to the right, while the marginal cost remains constant.

The first- and second-order conditions describe local maxima. A local maximum may not be a global maximum, however. Assume a student reads a textbook, starting from the beginning and continuing until the perceived marginal value of the next page falls below the marginal cost of reading it. This student is likely to stop too early. After a particularly tedious page, some more interesting or enlightening pages may follow. The local maximum need not be a global maximum.

5 C c Example: Choosing signal precision

The following example illustrates the optimization rule for the choice of the precision of a signal. Unlike the sequential decision of continuing to read a book or browse the internet, this is a one-shot *ex ante* decision.

Background: The Elementary Game

An individual is invited to play (or reject) the Elementary Game introduced in Chapter 4, and illustrated in Figure 4.1. Accepting the lottery, the individual can win X dollars with probability p or lose one dollar with probability $(1-p)$; rejection yields a payoff of zero. Before accepting or rejecting, the individual can observe a signal on the outcome of the lottery with precision q. Unlike in Section 4 C c,

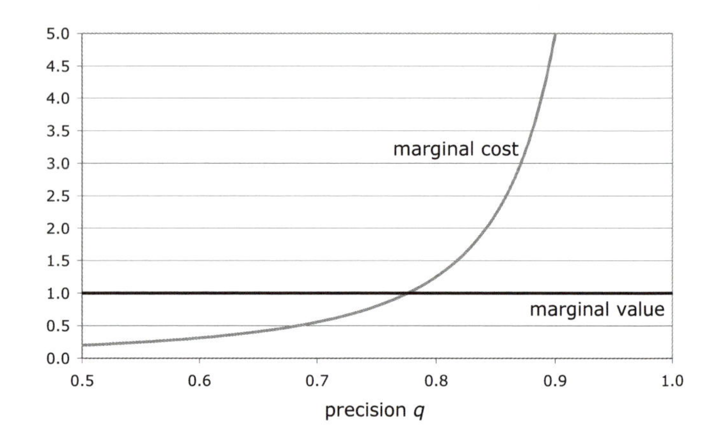

Figure 5.3 Marginal value and marginal cost of precision q (in the interval $0.5 \leq q \leq 1$) for the Elementary Game with $X = 1$ and cost function $c(q) = \beta\frac{q-\frac{1}{2}}{1-q}$ ($\beta = 0.1$). The optimal precision $q^* \approx 0.777$ is given by the intersection point (marginal value = marginal cost).

the precision q is not given exogenously. The individual can achieve any precision $0 < q < 1$ at cost:

$$c(q) = \beta\frac{q - \frac{1}{2}}{1 - q},$$

where β is a scaling factor.

Signal value and cost

The value of the signal, illustrated in Figure 4.4 on page 45, depends on whether X is smaller or larger than unity. By the logic introduced in Section 4 C c this is the maximum of zero (a case we can neglect here) or:

$$V(X < 1) = pqX - (1 - p)(1 - q)$$
$$V(X \geq 1) = (1 - p)q - p(1 - q)X$$

The marginal *value* of the signal as a function of q in both cases is:

$$\frac{\partial V}{\partial q} = pX + (1 - p).$$

The marginal *cost* of precision is:

$$\beta \frac{\partial c}{\partial q} = \frac{\beta}{2(1-q)^2}.$$

Optimal signal precision

The optimal precision, q^*, is the solution to the first-order condition (marginal value = marginal cost):

$$pX + (1-p) = \frac{\beta}{2(1-q)^2}.$$

This yields

$$q^* = 1 - \frac{\beta}{\sqrt{2(pX - p + 1)}}. \tag{5.1}$$

For $X = 1$, and $\beta = 0.1$ (5.1) yields an optimal precision of $q^* = 0.777$.

5 D Application: The central bank's inflation forecast

The importance of looking ahead

A central bank's main job is defending price stability. Central banks have a (mutually exclusive) choice among a set of policy variables (interest rates, the quantity of money, an exchange rate). Whatever instrument a central bank uses, the impact of its actions lies mainly in the future. Therefore, it is important for a central bank to know what state the economy (boom, recession, neutral) will be in, say, one year from now. The central bank has many indicators providing some information about the future state of the economy, but the future remains uncertain.

The cost of policy errors

Let us assume the central bank's ultimate target is a zero inflation rate. Its policy actions are a function, not of today's inflation rate, but of the inflation rate forecast for a year ahead (the forecast of the rate that would prevail *without* the bank's policy action). The central bank thinks it is important to forecast inflation correctly. Deviations from the true value (resulting in deviations of the actual rate from price stability) lead to economic damage which the central bank measures, as we assume, by the *square* of the forecast error.

Choosing optimal precision

The central bank can *choose* the precision of its inflation forecast. However, higher precision is costly. It requires larger computers, more (and better) economists, collecting more data, etc. The bank faces a trade-off between better forecasts and higher costs. What is the optimal level of precision the bank should choose for its inflation forecast?

To answer this question, we shall use the model in Radner and Stiglitz (1984), which describes the choice of the costly precision of a signal. In terms of this model, the true state of the economy, s, is the future level of inflation that would prevail in the absence of a policy action by the central bank. The central bank's policy action is denoted by a. This should be read as the action corresponding to a forecasted inflation level of a. The forecast error thus is $a - s$.

The central bank's utility from action a, if state s occurs is:

$$u(a) = -(a - s)^2. \tag{5.2}$$

Information quality …

The central bank must choose an action a *before* it knows the state of s. Before taking an action, the bank receives a signal y (the inflation forecast) about state s. Both y and s are random variables drawn from normal distributions with a mean of zero and unity variance:

$$E(s) = E(y) = 0,$$
$$Var(s) = Var(y) = 1.$$

The signal y is informative, if it is correlated with s. We denote the correlation coefficient by ρ:

$$Corr(y, s) = \rho.$$

The central bank can set the quality of the signal, measured by the correlation coefficient ρ, anywhere in the range between $\rho = 0$ (meaningless signal) and $\rho = 1$ (perfect signal).

… and information cost

Improving signal quality is costly. We denote the cost of signal quality by $C(\rho)$, with $C(0) = 0$ and $C' > 0$. A completely uninformed central bank ($\rho = 0$) incurs no cost of information, but remains with an expected squared policy error $Var(a - s) = 1$, as the optimal action in the absence of information is $a^*(0) = E(s) = 0$. The "reservation level" of utility, i.e. the expected utility the central bank obtains by investing nothing in the signal, is:

$$E(u(a, \rho = 0)) = -1.$$

The central bank's problem

What is the central bank's optimal choice of signal precision? The central bank maximizes expected utility (5.2). It has two choice variables: signal quality ρ action a. The bank solves:

$$\max_{a,\rho} E(u(a,\rho)) = E(-(a-s)^2 - C(\rho)).$$

The bank's decision is sequential. In real time, the bank first chooses ρ and then, after observing signal realization y, decides about action a. The solution of the decision problem, however, runs *backwards*. The bank first determines the action optimal for any signal source with precision ρ that has yielded a realization y. Second, the bank finds the precision ρ which is optimal under the assumption that the bank responds to any y with the optimal action.

Optimal action

The central bank now observes a signal y, knowing that its predictive power is given by the correlation ρ. The optimal action, $a^*(\rho)$, is simply the expectation of s, conditional on observing signal y with correlation ρ:

$$a^*(\rho, y) = E(s \mid y) = \rho y.$$

To see this, the predicted value of s (given the value of the signal) can be computed by means of a regression of s on the signal y. As the variances of the signal and of the state are equal, the regression coefficient for y is exactly the correlation between the two random variables.[10]

Optimal signal precision

The optimal precision, ρ^*, is found by balancing the cost of precision $C(\rho)$ against the cost of policy error. As a first step in the derivation, the expected policy error can be computed as follows:

$$\begin{aligned} E((a^*(\rho,y)-s)^2) &= E((\rho y - s)^2) \\ &= E((\rho y)^2 + s^2 - 2\rho y s) \\ &= \rho^2 + 1 - 2\rho \underbrace{Cov(y,s)}_{=\rho} = 1 - \rho^2 \end{aligned}$$

When the bank responds to any signal realization y with the optimal action $a^* = \rho y$, the expected cost of policy error is $1 - \rho^2$. The central bank's expected utility is:

$$Eu(\rho) = \rho^2 - 1 - C(\rho),$$

where ρ^2 can be interpreted as the (gross) value of a signal (an inflation forecasting model) having correlation ρ with the true state (the future inflation rate) s. It is the increase of the central bank's utility over the reservation utility -1 disregarding signal costs.

Taking into account that more precision is also costly, the optimal forecast quality is given by the first-order condition:

$$(Eu(\rho))' = 2\rho - C'(\rho) = 0.$$

If, for example, we specify the cost of signal quality as $C(\rho) = \rho$, the optimal value of ρ is $\rho^* = 1/2$.

5 E Application: Search

To shop or to stop?

Information often arrives piecemeal. This makes the question "How much information should I acquire?", a sequential decision problem. After each piece of information, the recipient can decide that he now knows enough or that he wishes to acquire more information. One example is optimal internet browsing (see Section 5 C a), which returns a sequence of pages. Shopping is another example. At any point on a shopping trip, a shopper can decide to buy, to stop without buying, or to visit one more shop (and then face a similar choice again).

Search models

The sequential arrival of information gives rise to *search models*. In some search models, the individual may return to a shop he has already visited. In other models, rejected opportunities are spurned forever. A search is typically costly. Going from one shop to another takes time and effort. Or, asking for a second medical opinion means duplication of the expense of a diagnosis. Search costs help to explain many institutional arrangements. Stock exchanges minimize the cost of a search for a matching buyer or seller. Money spares individuals searching for matching partners in each and every trading period. Search costs may also help to explain why in reality we find price dispersion. Even in a competitive environment, the "law of one price" often does not hold, at least not strictly, because it is costly to find out which supplier offers the lowest price. The search for the lowest price originally modelled by Stigler (1961) is discussed in Section 5 E a.

Waiting as a form of search

A particular form of acquiring information is *waiting*. If information arrives over time, without any particular effort on the part of the recipient, the only cost of information is the cost of waiting. Waiting may be costly because of impatience or because the given investment opportunity is a one-off, for instance. A model building on that motive is Cukierman (1980); we will discuss it in Section 5 E b.

One day, Alice hopes to find her Mister Right. The problem is that she cannot compare all potential partners. She can only look at different candidates one after the other. Once she has rejected a candidate, she can never go back. Alice's goal is to maximize the probability of finding the best among those N candidates she can seriously look at. Alice thinks that a search *per se* is costless, as the fun and the hassle balance each other. Yet, at one point she has to stop. When is the optimal point to stop?

As mathematicians have found out, Alice's task, in principle, is simple: Alice maximizes her chances by following the "37 per cent rule" (see Box 5.1). If she thinks she can examine 20 partners, she would not settle with any of the first seven ($7 \approx 0.37 \cdot 20$), but with the first candidate in the subsequent set who is better than the best from the first bunch. This way she maximizes her chances of finding the best partner—at some risk of ending up with the worst one (assuming she has to decide for one of them, rather than stay alone)!

Alice may be risk-averse or slightly less ambitious. In order to minimize the risk of ending up with a very bad partner, she may aim at one of the best ten per cent, rather than with the very best partner. In this case she would settle faster: She would not necessarily wait with her decision until after the seventh candidate (see Todd, 1997). But there is a further problem: Alice's stopping rule may identify Peter as the right one. Yet, Peter's own optimal stopping rule may or may not tell him to stop at Alice!

Box 5.1 The 37 per cent rule

A person has to find the best among N potential candidates (pick the envelope with the highest amount of money, find the best candidate for a job, the best partner to marry, etc.). The quality of individual candidates can be discovered by examining candidates one-by-one. However, a candidate, once rejected, cannot be chosen later on. The person who decides knows N, but has no idea about the properties of the actual distribution (like the range) of qualities. How many candidates should the person examine? What is the optimal rule with regard to stopping and accepting a particular candidate?

This optimal stopping problem was first presented in Martin Gardner's Mathematical Recreations column in the February 1960 issue of *Scientific American*. Gardner showed that the probability of finding the best among N potential candidates is maximized by following a simple rule: Review the first 37 per cent of candidates (without accepting any), then take the next best (the first who is better than the best of the first 37 per cent).

Why 37 per cent? The reason is mathematical. With N going to infinity, the optimal number of candidates to be examined before accepting goes towards $1/e = 0.36787\ldots$, i.e. towards approximately 37 per cent (see Todd, 1997). If an individual is risk-averse or if there are search costs, the optimal pre-decision sample is smaller than 37 per cent of available candidates. If an individual could reasonably evaluate between fifty and one hundred potential partners, "take a dozen" looks like a reasonable rule (Todd, 1997).

Later on, Gardner posed the optimal stopping problem under the name of *Googol.* This term, introduced by the mathematician Edward Kasner (1878–1955) in 1938, refers to a high number: a 1 followed by 100 zeros (a *googolplex* is 10^{googol}, a very large number by most standards). Legend has it that "googol" was the answer Kasner got when he asked his nine-year-old nephew, Milton Sirotta, what he should call a very large number. "Googol" became synonymous with the "search among a vast number of possibilities" and was the inspiration behind the name of the internet search engine *Google* (see: (http://www.google.com/corporate/history.html)).

Gardner's problem has since been discussed under various names ("the secretary problem", "the sultan's dowry problem", etc.) and in a number of slightly different variations. An excellent account (including some refinements) can be found in Todd (1997).

When both sides search . . .

While optimal search is already difficult, optimal bilateral search may seem to be a matter of pure good luck. However, Gale and Shapely (1962) have shown that there is more to the issue than pure good luck. The question is: Is there a stable pairwise allocation in a group with an equal number of men and women? "Stable" would mean that in the final allocation no individual could find a *better* mate *ready to leave* the allocated partner for the challenger. Such an allocation is Pareto optimal: It is impossible to improve one individual's lot without damaging the lot of another. Gale and Shapely have even formulated an algorithm that leads to stable allocation. The GS algorithm is successfully used in practice. For example, it used to match newly qualified physicians with available jobs in US hospitals. The GS algorithm makes people as happy as possible under the mutual consent constraint. Yet, even within the GS algorithm, an individual may not obtain his or her Mister or Miss Right, but only the best one available in the presence of competitors. This is reminiscent of the often told joke of the 90-year-old bachelor. The man was asked if he was still unmarried because he had never found the perfect woman. "Oh, I found her all right," said the bachelor, "but she was looking for the perfect man."[11]

5 E a Search for the lowest price

In the bazaar . . .

A leisurely visitor to an oriental bazaar rarely buys at one of the shops close to an entrance. Shops deeper in the bazaar may offer more attractive prices, shops close to the entrance often specializing on the more hurried customers (Arbatskaya, forthcoming). Similarly, in the bazaar most familiar among economists—the publishers' exhibit hall at the American Economic Association meetings—the first stand after the entrance is normally occupied by a seller of relatively pricey books.

. . . the law of one price may not hold

The first economist to model search and the impact of search cost in the price structure was George Stigler, the 1982 Nobel Laureate. Stigler claims that "some important aspects of economic organization take on a new meaning when they are considered from the viewpoint of the search of information" (Stigler 1961, p. 213). Stigler's main example is price dispersion observed for fairly homogeneous goods like "1959 Chevrolets available in Chicago" or "anthracite coal delivered April 1953 to Washington, D.C.". Such differences in a competitive market cast some doubt on the "law of one price". Stigler demonstrates why competition may not enforce perfect price convergence. In the presence of search cost (i) total market transparency is expensive, and (ii) at some point the incremental value of search falls with each additional seller visited. In the optimum, a buyer does not visit all sellers. Sellers are not forced to set the same price.

Modelling sequential search

The Stigler (1961) model starts from the following assumptions:

- Potential buyers try to find the cheapest seller of a good;
- Sellers' prices are drawn from the same distribution;
- Buyers learn a price by visiting the respective seller;
- After each seller a buyer can (i) buy from that seller, (ii) go to one more shop, or (iii) buy from a seller visited before.

Uniform distribution of prices

A buyer faces S potential sellers. Each seller has set a price. Prices are distributed uniformly (see Box 5.2):

$$p \sim U(0, 1).$$

The distribution of minimum prices with n searches (see Stigler, 1961) is:

$$n(1 - p)^{n-1},$$

and the average (i.e. expected) minimum price (see Box 5.2) is:

$$\frac{1}{(n+1)}.$$

Box 5.2 The Uniform distribution

Under a Uniform (or rectangular) distribution

$$k \sim U[\underline{X}, \overline{X}]$$

a *continuous* variable k, in an interval between $\underline{X}$ and $\overline{X}$ (the so-called support of k), has constant density $1/(\overline{X} - \underline{X})$. The cumulative distribution function (in the interval) is $F(k) = (k - \underline{X})/(\overline{X} - \underline{X})$. Variable k has mean $(\overline{X} - \underline{X})/2$ and variance $(\overline{X} - \underline{X})^2/12$.

The Uniform distribution also exists for *discrete* variables like the numbers shown on a dice. With n possible outcomes (for the numbers on a dice $n = 6$) the probability of each outcome is $1/n$. The mean is $(\overline{X} - \underline{X})/2$ and the variance is $(n^2 - 1)/12$. The sum of two Uniform distributions is a triangular distribution (as can be checked by adding the numbers thrown with two dice).

In many contexts, like auction theory, a result from *rank order statistics* is helpful: The expectation for the r-th highest value of n uniformly distributed valuations $k \sim U[0, 1]$ is:

$$E(r) = (n + 1 - r)/(n + 1).$$

If there are three bidders ($n = 3$) whose valuations for an object are drawn from an interval of between 0 and 100, the expected value of the highest valuation ($r = 1$) is 75.

The Uniform distribution is not commonly found in nature. Yet, it is arithmetically convenient and frequently used in economics.

The buyer's problem

Increased search yields diminishing returns as measured by the expected reduction in the minimum asking price. Assume that search cost, $c < 0.5$, is a constant amount per seller visited and that visiting no shop would be equivalent to paying a price of 1. If the buyer has to decide *ex ante* how many sellers to visit, he thus calculates:

$$\max_n -\frac{1}{(n+1)} - nc.$$

The first-order condition (marginal benefit equals marginal cost) for the optimal[12] number of buyers to visit, n^*, is

$$\frac{1}{(n^* + 1)^2} = c.$$

The lower the search cost, the higher the optimal amount of search, i.e. the number of sellers to visit.

Price dispersion . . .

Stigler (1961) concludes from his search model that the phenomenon of price dispersion in competitive markets can be explained by the existence of search costs. Price dispersion would diminish with reductions in search cost caused by, e.g., advertisement, reputation, specialized traders, or centralization of markets (as in medieval markets or on eBay).

. . . seems paradoxical . . .

Stigler mentions, but does not model the optimization problem of suppliers: A supplier faces a trade-off between charging a high price (and doing well if nobody searches) and a low price (doing well if everybody searches). Assume, for simplicity's sake, that there are only two sellers who consider setting a price somewhere between the monopoly price and the competitive price. With a strictly positive search cost, the unique equilibrium is both sellers setting the monopoly price (and nobody searching). When search costs are zero, the equilibrium price is the competitive price (still nobody searching). There is some critical level of search cost in between, where the equilibrium price switches from one extreme to the other. But, contrary to Stigler's intuition, in equilibrium there is neither price dispersion nor search. This result was discovered by Peter Diamond in 1971 and became known as the Diamond Paradox.

. . . but exists

Yet, price dispersion is well documented empirically. Subsequent research (see Arbatskaya, forthcoming) has looked for ways around the Diamond Paradox. It was found that price dispersion and search occur if one of the following assumptions hold:

- consumers are heterogeneous with respect to search cost;
- suppliers are heterogeneous with respect to producer costs;
- consumers search in an ordered (rather than in a random) sequence;
- there is multi-commodity search.

A model combining the first and the third of these assumptions (Arbatskaya, forthcoming) yields insights that are relevant in the age of the internet. In an internet search (see Section 5 C a), the search sequence is not randomly chosen by the customer. Search results normally come page by page, and in order of declining relevance, for example. The firms on top of the list are more likely to be visited (on the internet) by a customer; these tend to set the highest prices (just like the sellers next to the entrance of a bazaar). Some search engines thus auction off the advertising position of firms.

In practice, customers normally search for price *and* quality at the same time. Many goods are not homogeneous. Even if they were, they are sold with additional components, services—or with a smile (see Box 5.3).

Box 5.3 Antonio from Helvetia Square market

Antonio is a successful (and some would say the most handsome) vegetable and fruit seller on Helvetia square market in Zürich. His father, having immigrated from Italy in the 1960s, started a business with a market stall of barely 3 meters in length. Today, a 24-meter-long market stall, offering everything from *albicocche* to *zucchini*, employs the whole family. Every morning at 4.30 when the wholesale market opens its gates, Antonio and his father are there to buy their supply for the day. By 6 a.m. they have their stall ready for the first customers.

We wanted to know whether sellers take into account that customers may search among stalls for the best price, and whether this has an impact on price setting policies. “Absolutely!” says Antonio, “sellers look at each other’s prices.” However, he continues, he and his father are considered as market leaders, in particular with respect to quality. They rarely have to cut prices. Though, there was one exception recently, when a new competitor just opposite their stall slashed the prices of his strawberries.

Price setting is more tricky than just looking at competitors’ price tags. The first step is finding the right markup for each product. Domestic products, for example, command higher prices (it is Switzerland, after all), but tolerate a smaller profit margin than imported goods. The true art, Antonio adds, is not in pricing but in buying. When his father spent three weeks in hospital, Antonio had to buy on his own. After many years’ practice, he thought, this cannot be too difficult. Yet, by 10 a.m. he had already run out of asparagus (the main cash crop before the coming Easter weekend), but sat on other products he had bought too optimistically. “My father’s nose is still far superior in anticipating demand to any amount of information you could collect.” Buying and selling are also closely interrelated. To buy well, you must know the produce inside out. But this is exactly what also sells it, as customers frequently ask questions on how long the cherry season would last, how to prepare *barba frati*, or what was the difference in taste between *sarde* and *ramato* tomatoes. There is, we conclude, a lot of information processing going on in a market, but only the smaller part of it relates to prices; the larger part concerns quality. “Of course,” Antonio smiles.

5 E b The value of waiting

Waiting is a form of search

The logic of sequential search is also applicable to decision problems in which an agent does not actively search, but has to wait passively until more information arrives. Think of an entrepreneur who wants to build a new factory. The future demand for the factory's output—and hence the factory's optimal size—is not known with certainty. Yet, once the factory is built, its size is fixed and cannot be modified. The entrepreneur can build now or wait for more information about the demand to arrive. Postponing the decision improves its quality. Waiting, on the other hand, is costly. The information acquisition itself may consume resources, plus the entrepreneur may lose the market to a competitor. Such is the decision problem analyzed in Cukierman (1980). The author shows that (i) prior uncertainty deters investment and (ii) the precision of further information that becomes available through waiting has an ambiguous effect on the timing of investment. On the one hand, a decrease in precision makes waiting less attractive. On the other hand, with a less precise signal it takes more time to achieve the same level of information.

Sizing the factory

We present a modified version of Cukierman's model, expressed in terms of our above example (5 D) of a central bank's choice of the precision of a forecast (from Radner and Stiglitz (1984)). A risk-neutral entrepreneur considers building a factory of actual size a. The optimal size, given future demand, would be s. If the actual size does not perfectly match the optimal size the entrepreneur suffers a loss. The entrepreneur's payoff from size a, where size s would be optimal, is:

$$u = -(a - s)^2.$$

Wait and see

The entrepreneur has prior information about s. She knows that s is normally distributed and, after normalization, has a mean of zero and variance $\sigma^2 = 1$. In addition, she can receive a signal y which is correlated with s and has the same distribution $y \sim N(0, \sigma)$. The coefficient of correlation between s and y is ρ and is here called the *precision of the signal*. Unlike in the problem analyzed in (5 D), the entrepreneur cannot "buy" signal precision by incurring some direct financial cost. Instead, we assume with Cukierman (1980) that the entrepreneur can improve signal quality by waiting for more information. Thus ρ increases with time. We assume that:

$$\rho^2(t) = \frac{\lambda t}{1 + \lambda t}, \tag{5.3}$$

where t is the number of periods used to gather information and λ represents the effectiveness of waiting, the "leverage" of information collection, so to speak.

In the limit, ρ approaches unity, but its increment decreases the longer the entrepreneur waits. The higher λ, the sooner ρ approaches unity and the signal becomes close to perfect. As a proxy for ρ, λ is an indirect measure of information quality.

The (per period) cost of waiting is assumed to be constant:

$$C(t) = ct.$$

The impact of prior uncertainty

The entrepreneur solves:

$$\max_{a,t} u(s, a, t) = -(a - s)^2 - ct.$$

The optimal choice for a when signal realization y has been observed, is ρy (as $Ey = Es = 0$). The net benefit from investing after waiting t periods is:

$$V(t) = \rho^2(t) - \sigma^2 - ct. \tag{5.4}$$

This is the first of Cukierman's (1980) results: the detrimental effect of prior uncertainty (σ) on investment. If the effect is large enough, it may never pay to invest, even with optimal t. Prior uncertainty thus is bad for investment.

The impact of information quality

The impact of information quality is more tricky. Information quality is represented by λ, the effectiveness of waiting on signal precision ρ. Inserting (5.3) into (5.4) yields:

$$V(t) = \frac{\lambda t}{1 + \lambda t} - \sigma^2 - ct.$$

The first-order condition is:

$$\frac{\lambda}{(1 + \lambda t)^2} - c = 0.$$

Solving for t yields the optimal waiting period:

$$t^* = \frac{\sqrt{\lambda} - \sqrt{c}}{\lambda\sqrt{c}} = \frac{1}{\sqrt{\lambda c}} - \frac{1}{\sqrt{\lambda}}. \tag{5.5}$$

Obviously t^* decreases in c. The higher the cost of waiting, the shorter the optimal waiting period. The impact of λ, however, is ambiguous. An increase in λ, and hence in the precision of incoming information, may shorten or lengthen the optimal waiting period.[13] On the one hand, more precise information is attractive to wait for; on the other, more precise information which is obtained relatively quickly leads to some level of certainty. The first effect suggests postponing investment, the latter suggests accelerating it.

While higher information precision, as measured by λ, has an ambiguous impact on the *timing* of investment, it unambiguously improves the *quality* of the investment decision, measured by $-(a-s)^2$.

Inserting (5.5) into (5.3) yields

$$\rho(t^*) = \sqrt{1 - \frac{\sqrt{c}}{\sqrt{\lambda}}}.$$

The quality of the optimal amount of accumulated information (the correlation with the true state) thus increases with information leverage λ and decreases with information cost c.

Uncertainty and investment

The Cukierman (1980) model confirms the rule that "uncertainty is bad for investment", but it also shows that some qualification is necessary. Prior uncertainty is clearly bad for investment; Cukierman explicitly warns that "ambiguous and sometimes contradictory statements by government officials will have adverse effects on investment as well as on the acquisition of consumer durables" (p. 474). Improving the quality of incoming future information, like data arriving over time, may, however, lead investors to accelerate or to postpone decisions. Yet, better information leads to better decisions in both cases.

The option of waiting

The Cukierman (1980) model helps us to understand the impact of uncertainty and information on investment and on other decisions like the demand for liquidity. The idea that investment decisions do not boil down to "yes or no" questions but rather allow a choice between now, later, and never, has paved the way for models of waiting as an option, such as Dixit and Pindyck (1994).

5 F Conclusions and further reading

The optimal amount of information is determined by the same logic as the optimal amount of any other economic good: By balancing marginal benefit against marginal cost. In the context of information, this optimization has interesting applications, like the optimal precision of a forecast or optimal sequential search.

The optimal amount of information is an application of information valuation. For further reading we therefore recommend the works already quoted in the previous chapter, in particular Hirshleifer and Riley (1992, Ch. 5). The classical contribution on search is Stigler (1961). Optimal information acquisition plays a role in many economic situations. One example is the optimal timing of investment in Cukierman (1980), an article discussed above. Other applications of endogenous information are Doherty and Thistle (1996) on insurance, Grinblatt and Ross (1985) on security markets or Persico (2004) on committee decisions, to name just a few.

One assumption underlying the analysis in the present chapter was the absence of strategic interaction. Optimal information acquisition is a game of man against nature. The acquisition of information by one individual did not affect information or payoffs of another individual. There were no external effects, either positive or negative. (One single exception is the effect of a search by one buyer on a seller's pricing and thus on another buyer's payoff.) In reality, information acquisition has external effects: An individual may observe another individual's information and get a free ride. This leads to strategic issues of information production, which are analyzed in Chapter 6. Sometimes individuals cannot see each others' information directly, but only guess it from observing their actions. This leads to effects which we shall discuss in Chapter 10.

Checklist: Concepts introduced in this chapter

- Marginal value
- Marginal cost
- The precision of a continuous signal
- Sequential search
- The Diamond Paradox
- The value of waiting
- Uniform distribution

5 G Problem sets: Paying, searching, and waiting for information

For solutions see: www.alicebob.info.

Problem 5.1 Internet search
Assume you use an internet search engine to find instructions for rolling sushi. Entering "rolling sushi" returns around 2 million results. Assume the dollar value of looking at the 1st, 2nd, 3rd page and so on is $10/1$, $10/2$, $10/3$, etc. The cost of looking at a page is constant at $2.5 per page. How many pages do you visit?

Problem 5.2 Reading headlines
Newspapers and magazines almost without exception use headlines. How does the presence of headlines impact on the time you spend reading a newspaper?

Problem 5.3 Clustered search
Assume you can only work with one of two internet search engines. One returns pages with unstructured lists of results. The other returns pages with clustered

results. Assume the former engine returns more useless information per page than the latter.

(a) With which machine would you expect to find better results?
(b) With which machine would you spend more time searching?

Problem 5.4 Shopping around

A buyer looks for the lowest price for a given standard article. There is an unlimited number of shops. The price of the article at a shop is an independent draw from a Uniform distribution between \$0 and \$100. Visiting a shop for the first time costs \$1. The buyer can return to any previously visited shop at no cost.

(a) What is the optimal number of shops to visit if the buyer has to decide about the number *ex ante*?
(b) What is the expected number of shops visited if the buyer can stop at any time?

Problem 5.5 Waiting for information

An entrepreneur plans to build a new factory. The optimal size of the factory depends on the unknown demand for the good to be produced. The entrepreneur can learn about the true demand by collecting information over time. The longer the entrepreneur waits the more precisely does demand become known. What is the impact of the precision of incoming information on

(a) the optimal waiting period?
(b) the quality of the entrepreneur's decision?

Problem 5.6 Mediocrity

Why do teachers hate mediocre work by students?

Problem 5.7 Forensic evidence

A newspaper reports that "according to a recent study, the amount of evidence gathered by a court is not correlated with the quality of the court's decision". Would you agree with the conclusion that gathering evidence by courts is obviously a waste of taxpayers' money without any corresponding gain of justice?

Problem 5.8 Free to stop

Prove that in the Stigler model buyers search longer when they have to commit to a number of search steps (n) *ex ante* than when they are free to stop their search any time.

6 The production of information

> 1940: The year the U.S. Census Bureau eliminated the title 'inventor' as a job category.
>
> (Schwartz, 2004)

6 A Introduction

"Information is costly to *produce* but cheap to *reproduce*." (Shapiro and Varian, 1999). The fact that information can be produced as well as reproduced—often at low cost—is the topic of the present chapter. This fact adds a strategic dimension to the question of optimal information acquisition dealt with in Chapter 5: Information cannot only be extracted from nature—through research, search or waiting—but it can also be bought or copied from others.

The hope to free-ride on others' information and the corresponding fear of being free-ridden by others can have serious consequences. It can lead to outcomes anywhere between the extremes of (i) no information production at all, and (ii) hectic patent races.

None of these extremes are optimal outcomes from the point of view of society as a whole. In one, there is too little research, in the other there is too much. Such deviations of information production or research and development from their optimal levels may hurt economic growth. The production of information, including research and development, is a key factor behind economic growth. In 1957 Robert Solow discovered that only a small fraction of per capita growth is due to an increase in the capital to labor ratio; more important is technical progress along with the improved education of labor (see Tirole, 1988, p. 389). Unlike in the Solow model, innovations do not fall from the sky. Innovation, like any other economic activity, is motivated by profit opportunities and thus is endogenous (Romer, 1990).

Technical progress is more than invention. Inventions only become effective if they are implemented. The discovery of electromagnetism was one thing, but the introduction of the tape recorder was another. Here is a role for a person traditionally called the entrepreneur. The entrepreneur is at the core of a process of "creative destruction", a term introduced by Joseph A. Schumpeter (1883–1950) in 1942. Creative destruction explains the dynamics of economic progress, including the ruins of disused businesses that were driven out of the market by new technologies. We introduce Schumpeter's vision as an application of the theory of information production.

With the present chapter we leave the dull world of the lonely Robinson Crusoe before he met Man Friday. From now on, the payoffs to one individual's actions will depend on another's actions. Decision problems, in other words, become strategic.

Strategic means that Alice's best choice from a number of possible actions depends on how Bob decides, and vice versa. If both read the weather report, they duplicate their effort. If both think to ask the other about the weather forecast, they end up with too little information.

Due to the strategic dimension, the question "make or buy?" becomes difficult in the context of information. Individuals will exert too little research effort from a social point of view, either hoping to free-ride on others' information or because they fear being copied. Or, individuals will spend too much on research from a social point of view, trying to "snap away" profitable discoveries from their competitors.

An example of the economics of producing and selling information is the credit ratings industry, which we will discuss in another Application. A failure in information production may also explain why banks in most countries are supervised, as we will argue in a third Application section. Another failure in the use of information occurs when individuals try to free-ride on each others' information by looking at how others decide. Such observational learning (in a sequential setting) results in "cascades" and is discussed in Chapter 10

Finally, the make or buy (produce or reproduce) decision very much depends on the allocation of property rights. The main instruments of allocating property rights in terms of information are patents and copyrights. We will not discuss the role of intellectual property in depth. But we will take a brief look at a potentially important—and strongly debated—development: Patenting business methods, financial instruments and software (see Box 6.1).

Box 6.1 Business patents

On 10 April 1790, President Washington signed the first patent statute passed by the Congress of the (twelve) United States. On 31 July, the "Commissioners for the Promotion of the Useful Arts," chaired by Secretary of State Thomas Jefferson, granted the first US patent to a chemical method for making potash and pearl ash. The first financial patent was granted on 19 March 1799, to Jacob Perkins of Massachusetts for an invention "Detecting of Counterfeit Notes." Unfortunately, all details of Mr. Perkins' invention were lost in the great Patent Office fire of 1836.

Business patents continued to be granted for methods ranging from interest calculation tables to lotteries during the following decades. On 8 January 1889, three patents were granted to the inventor-entrepreneur

Herman Hollerith for an apparatus built for "the art of compiling statistics". The business data processing method patent was born. The protection of his invention saved Mr. Hollerith's fledgling Tabulating Machine Company which, in 1924, changed its name to International Business Machine Corporation (The above account is based on http://www.uspto.gov/web/menu/busmethp/index.html).

In the following decades, electricity widely replaced the electrical-mechanical devices for business data processing that Hollerith and his successors had invented. Business methods became pure algorithms without any physical or technical consequences, beyond the computation of, for example, a price. The dominant "business methods exception" view held that pure business methods without a tangible result, such as a bookkeeping system to prevent embezzlement by waiters, were *not* patentable. Nevertheless, business methods, including software, continued to be patented occasionally. British Telecom, for example, took out a patent on the internet hyperlink in 1976 in the UK (expired in 2001) and in 1989 in the US (rejected in court in 2002).

In 1998, the *State Street* decision, however, turned the tables (State Street Bank & Trust Co. v. Signature Financial Group, 149 F.3d 1368; Fed. Cir. 23 July 1998). In question was a patent for a method of administrating an investment portfolio of pooled assets owned by mutual funds. The court held that the production of "a final share price momentarily fixed for recording and reporting purposes and even accepted and relied upon by regulatory authorities and in subsequent trades," is indeed the production of a useful, concrete and tangible result (State Street, 149 F.3d at 1373).

The State Street decision led to an explosion of *applications* for business patents in the following years as well as to an increase in the number of *granted* patents: (Source: http://www.uspto.gov/web/menu/pbmethod/applicationfiling.htm)

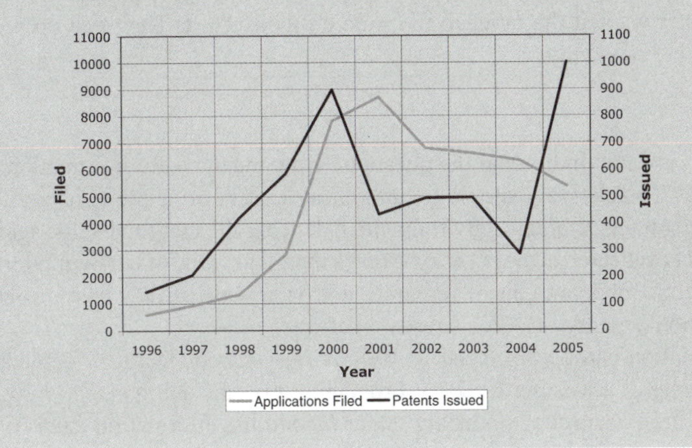

Patented inventions include trading systems, payment and transaction systems, portfolio and risk management methods as well as straightforward software. The proliferation of software patents led to a number of legal disputes. The wider public became aware of the issues when, in early 2006, the producer of *Blackberry* (a hand-held tool particularly popular among managers) only succeeded in settling a patent dispute shortly before a court decision.

The pros and cons of patentability of business methods and of software remain highly disputed. Indeed, it is difficult to see what the financial world would look like if Harry Markowitz and William Sharpe had patented the capital asset pricing model (CAPM), or if Fisher Black and Myron Scholes had patented their option valuation formula. Some claim that patentability remains the most powerful incentive for research. Others claim that innovation would have come to a standstill. Among the most ardent critics are the adherents of the Open Source Software movement (Von Krogh and Von Hippel, 2003). In this area, academic research, itself very much "open source", still has a number of issues to resolve.

6 B Main Ideas: Too little research or too much?

6 B a Free-riding on information

Alice and Bob plan to go to the movies. They both like movies with a good plot and a lot of suspense. But they know nothing about the films presently on show. They decide that by the next day they will both try to find out some information. As it turns out, though, neither made much of an effort. Both hoped the other might; and neither wanted the other to free-ride on their effort. They end up picking a movie quite randomly.

For an isolated individual the question "How much costly information should I acquire?" is relatively simple to answer, at least in principle (Chapter 5). In a multi-person, i.e. strategic, environment, balancing the marginal value against the marginal cost becomes more tricky. Alice's marginal value of her own information depends on Bob's amount of research, and vice versa. Bob's cost of obtaining information depends on Alice's research efforts.

Information can be *produced* as well as *reproduced* (learnt or copied). Take the example of a weather forecast: Producing it means doing *research* (buying a jar and a frog); reproducing means either *buying* the information (subscribing to a weather service) or copying or free-riding on someone else's information (by observing whether people on the street are carrying umbrellas).

The possibility of free-riding reduces the incentive on both sides to produce information. It lowers the cost to the free-rider and reduces the profit from information production of the "free-ridden". In the equilibrium, less information is produced than would be optimal from a social point of view. Information is under-produced because it shares the characteristics of a public good: (i) the same piece of information can be used by more than one person (non-rival use); (ii) it is difficult to exclude others from obtaining a piece of information and from using it (non-excludability).

One remedy against underproduction of knowledge is privatization (establishing excludability). Researchers may try to keep their findings secret. The ancient Greeks used the word sycophant ("fig-talker") to refer a person kind (or stupid) enough to reveal where he had discovered the highly esteemed fruit. Secrecy has one big drawback from a social point of view: It prevents the dissemination of knowledge and thus the efficient use of a public good.

An attempt to cure the disease (of underproduction) without killing the patient (efficient use) is patenting. A patent gives the inventor the right to the exclusive use of the invention; at the same time it lays the invention open ("patent" comes from Latin *patere*, "to be open"). Under a patent the sycophant discloses the location of the tree but retains the right to keep the fruit.

6 B b Racing for information

Privatization of the use of information by secrecy or patents prevents free-riding. Yet, the "winner-takes-all" situation causes another problem: Potential inventors face the danger that a competitor is faster and makes their own research efforts obsolete.

This may lead to overproduction of information. One example is patent races. Another are trials in court, where lawyers' fees play the role of research expenditure, in that they increase the probability of a party winning over the judges with their arguments. In both instances, one party takes into account that their own expenditure increases their chance of winning, but not the corresponding reduction in the other party's chance. The parties ignore the fact that their efforts impose a negative externality on their competitors.

Alice and Bob again make a date to go to the movies. They still like suspense. But this time, Bob wants the movie to have a happy ending, while Alice would still prefer something darker.

Both know of each other's preferences. They also know that they have a good chance of changing the other's mind if they can provide sufficient evidence that their favorite choice has more suspense.

In fact, they manage to agree on a movie. They both like it. But, leaving the theater they both admit: I should have accepted your proposal blindly, rather than having spent a day reading movie critics.

Why is it that, from a social point of view, information is sometimes under-produced and sometimes over-produced? The answer is: There are two different public goods involved. One is the stock of (publicly accessible) information *already discovered*. The other is the pool of knowledge *yet to be discovered*. Hirshleifer and Riley (1992, p. 259) distinguish the right *to* fish (as a verb) from the right *in* fish (as a noun). If the stock of fish caught is public property, too little fishing will be undertaken. If the pool of fish yet to be caught is public property, but caught fish is private property, there will be excessive fishing.

Similarly, if information is a public good, there will be *too little* private research. If there is a limited public pool of ideas yet to be discovered, research is likely to be *excessive*, from a social point of view.

In a dynamic setting, innovations build on each other. Assume that one firm's technology is superseded by another firm's more efficient technology. The innovator may drive the incumbent out of the market (the incumbent being the losing "third party"), only to be replaced soon after by a competitor with an even better technology, *ad infinitum*. This leads to "creative destruction" in the sense of Schumpeter. Creative destruction may or may not give rise to excessive research. But beyond optimality considerations, creative destruction is a very interesting concept with important macro-economic consequences.

6 C Theory: The incentive to innovate

6 C a How information is produced

> Pigmies placed on the shoulders of giants see more than the giants themselves.[14]

The information production function

One element in the optimality criterion "marginal value = marginal cost" introduced in Chapter 5—the cost of information—crucially depends on the underlying *production function* of information. Here we present and discuss a general production function, which can capture important aspects of reality: It recognizes that innovations may build on the stock of a previously discovered piece of information or that—conversely—with each invention it may become more difficult to find radically new ideas. The production function also captures the fact that research groups can be too small or too large.

Let Q denote the level of knowledge, which is equivalent to the stock of previously discovered information. ΔQ is then the amount of new ideas discovered by research. If N_R is the number of research workers, we can write a general production function f of new information as follows:

$$\Delta Q = f(Q, N_R)$$

For expositional reasons, let us postulate a concrete production function as:

$$\Delta Q = \gamma N_R^{\lambda} Q^{\phi}, \tag{6.1}$$

where γ, ϕ and λ are constant parameters.[15] While γ is just a scaling parameter, the other two are economically important: ϕ describes the role of *the stock of ideas*, λ the impact of *the productivity of researchers*.

The stock of ideas

Let us first dwell on the role of the previously accumulated stock of information or ideas Q. We can distinguish three cases: If $\phi > 0$, research is more productive the higher the stock of previously discovered ideas. This positive externality is associated with the "standing on shoulders" effect, alluded to in the quotation at the beginning of this section. To take an example from the financial world: Assessing the quality of one debtor usually becomes easier with each debtor that has been examined beforehand.

On the other hand, $\phi < 0$ corresponds to the idea that the stock of potential innovations may be limited, or that the probability of discovering a new idea gets smaller with each discovery made. The higher Q is, the lower the discovery rate of new ideas, equivalent to a classic fishing-out case, in which the fish get harder to catch over time. Finding new prime numbers, for example, becomes increasingly difficult. For $\phi = 0$ the discovery rate is independent of the stock of previously accumulated information. This case can also be viewed as a combination of the previous two effects, which may offset each other perfectly. While it is easier to produce the most obvious pieces of information, the accumulation of this knowledge can also facilitate the search for more sophisticated ideas.

The productivity of researchers

The second important determinant in the production function of information is the number of workers allocated to research, N_R. If $\lambda = 1$, the amount of new information is proportional to the number of researchers. For $\lambda > 1$ there are increasing returns to scale, making two researchers more than twice as productive than one. On the other hand, if $\lambda < 1$, an increase in the number of researchers creates a negative externality often referred to as the "stepping on each others' toes" effect. It is equivalent to congestion on highways, for example.

In reality, the impact of the number of researchers on the productivity of research is usually not constant. There may be too few researchers until a critical mass is reached ($\lambda > 1$), but too many after a certain number ($\lambda < 1$). The high fraction of co-authored research papers in science and economics, for example, seems to suggest that there are increasing returns to scale for a small number of researchers, turning to congestion if too many work on the same paper.

Note that the current state of knowledge Q is usually given and can only be increased by producing new information ΔQ in the future. The number of researchers N_R, on the other hand, can be freely chosen. As we have seen before, optimality requires that the value of research equals its marginal cost. If workers can be employed both in research and in production, this also implies that the optimal number of researchers is reached when their marginal product in research equals their wage in the production sector.

6 C b Production and reproduction cost

The quotation from Shapiro and Varian (1999) at the beginning of this chapter reminds us of the importance of distinguishing between the *production* of new information and the *reproduction* of already existing information.

The cost of production

The *production* of new information (the exploration of an oilfield, the forecast of demand for a new product, the examination of a credit application) was discussed in the previous section. In reality it is often subject to some *fixed cost*. For example, $\hat{N}$ researchers may be required to build a laboratory, before research can actually start. In (6.1), the number of researchers in that case would enter as $N_R - \hat{N}$. Another reason for a fixed cost is the need for elaborate technical equipment (not included in the simple production function (6.1)). The fixed cost may be high or low, depending on the circumstances. Information about the dark side of the moon, for example, is not available without a considerable amount of prior investment. By contrast, a first approximate weather forecast is available from a look through the window. The *marginal cost* of producing new information is normally positive (except for those researchers who derive so much fun from thinking that they would even pay to do it). It may increase or decrease in the amount of information produced as we have outlined above.

As with other goods, the cost structure shapes the market structure. In particular, decreasing marginal cost and (by consequence) decreasing average cost lead to so-called natural monopolies. The high degree of concentration in the credit ratings industry may be the consequence of marginal cost decreasing over a large number of rated firms; the fact that there is more than one agency suggests that, after some point, the marginal cost may increase, thus leading to a "natural oligopoly" (see 6 E).

The example of the ratings industry also illustrates that the notion of "more" information may have different dimensions. It is not the same if a rating agency issues a rating for an additional company or if it improves the precision of an existing rating. The marginal cost of rating one more company is likely to decrease; improving the precision of an existing rating is likely to be subject to an increasing marginal cost.

The cost of reproduction

The *reproduction* of information, as distinct from its production, has some very special features. The existence of a *fixed cost* is nothing extraordinary, but the nature of these fixed costs deserves a closer look. The reproduction of information (included its transmission) requires a physical and a virtual infrastructure, namely: (i) a copy and transmission mechanism (copy plus fax machine plus telephone lines; the internet, i.e. cables, servers, numerous local computers, and last but not least a bunch of satellites) as well as (ii) a language (like English) plus a protocol (like the Hyper Text Transfer Protocol), shared by both sender and recipient alike.

The *marginal cost* of reproducing information is close to zero. Once a network in its physical and virtual dimension is set up, the marginal cost of reproducing information is very low. This is true even for very large numbers of copies. Today, no other good is as easy and cheap to reproduce as information.

The impact of low marginal cost

This has important consequences. First, marginal cost pricing will not work for information, just as it does not work in pricing seats on a train. A producer charging a marginal cost would make a loss, since the price would reflect the cost of the "last" unit, which is zero. Information or information goods thus can only be profitably supplied (i) by a monopoly or a firm dominant enough to earn some monopoly rents like Microsoft, or (ii) as a complement to some other good which exhibits a more "normal" cost function. If none of these cases prevails, an "alternative" funding for information production must be found, like letting the debt issuer pay for a credit rating or allowing advertisements on one's home page.

Second, the market price of a piece of information falls to zero very quickly. Anybody with a piece of information would try to sell it quickly, before others would offer it at an even lower price. That is why gossip travels so fast. And, that is why there are patent and copyright laws which try to define property rights for information. However, property rights for information are hard to define, as neither information itself nor its transfer are visible. The internet, which can be seen as the most efficient copy machine ever, is the stage for a continuing race between copyright law and defensive technology (like the self-destructing disc) on one side, and copying technology on the other. Even the suppliers of software for illegal copying are using ingenious devices to protect *their* software from being copied.

Third, information is not only cheap to reproduce, it can also be used by many people at the same time. A weather forecast, for example, reminds me to take my umbrella, just as it reminds other people to take theirs. If I do not suffer from other people carrying umbrellas (I might even benefit from a reduced risk of umbrella theft!), information becomes a public good. This will be discussed in the next section.

6 C c Information as a public good

The properties of a public good

Information is not only cheap to reproduce. The "owner" of a piece of information often finds it difficult or impossible to exclude others from acquiring, reproducing and disseminating it. Information thus shares (to some degree) one of the two properties that define a public good: *non-excludability*. The other property of a public goods is *non-rival use*. It may be consumed or used by one person without preventing another person from also using it at the same time (a broadcast radio program, for example). Information does not necessarily share this second property. True, several people may laugh about the same joke at the same time; but two competitors in the product market cannot make use of an innovation, like a new product, without diminishing each other's profits.

The underproduction of information as a public good

Even if the use of a piece of information does not hurt other users, the private production of information may fall short of its socially optimal level. The reason is the non-excludability of information; it may be cheaper to free-ride on others' information production than to incur the costs of producing it. The problem is that everyone may think alike. As a consequence, the good will not be provided in the quantity everyone would agree to if people could commit. The tendency towards underproduction increases with the number of potential contributors. With a large number of potential contributors, each individual's contribution does not raise the level of the supply of a public good by more than a marginal amount. The incentive to contribute thus is minimized.

A simple model

The following model formalizes the underprovision of information as a public good. Assume that there is a given number $N = \bar{N}$ of individuals who independently produce information by engaging in research. The information production is a special (static) case of (6.1) with $\phi = 0$ and $\gamma = 1$. We interpret the term N_R in (6.1) as research effort (like research hours), rather than as the number of researchers. If all N researchers were to exert the same effort, we could write total effort as $N_R = Ne$, thus splitting total effort N_R into (i) effort per head e and (ii) the number of individuals in research N. If efforts may differ across individuals, aggregate effort has to be written as $N_R = \sum e_i$.

The aggregate effort produces new knowledge q (shorthand for ΔQ). New knowledge q is a public good which benefits everyone. Individual utility is the difference between new knowledge, q, and the disutility from individual effort. We assume that disutility increases with the square of effort and is scaled with a constant factor θ; thus:

$$q = \sum_{j=1}^{N} e_j$$

$$U_i = q - \frac{1}{2}\theta e_i^2$$

It is important to be precise about the relative roles of an individual's decisions and everybody else's decisions. In the individual utility function, the positive term is the sum of knowledge, i.e. the sum of all individual research efforts. This term reflects the public good character of information: Once gathered, it benefits everyone. The subtracted term is the individual cost of research reflecting only one individual's effort (not everyone else's effort).

We are now ready to compare the social optimum with the individual optimum. The social optimum is the choice of effort level(s) that would result if a social planner could decide in the place of individuals. The private optimum is the effort level of an individual that results from individual choice, given that all other individuals chose their individually optimal level.

The social optimum

We first look at the *social optimum* as a benchmark against which the private optimum can be measured. What effort level(s) would a social planner chose? As individuals are identical, we restrict our attention to symmetrical solutions in which all individuals contribute the same level of effort $e_i = e$. Let us further assume that the planner defines social welfare as the sum of individual utilities. Maximizing social welfare under these assumptions is equivalent to maximizing welfare *per head*, or—equivalently—of a *representative individual*. The planner thus chooses e to maximize per head utility U_s:

$$\max_{e} U_s = Ne - \frac{1}{2}\theta e^2.$$

Taking derivatives with respect to e yields the first-order condition for a maximum:[16]

$$\frac{\partial U_s}{\partial e} = N - \theta e = 0.$$

The optimal level of effort from a social point of view, e^*, solves:

$$e^* = \frac{N}{\theta}.$$

The socially optimal effort level reflects the fact that an increase in effort raises the welfare per head N times, as e refers to the research effort of *all* individuals.

The privately optimal research effort

The individual, deciding simultaneously and in isolation from other individuals, only cares about his own effort. The individual solves:

$$\max_{e_i} U_i = \sum_{j=1}^{N} e_j - \frac{1}{2}\theta e_i^2.$$

The first-order condition for a maximum:

$$\frac{\partial U_i}{\partial e_i} = 1 - \theta e = 0,$$

reflects the optimality criterion marginal value = marginal cost, or $1 = \theta e$, which yields the optimal level of effort from the individual's private perspective:

$$e^{**} = \frac{1}{\theta}.$$

Obviously the privately optimal effort falls short of the social optimum:

$$\frac{1}{\theta} = e^{**} \ll e^{*} = \frac{N}{\theta}.$$

With an increasing number of individuals, N, the difference between the private and the social optimum becomes considerable. As mentioned above, this difference reflects the fact that the individual only takes into account his own impact on the pool of knowledge (available to him), whereas the social planner takes into account all individuals' contributions to the pool of knowledge (available to everybody).

Reaction functions

In the present case, the solution for the optimal individual effort is drastically simplified by the *additivity* of efforts e_j. The optimal effort of one individual is *independent* of the effort levels chosen by others. The function describing the optimal response (here: choice of effort level) of one individual to any action (again: choice of effort level) of another is normally called the individual's *reaction function*. In the present case, the reaction functions—for example $e_1^*(e_2)$ in the 2-person case—are straight lines parallel to the axes. Under alternative assumptions regarding the interaction of individual efforts, the problem becomes more difficult. For example, convex reaction functions may lead to multiple equilibria, i.e. to coordination problems (Chapter 9).

6 C d Research as a public right

An exhaustible pool of inventions

Here we will formalize the idea that there may be excessive information production, because everybody has the right to research. In the language of Hirshleifer and Riley (1992) this is the right to fish (as a verb), as distinct from the right in (caught) fish. A pool of undiscovered ideas thus is like a pool full of fish, and research is an attempt to catch one after the other. The assumption behind this view is that fish caught becomes private property. The inventor owns an invention, either because it can be kept secret, or because it can be protected by a patent. Whatever one inventor "catches" reduces the available inventory of ideas that can be appropriated by other inventors.

It is easy to see why under these circumstances there is an excessive incentive to fish in the common pool of undiscovered ideas. The over-use of a public resource is the mirror image of the under-production of a public good. An individual, facing the choice to enter the race for innovations or not, only looks at her private benefits. These are equal to the *average* benefits of all other participants (assuming these are all identical). Their *marginal* benefit from one additional participant, however, is negative since any additional participant imposes a negative externality to all others.

A simple numerical example may illustrate the negative externality.

A new local restaurant offers a luxurious free dinner worth $100 to the first guest. If only Alice and Bob compete to be first, they each have a chance of 1/2 to win the free meal. If Clara also decides to join the competition, the individual odds fall to 1/3. Clara's benefit from participating thus is equal to the *average* expected benefit for all participants of $33.33. . .. What Clara disregards is that from a social point of view her individual (marginal) benefit from joining is compensated by a reduction in Alice's and Bob's expected benefits (by $16.66. . . for each).

In a competition in which participants hope to appropriate a public resource privately, their marginal benefit from a further participant thus is equal to the average benefit for all participants, not to the (smaller) marginal social benefit. Sending one more sheep to the commons pays off as long as the animal finds something to eat (the quantity being equal to the average benefit for all participants), even though the total amount of grass remains constant at best (the marginal social benefit being zero).

The tragedy of the commons

The over-consumption of a public good is the mirror image of its underproduction. It leads to the "commons effect", named after the famous "tragedy of the commons" in nineteenth-century Britain.[17] The historical commons problem was particularly severe, because with each new entrant (grazing sheep) the social returns from the commons decreased in absolute terms. Our free dinner example is a mild case of a commons problem. Entry of further participants does not change the value of the common good, it only redistributes it among more people.

Hirshleifer and Riley (1992) distinguish two cases of such fishing-like competitions for new ideas: (i) searching among a large number of small ideas and (ii) "the Quest for the Holy Grail" in the sense of an attempt to find one big innovation. The conceptual difference between the two cases is small, though. We will try to treat them together.

Fishing in the pool of inventions

To keep things simple we look at a benchmark case. Assume that there is a pool of potential inventions which will become known some day. Any research effort is lost from society's long-term point of view (with a positive rate of discount, society would benefit from early discovery). But from an individual point of view, research efforts are not lost: Inventors appropriate inventions on a first-come-first-served basis, by taking out a patent for each invention, for instance.

Again we use a special static version of the information production function. The main difference to (6.1) is that both the cost and the benefit of research accrue on an individual basis. A researcher's effort benefits only himself, but hurts others (by "stealing" inventions). As an increase in the stock of knowledge only accrues to the inventor i (rather than to the public), we can write the additional knowledge ΔQ as q_i. Further, we use the researcher's own effort e_i as the individual counterpart for the aggregate number of researchers or research hours N_R in (6.1). The previous stock of knowledge Q^{ϕ}—the sum $\sum e$ of innovations already made—here has a negative impact on further innovation (the more innovations are already made, the less remain in the pool of undiscovered ideas).

Writing the production function (6.1) with these adaptations and setting $\gamma = 1$, $\lambda = 1$ and $\phi = -1$ yields the following specification:

$$q_i = e_i \left(\sum e_j\right)^{-1} = \frac{e_i}{\sum e_j}.$$

The inventions that a researcher i finds are a reflection of her share in the total research effort. Individual utility is equal to inventions found minus the research cost. We assume that the research cost, θe_i, is linear with respect to effort. Therefore:

$$U_i = q_i - \theta e_i.$$

The social optimum

In the social optimum, by assumption, nobody is doing any research at all. A social planner would maximize per capita utility:

$$U_s = \frac{e}{\sum e} - \theta e = \frac{e}{Ne} - \theta e = \frac{1}{N} - \theta e,$$

which is obviously maximized by setting $e = 0$.

Although research has no productive role here, its redistributive effect creates an incentive for positive individual research efforts.

The private optimum

Individual researchers, indifferent to social welfare, maximize:

$$U_i = \frac{e_i}{\sum e_j} - \theta e_i.$$

The first-order condition for an optimum is:

$$\frac{\partial U_i}{\partial e_i} = \frac{1}{\sum e_j} - \frac{e_i}{(\sum e_j)^2} - \theta = 0.$$

In a symmetric equilibrium $e_i = e$ and $\sum e_j = Ne$. This leads to the solution for privately optimal effort:

$$e^{**} = \frac{1}{\theta}\frac{N-1}{N^2}.$$

An individual's privately optimal research effort decreases with the number of competing inventors (if $N > 1$). Aggregate research effort Ne^{**}, however, which measures the aggregate social loss, increases with the number of competing inventors:

$$Ne^{**} = \frac{1}{\theta}\left(1 - \frac{1}{N}\right).$$

Research: production or stealing?

The above model describes a benchmark case, as we mentioned at the outset. Individual efforts of competing inventors are lost for society as their impact is canceled out in the aggregate. Admittedly, this is not very realistic. In reality, individual research efforts have two effects. They may increase the pool of knowledge (production) and they may "steal" the right to use knowledge from competitors who come too late to win the patent (redistribution). Which one of the two effects dominates (net of research costs) cannot be decided a priori.

Hirshleifer (1971) distinguishes between the discovery of technological processes (like supra-conductivity at room temperature) and foreknowledge of future states of nature (like the winner of the next Melbourne Cup). He claims that the first is relevant for production, while the latter is only relevant for (re)distribution. In Hirshleifer and Riley (1992, p. 260) the redistribution effect is called the "speculative" effect.

6 D Application: Creative destruction

> But in capitalist reality as distinguished from its textbook picture, it is not that kind of competition which counts but the competition from the new commodity, the new technology, the new source of supply, the new type of organization (the largest-scale unit of control for instance)—competition which commands a decisive cost or quality advantage and which strikes not at the margins of the profits and the outputs of the existing firms but at their foundations and their very lives (Schumpeter 1947, p. 84).

The entrepreneur: creator and destroyer

Joseph A. Schumpeter developed an evolutionary theory of capitalism, marked by periodic change driven by technological innovation. The replacement of obsolete production technologies by new ideas led to the best known theory by Schumpeter: The process of creative destruction. New information is destructive as it destroys the value of the previous information for some firms, but it is creative in as far as it generates economic growth. The (capitalist) entrepreneur is always at the center of Schumpeter's analysis. She decides optimally on the level of research and the production of final goods.

Schumpeter's original argument is purely verbal and expressed in a manner that is not easily accessible to modern students of economics. This may also be the reason why it lay dormant for so long. Pioneering work by Aghion and Howitt (1992) resuscitated Schumpeter's ideas[18] and transposed them using a modern formal approach. As the latter is beyond the scope of this book—mainly because intertemporal issues tend to become complicated very quickly—, we attempt to take an intermediate approach and formalize as much as we can to make the argument clear, without introducing too much technique.

Building blocks

The main building blocks of the Aghion and Howitt's model, but also in Schumpeter's original work, are the following:

- Innovations are not gradual, but occur in steps. New information can be perceived as the base for an all-purpose technology used for the production of final consumption goods.
- Innovations are additive in the sense that discovery $n+1$ can only be made after innovation n, but not directly after $n-1$. This also implies that an innovation stays valuable forever, as it serves as a stepping stone for all subsequent innovations.
- The probability of discovering a new technology is proportional to the amount of research activity.[19] The intervals between two successive innovations are thus stochastic.
- Labor can be used for both research and the production of the monopolist good. The labor market is competitive.
- The discoverer of a new innovation takes out a patent for exploitation in a monopolist way. The duration of the patent is not limited, but to the entrepreneur the patent is valid only as long as it is not replaced by a better technology. As in real life, many patents become obsolete long before the official patent period expires.
- A patent implies the disclosure of the relevant information. The information can be used immediately by researchers for further research, but the patent holder has sole production use.

Equilibrium conditions

Aghion and Howitt nicely demonstrate that these assumptions result in two simple equilibrium conditions:[20]

1. *Total labor force = Number of researchers + Number of production workers*

   As workers are *ex ante* identical and the labor market is competitive, this means that the wages in research and production have to be equal in equilibrium.
2. *Marginal research costs = Expected future marginal profits from an innovation*

   If the research sector is competitive, the hiring of an additional researcher is profitable as long as the expected return of the research exceeds his cost. Equivalently, the marginal return to an additional researcher has to be equal to his wage.

The need for capitalists

Let us look again at the second condition for a moment. An entrepreneur has to trade off the cost and benefits of research efforts. The optimizing calculus of the researching entrepreneur thus reads as follows: The cost of research (= labor cost) must equal the expected benefit from being granted the patent for the new innovation. The size of the expected payoff from research, in turn, is driven by the probability of success, the size of the anticipated monopoly profit in each period, and the expected length of the patent, i.e. the expected time until the patent becomes obsolete by the discovery of a new innovation. There are two important implications of this optimization. First, at any point in time, an entrepreneur either produces or engages in research. The incumbent monopolist does not do research, as we will argue below. Second the cost of an innovation accrues before its benefits materialize. A researching entrepreneur either has deep pockets or access to a perfect capital market. Schumpeter (1939, p. 223) describes this implication as follows:

> We have to define that word which good economists always try to avoid: Capitalism is that form of private property economy in which innovations are carried out by means of borrowed money.

For queens or girls?

For the sake of argument, let us assume that the relevant process innovation is the marginal cost (other than wage costs) of output production. New information means the discovery of a lower production cost, $C_{n+1} = C_n - \Delta C$, for the same good, a notion very close to Schumpeter's original perception.

> Queen Elizabeth owned silk stockings. The capitalist achievement does not typically consist in providing more silk stockings for queens but in bringing

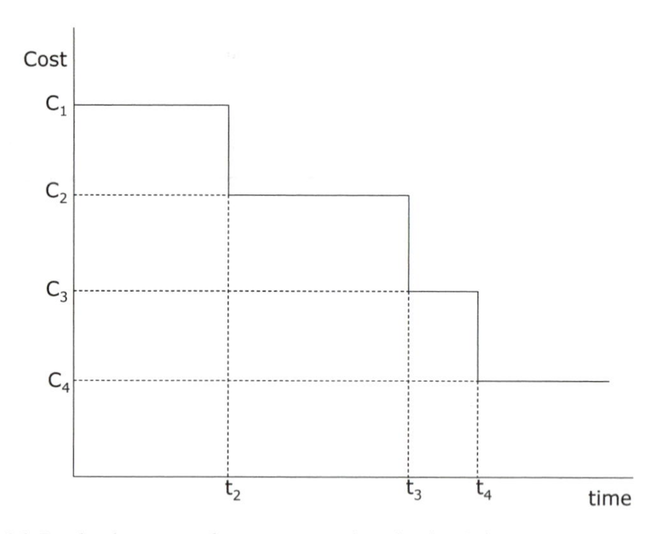

Figure 6.1 Production costs decrease over time in the Schumpeter creative destruction model with process innovations at t_2, t_3, and t_4.

> them within the reach of factory girls in return for steadily decreasing amounts of effort.
>
> (Schumpeter 1947, p. 67)

The possible evolution of production costs over time is represented in Figure 6.1. Note that the intervals between two innovations are not necessarily equal, but depend on the amount of research and stochastic factors such as luck.[21]

Market structure

Recall that the researcher who discovers the new information is granted a patent, establishing a (temporary) monopoly on the production of the relevant good. The monopoly is a source of profit. The famous profit maximizing combination of price and quantity, known as the Cournot solution to monopoly pricing, is derived in Box 6.2.

Box 6.2 Maximizing monopoly profit

In the figure, a supplier of a good with constant marginal production cost c faces a downward sloping aggregate demand function $p(X)$ represented by the solid line in the figure below. Under competition, the supplier would charge a price equal to marginal cost, $p_c = c$. At such a price, the supplier earns zero profit, but welfare—the sum of consumers' and producers' rent—is maximized.

A monopolistic supplier maximizes profits by selling a quantity at which marginal revenue ($\partial p(X)X/\partial X$) is equal to marginal cost. In the figure, this condition is represented by point A. Monopoly thus leads to a higher price and to a smaller quantity than competition. The reason is that a monopolist, unlike a competitive supplier, takes into account that an increase in the quantity sold brings down the price.

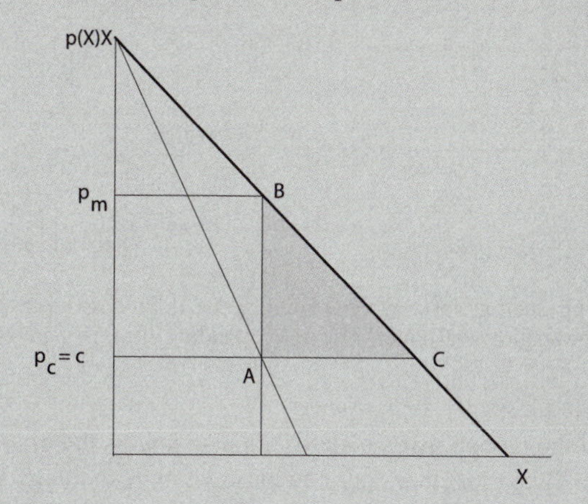

Monopoly pricing implies an inefficiency. The potential gains from trade are the sum of producers' and consumers' rent under competition, i.e. the triangle under the solid line above point C. Under competition the aggregate rent is earned by consumers. Under a monopoly consumers keep the triangle above point B, while the monopolist appropriates the rectangle left of A–B. The shaded triangle is lost. Some consumers do not get the product, even though they would have paid more than the marginal cost necessary to produce it. The monopolist dissipates a part of the consumers' rent in order to appropriate a higher share of the total surplus. This is the deadweight cost of monopoly.

Figure 6.2 depicts the development of the price–quantity policy by a monopolist using the latest innovation in the production of some good. A newer (better) technology leads to a fall in marginal cost; over time, marginal cost falls in three steps from c_1 to c_4. The quantity produced by the monopolist can be read off at the intersection of the marginal revenue curve and the marginal cost curve relevant at the time. The corresponding price is the price that equalizes supply and demand at the monopoly price. The shaded rectangles in Figure 6.2 represent the surplus (net benefit) earned by the monopolist at different levels of marginal cost. An innovation, i.e. a downward movement of the marginal cost function,

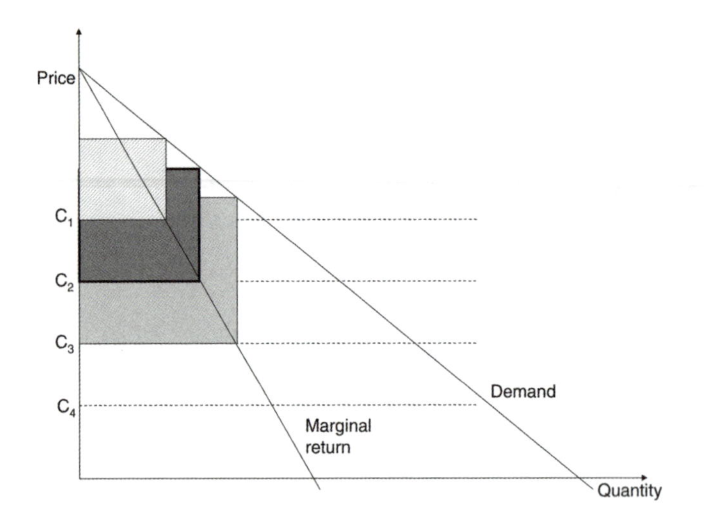

Figure 6.2 As the marginal production costs become smaller due to process innovations, the monopoly power grows stronger. The corresponding surpluses are shown by the squares.

increases the quantity and the monopolist's benefit, but also lowers the price and increases consumers' rent. Taken together, any new production technology leads to economic growth.

An example

To give a concrete example, let the demand curve be represented by:

$$p(X) = 1 - X.$$

The equilibrium monopoly price and quantity are therefore (see Box 6.2) $p = (1 + C)/2$, and $X = (1 - C)/2$, respectively, and the monopolist's profit is $(p - C)X = \{(1 - C)/2\}^2$. The anticipated monopoly profit is the reason why an entrepreneur without a valid patent will want to carry out research. There is also a direct benefit from an innovation to consumers, although it is arguably smaller than under perfect competition. The consumer surplus is the triangle just above the monopolist's surplus. In our example, it amounts to half the monopolist's surplus. Production innovation also leads to a higher consumer surplus, an effect that is not taken into account by the optimizing researcher, though.

A technical issue

There is a somewhat tricky issue, as in our simplified example, that it may still be profitable to produce for the previous monopolist. We can assume that other forces drive the previous incumbent out of the market. Or one can think of an alternative market structure induced by a regulated monopoly, for example. Assume that the

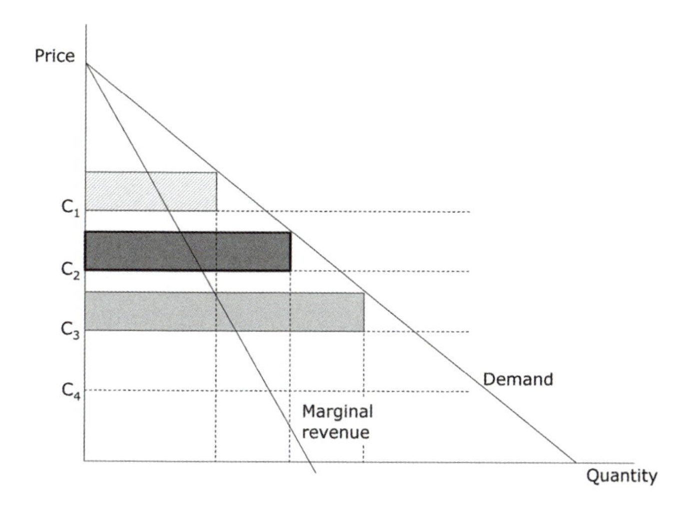

Figure 6.3 In a regulated monopoly, the monopolist does not receive the whole surplus but only a markup yielding horizontal bars instead of squares.

holder of the patent is granted a constant mark-up $\mu < \Delta C$ over the production cost C, but not a full monopoly as before. Instead of the relatively large monopoly squares in Figure 6.2, we would have smaller profit rectangles, as represented in Figure 6.3. In this case, the incumbent producer would be driven out of the market automatically. But as above, an innovation leads to an increase in quantity and in the monopolist's profits, but also to lower prices and to an increase in overall welfare.

Lazy monopolists . . .

An important implication of the Creative Destruction model is that the incumbent monopolist has no incentive to do research. Let us see why. Assume we are in regime C_1, meaning that the current monopolist earns a gross profit (before wage payments) of $\{(1 - C_1)/2\}^2$ in each period. If she reallocates one worker to do research, she would forego a fraction of this monopoly benefit in exchange for a share of the future innovation's expected value $\{(1 - C_2)/2\}^2$. In contrast, entrepreneurs who do not hold the current patent only "pay" the wage w for the expected future profit $\{(1 - C_2)/2\}^2$. If the labor market is competitive, the expected future benefit is equal to the labor cost of research, implying that the incumbent monopolist would lose out by doing research.

. . . but motivating monopoly

A second implication is that product market competition is unambiguously bad for research incentives and for economic growth. The smaller the anticipated return from a patent, the less profitable the research and, as a consequence, the less manpower is devoted to the discovery of new information. Comparing the two market structures depicted in Figures 6.2 and 6.3, entrepreneurs would invest more effort in research in the former due to the higher expected payoff in the full monopoly case. In the long run, the higher distortions in the former case will be compensated by faster technological progress. Nonetheless, the impact on society's welfare is a priori ambiguous, as a higher growth rate also entails a greater loss for incumbent producers.

Welfare considerations

The key question then is whether the amount of research carried out in the economy is optimal from society's point of view. Let us revisit the different effects analyzed in this chapter:[22]

Knowledge spillover effect: When an entrepreneur considers doing research, she trades off the cost of research with the expected benefit from obtaining the patent. Here, she only considers the expected length the patent can be exploited for, before a new innovation is made. A social planner, on the other hand, would recognize that any new innovation increases the social welfare for all future points in time, as it lays a new foundation on which other discoveries can be made. For society as a whole, new information stays valuable forever, as it provides the basis for future innovations. As the entrepreneur does not take the spillovers of her research into account, too little research is carried out in equilibrium.

Appropriability or consumer-surplus effect: A new innovation not only results in a higher profit for the patent holder than for the previous innovator, but it also leads to a higher consumer surplus.[23] The total increase in societal welfare is thus greater than the benefit of the monopolist. A potential innovator does not take this into account. As before, *laissez faire* results in too little research. The more product market competition depresses the patent holder's surplus, the more the incentive to innovate and, as a consequence, economic growth is reduced.

Business-stealing effect: We have seen above that the incumbent monopolist does not engage in research. The expected return from one unit of research is much smaller to her than to a research competitor without a valuable patent. The incentive to do research lies solely in the value of the future patent (the future monopolist's rectangle), and does not internalize that an innovation also results in a loss to the incumbent monopolist. From a societal point of view, stealing business results in too much research.

Taken together, the net effect of these three forces is unclear. While the spillover and appropriability effects lead to a suboptimally low level of research, business stealing alone leads to an overprovision of research efforts. The optimal policy of a government can thus either be a subsidy for research (underprovision) or a tax on research activities (or the monopolist's profits). Most empirical studies find that the business-stealing effect is too small to offset the spillover and the appropriability effect. The main reason for this finding is that business stealing is rarely perfect in reality. The holder of an outdated product still has some market share, and is not totally replaced as in our stylized model.

Two faces of monopoly

The most important message of the setup is—as Schumpeter clearly pointed out—that short-term market distortions due to patent monopolies may be more than offset by the beneficial impact on economic growth in the long term. A good example is the production of medicines against the many illnesses still around. While it is tempting to regulate the price of the few drugs that are effective against avian flu in the wake of a potential pandemic, such a measure may lessen the incentive for the chemical industry to find effective vaccines against a future virus. In the same spirit, the desperate efforts of some developing countries to get a grip on AIDS by copying drugs designed to treat this devastating illness may seriously harm the research of pharmaceutical firms to develop new, and potentially cheaper, cures. A currently used strategy to serve the needs of the poor without harming the research efforts too much is price discrimination between countries.

6 E Application: Rating agencies

Rating agencies are important producers and disseminators of financial market information. The amount of debt outstanding rated by Moody's, Standard&Poor's, Fitch and some 150 smaller competitors according to most estimates is close to 50 trillion US-Dollars.[24]

Valuable information . . .

The ratings industry is a perfect illustration of the economics of information production and dissemination. Ratings information is potentially valuable. These days, few investors subscribe to a bond issue before checking that it has a satisfactory rating. Many mutual funds and pension schemes, by choice or by law, restrict their investments to bonds of "investment grade", i.e. bonds with a minimal rating such as a *BB* from Standard&Poor's. The Basel Committee on Banking Supervision (a body of central bankers and bank supervisors meeting at the Bank of International Settlements in Basel) in its new Capital Accord recommends that banks set aside more capital against credits to borrowers with low or no ratings than for top-rated borrowers (see BCBS, 2002).

. . . but hard to sell

Credit ratings, despite providing valuable information, are hard to sell. Typically, the large ratings firms do not, and probably could not sell their information to investors. Why not? The reason is that the equilibrium selling price for rating information is zero! Assume that rating agencies would indeed sell their ratings to investors. Each recipient of rating information could secretly sell the information to other investors for a somewhat lower price, and so on. The original producer of information thus would see his price underbid by the first purchasers, and even more aggressively by further purchaser. The price for the rating information would quickly fall to zero. The ratings industry is in a similar position as the music industry would be in the absence of copyright protection.

Fortunately for the rating firms, bond issuers are ready to pay for their ratings. A borrower of good quality is eager to signal his quality to the market by acquiring a rating. Unrated debt thus is suspected of being of inferior quality. This forces companies with less than top grade debt to obtain a rating, too. At the limit, all but the worst junk debt would be rated.[25]

The impact of technology

For the reasons explained, ratings firms bill the issuers, rather than the investors. This has not always been the case, though. In 1909, when John Moody first issued a book of annual railroad-bond-ratings, subscribers, not borrowers, paid for the ratings. What has changed since 1909? First and foremost, communication technology. The first ratings came as heavy books full of dense print. Photocopy machines did not exist, and the most efficient means of communication, the telegraph, was fast, but only for relatively small amounts of data. Today, information is copied with two clicks of a mouse and travels over the internet at great speed and in large amounts. Unlike in 1909, when the subscriber found it almost impossible to pass on ratings information, ratings today are public knowledge immediately upon their release, whether the raters like it or not.

But even today, there are a number of firms, mainly smaller ones, that sell debtor assessments directly to investors.[26] The public good problem may thus not always prevent debtor assessment specialists from selling their information. Assessments sold to investors often come with buy and sell recommendations. Investors may be afraid of reselling information to competitors who might snap away profitable investment opportunities.

Notwithstanding the fundamental difficulties of selling a public good, the ratings industry is thriving. Their aggregate gross revenue is an estimated 5 billion US-Dollars per year. One reason for the raters' profitability is the issuers' strong demand for ratings. Yet there is another reason, again rooted in the economics of information production.

Market structure

The ratings industry is heavily concentrated. The market is dominated by three big firms: Moody's, Standard & Poor's (S&P), and Fitch. More recently, a number of smaller contestants mostly on a national level have emerged, without really challenging the big three's oligopoly. This oligopoly reflects economies of scale in information production. First, a borrower assessment is much more efficiently done by one specialized agent who shares the result with many investors than by individual investors separately duplicating each other's efforts. Second, it is very difficult, if not impossible, to assess the quality of one single bond issue. A rater who has followed an issuer over time has a natural advantage of rating just one more issue by the same debtor. In addition to these intertemporal economies of scale there are cross-sectional economies of scale. It is much easier to assess the quality of a chemical firm after assessing some of its competitors beforehand. Finally, rating a chemical firm may provide some relevant experience for rating a car producer or even a bank. Rating thus comes close to a "natural monopoly", or at least a "natural oligopoly". Under the prevailing oligopolistic structure, competition does not seem to be cut-throat (although in principle, even with only two competitors, it might be); rating firms thus have lucrative franchises.

High or low profits?

Or so it may seem. The industry's profits are less impressive when compared to the aggregate value of rated bonds. Even given generous assumptions, the rating agencies' aggregate gross revenue does not exceed one basis point of the amount of rated debt. This is an almost negligible fraction of the aggregate interest a company pays on a bond (e.g., 0.2 per mil of the aggregate interest on a ten-year bond paying 5 per cent p.a.). Is it plausible that rating information is only worth such a tiny fraction of the interest cost of a bond?

There are three possible answers: (1) the rating industry's revenue may underestimate the value of rating information, as competition among the few big players is fiercer than many observers think; (2) revenues only include the value of a rating to the rated company, while the value to investors (who get ratings for free) is not priced; (3) the predictive power of ratings and hence their value may be weak. The latter hypothesis is discussed in White (2001). Indeed, ratings are well correlated with average default rates. However, this does not mean that they convey any information not yet available to the market in the form of, say, interest spreads on the rated bonds. Although there are indications that rating changes provide some additional information, it is not clear how valuable this information really is (White, 2001).

6 F Application: Why are banks supervised?

In most countries banks are subject to special regulations and are supervised by a special government agency. Why? Dewatripont and Tirole (1994) start their economic analysis of prudential regulation of banks with this basic question: What makes banks so special that they need to be regulated by law and monitored by a special supervisor?

Traditional arguments

First, Dewatripont and Tirole (1994) list the arguments in favor of the regulation of banks (and similar intermediaries like insurance companies or pension funds) found in the literature:

- Banks' transformation function (borrowing short, lending long);
- Banks' role in payments systems;

- Banks' high leverage;
- The existence of deposit insurance.

Then, Dewatripont and Tirole (1994) discard each one of these arguments after the other. The transformation function is not specific to banks. A company that finances long-term research projects with short-term debt also transforms maturities and risks. Payment systems may need protection (for example, in the form of a central bank providing liquidity at peak times); but they are no reason to protect holders of banks' time deposits. Banks' high leverage is endogenous, perhaps reflecting banks' specific assets. Deposit insurance is one form of regulation, not a rationale for regulation.

The representation hypothesis

Finally, they present their own—the only surviving—rationale for bank regulation and supervision. Banks, to begin with, are companies. As such, they are subject to substantial problems of adverse selection (Chapter 13) and moral hazard (Chapter 16). Normally, investors (shareholders and creditors) perform various monitoring and corporate control functions. Not so with banks:

> Bank debt is primarily held by small depositors. Such depositors are often unsophisticated, in that they are unable to understand the intricacies of balance and off-balance sheet activities. More fundamentally the thousands of customers of a bank have little individual incentive to perform the various monitoring functions. This free-riding gives rise to a need for private or public representatives of depositors.
>
> (Dewatripont and Tirole 1994, pp. 31–32)

This is what Dewatripont and Tirole (1994) call the *representation hypothesis*.

Their argument exactly reflects the above argument (modeled in Section 6 C c), namely that information (here the insight from bank monitoring) is a public good. This public good characteristic leads to free-riding and underprovision. Bank regulation thus tries to correct a failure of corporate control: A serious lack of creditor discipline.

Of course, Dewatripont and Tirole (1994) are aware that the "nature" of banks—being companies with a large number of unsophisticated and small depositors—itself is endogenous. They cite three reasons why so many people hold bank deposits, rather than, for instance, bank shares (Dewatripont and Tirole 1994, pp. 34):

- There is a pyramid of "delegated monitoring" (Diamond, 1984): Non-financial firms selecting and monitoring projects, banks monitoring non-financial firms, bank depositors (read: their representative) monitoring banks.
- Bank deposits insure households against unexpected liquidity needs. Having a fixed value, deposits are safe from the risk of dealing with an insider.

- Deposits (like banknotes) are efficient means of payment, since the recipient does not need to assess their value before accepting payment.

Assumptions

The Dewatripont and Tirole representation hypothesis rests on the assumption that individual bank monitoring is either completely useless or at least undersupplied because it provides a public good. The public good assumption is plausible if at least one of two assumptions holds: (i) depositors can mutually observe the result of their individual monitoring activities; and (ii) monitoring efforts have a positive impact on a bank's solvency. For example, bank managers are more careful in their lending decisions when they feel that depositors are looking over their shoulders.

Yet, information gathered by one depositor, to some degree, may also be a private good which does not automatically benefit all other depositors. Private information of one depositor might even hurt others. If an informed depositor is first to withdraw money upon bad news, less money is left in the bank for all others. This may even lead to excess monitoring (Section 6 C d) or to bank runs (Section 9 D). An interesting intermediate case occurs when depositors cannot observe the result of each other's monitoring efforts, i.e. each other's information, but only each other's actions: Whether others stand in line to withdraw or not. This may lead to a cascade (Chapter 10).

Finally, individual monitoring efforts may produce a private good to the respective investors but still become available to all others. This is the case when individual pieces of information are aggregated in the market. The successful finder of new information is rewarded with a trading profit; all others benefit from asset prices reflecting all available information (Chapter 7).

While some assumptions behind the Dewatripont and Tirole representation hypothesis may need further discussion, the hypothesis is appealing as a simple and economically sound argument in favor of bank regulation and supervision.

6 G Conclusions and further reading

It is hard to say whether in the "information" society we suffer from too much or too little information—or from too much lousy information. There is a tension between the production and the reproduction or dissemination that may tilt the outcome towards either side. As information can be reproduced at low or zero cost, each piece of information should be produced only once by a specialist, who then distributes it freely to somebody else. The only problem is that the specialist may not have any incentive to incur some costs in order to produce a gift to society. Creating such incentives is possible through patents, for example. The race for patents may, however, lead to too much research (and to too little dissemination). Alternatives to patents in stimulating inventions are government grants, research awards, as well as procurement or contractual mechanisms (Scotchmer, 2004).

We have not covered all aspects of information production and dissemination in the present chapter. Important issues we could not discuss sufficiently are industrial organization aspects and network effects. For further reading we recommend the following texts. A standard text on the economics of information production is Hirshleifer and Riley (1992, Ch. 7). Industrial organization aspects are covered in Tirole (1988). A sound business economics introduction to strategic issues in markets for information goods is Shapiro and Varian (1999) who also refer to network effects, an important topic not covered in the present volume (except as a special case of strategic complementarities, see Chapter 9). A more specialized introduction to network economics (covering much more than informational aspects) is Shy (2001). An overview on the economics of patents and intellectual property rights in general can be found in Scotchmer (2004). The economics of open source software production, finally, are discussed in Myatt and Wallace (2002) as well as in Von Krogh and Von Hippel (2003).

Checklist: Concepts introduced in this chapter

- The information production function
- Production versus reproduction of information
- Average and marginal cost of information (re)production
- Information as a public good
- Underproduction of information
- Overproduction of information
- Private versus social optimum
- Creative destruction
- Rating agencies and the market price of rating information
- Bank supervision and the representation hypothesis

6 H Problem sets: Produce or copy—sell or give away?

For solutions see: www.alicebob.info.

Problem 6.1 A joke shortage?
Are there too many jokes or too few?

Problem 6.2 The equilibrium price of information
Assume that the transmission of information from one party to another is (i) costless and (ii) unobservable to third parties. What is the equilibrium price of information under these assumptions?

Problem 6.3 Is information a public good?
Is information a pure public good?

(a) Is information a pure public good?
(b) A lighthouse is an information good. Is it a pure public good?
(c) What problems arise in provision of information due to its characteristics?

Problem 6.4 Bank monitoring

This problem is a simplified version of Freixas and Rochet (1997, Problem 2.6.2) Depositor A and Depositor B deposit money in a bank. They are the only two depositors. The quality of the bank is $q = e_1 + e_2$, where e_i is monitoring effort provided by depositor i. Both, Alice and Bob, have utility $U(i) = q - e_i^2/2$.

(a) Draw a graph showing what effort levels Depositor A and B would choose as a response to any effort level chosen by the other (so-called reaction functions);
(b) Compare monitoring efforts in the private and the social optimum. (Hint: think of the social optimum as the effort level the two agents would agree on if they could credibly commit to it.)
(c) Is the assumption reasonable that the depositors, by monitoring the bank, benefit rather than hurt each other;
(d) What is the private optimum and the social optimum (in the additive case) if the number of depositors is n?

Problem 6.5 Cooperation with unequal talent

Assume that in the previous problem bank quality were:

$$q = e_1 + 2e_2.$$

(a) What are the monitoring levels depositor A and B would choose in isolation;
(b) What level do they choose if for fairness reasons they agree on the same level for both;
(c) What level maximizes their aggregate wealth?
(d) Compare their utility levels in the different scenarios.

Problem 6.6 Research teams

Assume now, that individual monitoring efforts have a multiplicative, instead of an additive impact on bank quality. The quality of the bank is $q = e_1^{1/2} e_2^{1/2}$.

(a) Calculate the equilibrium effort levels Alice and Bob will choose in this setup. Sketch a graph of the reaction functions.
(b) What are the differences to the case with an additive impact of monitoring efforts on bank quality?
(d) How would the results change if the quality of the bank were given by $q = e_1 e_2$?

Problem 6.7 Open source software

In the last decade open source software projects have spread widely. A great number of programmers participate in these projects and up to millions use the developed software. Prominent examples include the Linux computer operating system, Apache server software and the Perl programming language.

(a) Why is the open source movement successful? Is this a counterexample of the underprovision of public goods?

(b) Apart from the price difference, what may be reasons for or against using open source software rather than commercially produced software?

Problem 6.8 A patent buyout
Please read the following text:[27]

> In 1837 Louis Jacques Mande Daguerre invented photography by developing the Daguerreo-type process. He exhibited images created using the process, and offered to sell detailed instructions to a single buyer for 200,000 francs or to 100 to 400 subscribers at 1,000 francs each. Daguerre was not able to find a buyer, but obtained the backing of François Arago, a politician and member of the Académie des Sciences, who argued that it was "... indispensable that the government should compensate M. Daguerre direct, and that France should then nobly give to the whole world this discovery which could contribute so much to the progress of art and science." In July 1839 the French government purchased the patent in exchange for pensions of 6000 francs per year to Daguerre, 4000 francs to his partner, and half that amount to their widows upon their death. The French government then put the rights to Daguerre's patent in the public domain (except in England, where the French government allowed Daguerre's original patent to remain in force). The invention was rapidly adopted and subjected to technological improvements. Within months, Daguerre's instruction manual was translated into a dozen languages. Many complementary inventions improved the chemistry and lenses used in Daguerre's process.
>
> In England, William Fox Talbot had developed the calotype process independently, and when he heard of Daguerre's process, he patented his own system in 1841. The Daguerre process became the standard, while the English process was abandoned, perhaps in part because Talbot charged high fees for use of his process. However, twenty years later a new process was developed, which also involved making prints from negatives, as had Talbot's process. The subsequent development of photography followed this colloid type process.[28]

(a) Comment on the decision of the French government in light of Schumpeter's model of creative destruction. Did the decision increase welfare?
(b) Kremer writes further about the "successful" inventor of the Cotton Gin, Eli Whitney: *"Like Daguerre, Eli Whitney was unable to make much money from his patent"*. Give two possible explanations.
(c) From a social point of view: Is the amount of research in the model of creative destruction always optimal in a decentralized economy?
(d) What problems does the government face if it tries to implement the socially optimal level of research?

Problem 6.9 Myratings Inc.
After college you start your own rating agency.

(a) Would you sell the ratings to investors or to the issuers of rated debt? Explain a few reasons for and against each option.
(b) How might your policy described under (a) change over time?

Problem 6.10 Knowledge banks
Financial intermediaries (banks, pension funds, insurance companies, hedge funds, etc.) collect considerable amounts of information. Some also make information available to their clients.

(a) Do financial intermediaries *produce* information?
(b) Do they *re-produce* information?
(c) Do you find any inefficiencies in the production or reproduction of information in the financial sector?

Problem 6.11 Information acquisition in teams
In many situations, a decision is improved when a number of individuals with different information participate in the decision. Examples are juries, committees, or public votes. Let us assume that a committee has to decide about either keeping or rejecting the status quo. Members' incentive to gather information is provided by the cost of either rejecting the status quo when it would be preferable to keep it or sticking to it when, in fact, a change would be called for. Agents cannot influence the cost of a wrong decision but they can change the probability of influencing the decision (in the right direction). A formal model can be found in Persico (2004).

(a) What decision rule would you expect to lead to more information gathering by committee members: (i) unanimity rule or (ii) majority rule?
(b) Assume that among two otherwise identical committees Committee A starts from relatively noisy information compared to Committee B. In which of the two committees would you expect members to gather more information (distinguish between decision rules)?
(c) How does the incentive to gather information depend on group size (distinguish between decision rules)?

Part II

How the market aggregates information

7 From information to prices

7 A Introduction

In Friedrich Dürrenmatt's play "The Visit of the Old Lady", Claire Zachanassian, old and immensely rich, returns to her home town of Güllen and shocks citizens of this sleazy place with an unsavory offer. She publicly promises a bounty of one billion to be divided among the town's treasury and its citizens under the condition that Alfred Ill dies. Alfred was her early love, but he let her down when she became pregnant, and Claire finally left Güllen. Now, after surviving eight rich husbands, the old lady exacts her revenge.

What happens after the offer? Of course, officials reject the offer in disgust, as do the citizens. They also try to calm Alfred. Yet, to his mortification, his fellow citizens soon start to appear with new shoes. Other purchases follow, more and more of them on credit. Poor Alfred can tell from observed market transactions that he has no chance. Eventually Alfred meets his deadly fate.

In the play, the market very efficiently aggregates and reveals citizens' true thoughts and plans. This may work in drama or fiction, but what about reality? In July 2003, the US Department of Defense (also known as the Pentagon) disclosed a scheme for wagering on terrorist events. Prices of individual contracts were thought to reveal information about the probability of underlying events, such as a missile strike from an enemy country. Yet, the project caused an immediate stir among politicians and the press. The official in charge, John Poindexter, was eventually ousted. Most ironically, he could read his imminent fate from the continuous increase in the price of a security paying 100 US-Dollars upon his dismissal. This contingent claim had been quickly introduced by an internet betting exchange.

Market signals can be powerful purveyors of information. Such signals can come in the form of quantities as in the case of Alfred Ill, or of prices as in the case of John Poindexter. The price system performs the formidable task of aggregating "the dispersed bits of incomplete and frequently contradictory knowledge which all the separate individuals possess", as F. A. Von Hayek (1945) put it.

In this chapter, we look at how individuals can pool their information when they come together to meet in the market. We try to explain the basic mechanism by which the market as a "super-computer" feeds dispersed and private information into prices. We will also discuss the idea of market efficiency as well as some paradoxes that lurk behind the *Efficient Market Hypothesis*. Within two Application sections we examine the use of prices as early warning signals and as a basis for action by authorities.

7 B Main ideas: Revealing information through prices

In the summer of 1968, French *chansonnier* Jacques Dutronc had his greatest hit with "Il est cinq heures, Paris s'éveille". The song tells of tired lovers, redressed strip-teasers, and the Eiffel Tower having cold feet, but also of the trucks full of milk, the coffee returning to cups and the bakers making their *bâtards*. Indeed, it may be the bakers who make a Paris morning an unforgettable event. In the early Paris morning, the scent of *croissants au beurre* manages to prevail, albeit fleetingly, over traffic fumes. Suddenly, the *croissants* are there; by the evening they are gone again.

The miracle of the daily flow and ebb of the *croissants* is an example of the working of the famous "invisible hand", a metaphor used by Adam Smith (1723–1790) in his *Wealth of Nations* (1776). Smith stressed the role of individual self-interest behind the socially beneficial working of the market. But there is another miracle involved: The coordinating hand is not only invisible but also *ignorant*. No visitor of the many Parisian bistros ever pre-orders a *croissant* for the next morning; hardly any *croissants* are left over by the evening, and few people complain that they wanted some, but could not get any. The market, it seems, is able to *extract* all relevant information (who will go to which bar; what is the price of flour, etc.) from individuals and to *aggregate* it into prices and quantities. This is the vision Von Hayek had of the market economy: A giant processing machine for decentralized information. The workings of the invisible and ignorant hand seem even more spectacular given that they have never been invented by anyone, but have rather developed spontaneously.

Notwithstanding some imperfections, the market does a remarkable job. The market's performance is even more impressive when we take into account that information is not only decentralized, but also private. Although we want the baker to anticipate our demand for a morning croissant, once at the shop we would not want to disclose how urgently we need it for fear that she might increase the price. We have an incentive to keep our preferences and our resources secret, but still the market extracts the relevant information.

What is the "trick" by which the market pools information? As Adam Smith pointed out, the driving force is the profit motive. The market provides individuals who have knowledge not yet incorporated in market prices with considerable profit opportunities. By realizing these opportunities, informed individuals move prices and thus reveal their information, at least indirectly. The market thus is a mechanism to "bribe" individuals to reveal private information.

Hedgeville municipality has issued a perpetual bond which is traded daily on an exchange. Every day, before trading starts, one individual receives information with potential relevance for the value of Hedgeville bonds. This individual knows that the information (s)he obtains is (i) new and (ii) private, i.e. not known to anybody else. Other individuals do not get any information on the same day but have to wait for their respective "information day". To keep the example as simple as possible, we also assume that individuals who get no information on a particular day are naïve, in the sense that they do not suspect that anyone else might have private information.

On Day 0, Hedgeville perpetuals closed at 100. On Day 1 (early morning), Alice learns that the interest rate level will fall from 5 per cent to 4 per cent that day. Alice, assuming that the bonds had been adequately priced until yesterday, calculates that their value is 125 ($= 100 \times 5/4$) at the lower interest rate (the coupon on the perpetual remaining the same). She starts to buy and continues to do so until the price of the bond has reached the new equilibrium value. All other investors (who have not received any information today) are agnostic as to the bond's actual value and sell what Alice wants to buy. The increase in demand, however, leads to increasingly higher prices (for example because the other investors' portfolios cease to be balanced). The closing price of Day 1 thus reflects Alice's information.

On Day 2, Bob learns that Hedgeville will cut taxes. He calculates that the municipality's probability of failure increases by two per cent a year. He concludes that a risk-neutral investor would demand a risk premium higher by the same amount. He thus sells Hedgeville perpetuals until they fall to 83.33 ($= 125 \times 4/6$). The next day Clara learns that thanks to the tax cut announced yesterday Pumpkin Chips Inc. will announce plans to move corporate headquarters to Hedgeville. This will generate so much tax revenue as to make Hedgeville a wealthy community and outstanding bonds almost riskless. She buys the perpetuals until they close near their risk-free level of 125 (since the fall in interest rates), at 123.6, say. And so on until Yvette learns that demographic change will heavily increase Hedgeville's latent pension liabilities towards municipal employees, followed by Zeno, coach of the Hedgeville Hedgehogs, who had a dream last night foretelling that his team will qualify for the Champions' League this time round.

In the course of time, all new information is built into the market price. Information is extracted from its holders by the market mechanism. The more spectacular a piece of information is (i.e. the greater the price change it suggests), the higher the profit the informed insider can make and the stronger the incentive is to trade. Of course, the price is never quite correct in the light of *future* information, but it always reflects past information, i.e. the information known to date. (Alice's assumption that the price on Day 0 was about appropriate, given the best information at that time, was thus reasonable). When Zeno observes a closing price of, say, 103.2 on Day 25, and concludes, based on the information he received on Day 26, that the price is too low by 7 per cent, he does not need to know why it was 103.2 the day before (as long as he is sure his information is truly new). To put it simply, the price is a sum of all previous information. Furthermore, we have assumed that information available on any single day is fully incorporated in the price on the very same day. The price is, to use some jargon, a sufficient statistic for all relevant information.

A market like the one in Hedgeville perpetual bonds described above is *informationally efficient* in a theoretical sense. Finance literature uses the concept of informational efficiency in a more empirical sense, efficiency being defined as an absence of profit opportunities. Notwithstanding innumerable empirical studies, the efficiency of financial markets remains a contested issue. We will not participate in this discussion, but we will briefly introduce the related concepts (strong, semi-strong, and weak efficiency hypotheses) in the Theory section below. For readers who choose to skip that section, let us stress here that informational efficiency is not the same as allocative efficiency. Under perfect information (informational efficiency), for example, we would be unable to insure (allocative inefficiency), as we will demonstrate in Chapter 13 D.

The above example of a day-by-day market was made simple by some very restrictive informational assumptions:

- Individuals are completely ignorant until the day they receive their information.
- On the day individuals receive their information, they are omniscient: An individual receives the actual information available on the day, plus a summary of all previous information in the form of yesterday's observed closing price.
- The following day, an individual is ignorant again.
- Individuals know that, on their information day, no one else has any news. On every other day they ignore or disregard the fact that an insider (an individual with news) may be around.

We will relax these assumptions in later chapters. Here we will take a closer look at some of them.

First, informed people know they have an advantage and do indeed trade in order to exploit it. But imagine that Fatima is not quite sure whether the information she received (on Day 6) is really new. If she comes to the conclusion that the information is already reflected in the price, she has no incentive to trade. Yet, if all individuals think alike, nobody trades and information never becomes built into prices in the first place. This is a paradox we will discuss in the Theory section.

Second, uninformed individuals do trade. However, individuals who did not receive any information on a particular day know that with some probability they will trade against that day's insider. In expected terms they cannot but lose money. So they should not trade at all. Again, the available information might not be reflected in the prices.

Third, individuals in our example either are informed (about the day's news) or completely naïve. In particular, they do not speculate on what the majority believes the majority believes. This avoids biases in information aggregation that may become relevant in a more realistic setting. We will discuss such biases in Chapter 8.

One thing should now be clear. Once one looks into the details of information and market structures, including the timing of information and actions, the market's ability to aggregate information and to make it public via prices is all but obvious. One needs a sizeable assumptional structure to understand how exactly information is incorporated into market prices or, for that matter, into quantities. Maloney and Mulherin (2003) conclude that "while markets appear to work in practice, we are not sure how they work in theory." The Paris croissants thus are a miracle, not only from a culinary point of view!

7 C Theory: The market as an information processor

7 C a Contingent claims and market completeness

Spot markets

In a timeless economy, there are only physical goods or services for immediate consumption. Markets in such an economy are "spot" markets. In an economy that has a time dimension ($t = 0, 1$) three kinds of markets become possible: spot markets (markets operating now for goods available now), futures markets (markets operating now for goods available tomorrow) and future spot markets (markets operating tomorrow for goods available tomorrow). If there are more than two periods ($t = 0, 1, 2$) we have to add a fourth category: Future futures market. These are markets operating tomorrow for goods available the day after tomorrow.

Contingent markets

If we also add uncertainty, contingent markets enter the scene. Contingent markets are markets for assets whose payoff depends on the future state of nature. The typical example is a bond which pays the nominal value if the debtor is solvent, and a bankruptcy dividend otherwise. Contingent markets do not add an additional market category to the previously mentioned spot, futures, future spot and future futures markets. Rather, contingent claims can be seen as additional goods. The distinction of goods according to the state of nature in which they are available is comparable to the distinction according to location or time of availability. For example, an umbrella on a rainy day is a different good from an umbrella on a sunny day, and vice versa. We can thus distinguish between contingent and non-contingent claims. A swap contract, for example, is non-contingent (abstracting from the possibility that one of the parties may default), while an option contract is contingent, its payoff depending on the realization of some underlying variable, such as the weather in the umbrella example. Contingent claims are only traded on futures markets; contingent spot markets do not exist, because we cannot live simultaneously in more than one state of nature.

Futures markets

Futures markets, including the markets for contingent claims, are often called financial markets. This is because they normally do not exist for all possible pairs of goods, like apples for oranges, but only for goods against money or for money (today) against money (tomorrow). (The restriction on money prices does not imply any loss of generality, as the price of apples in oranges, that is the value of apples measured in units of oranges, is already defined through the money prices of the two commodities.)

Assets

The "goods" traded on futures markets are essentially promises. Economists call these promises "securities" or "financial assets". An asset is defined by the gross return r it yields in every different state of nature $\Omega = \{1, 2, \cdots, \omega\}$. If we denote a state i by subscripts, an asset J can be written as:

$$\mathbf{R}^J = \begin{bmatrix} r_1^J \\ r_2^J \\ \vdots \\ r_\omega^J \end{bmatrix}$$

The financial market, consisting of J assets with payoffs R^J conditional on states Ω, can be represented by the matrix:

$$\mathbf{R} = \begin{bmatrix} r_1^1 & \cdots & r_1^j & \cdots & r_1^J \\ r_2^1 & \cdots & r_2^j & \cdots & r_2^J \\ \vdots & \ddots & \vdots & & \vdots \\ r_i^1 & \cdots & r_i^j & \cdots & r_2^J \\ \vdots & & \vdots & \ddots & \vdots \\ r_\omega^1 & \cdots & r_\omega^j & \cdots & r_\omega^J \end{bmatrix}$$

Arrow securities

All assets can be thought of as bundles of some elementary assets, paying one unit of purchasing power in exactly one state of the world and nothing in all others.

Let us consider an asset $\mathbf{e}_i$ that pays one in the state $i \in \Omega$, and 0 otherwise.

$$\mathbf{e}_{i \in \Omega} = \begin{bmatrix} 0 \\ \vdots \\ 0 \\ 1 \\ 0 \\ \vdots \\ 0 \end{bmatrix}$$

Such assets are called *Arrow securities*. A portfolio with one Arrow security for each possible state is represented by a matrix with ω rows (one for each state) and J columns (one for each security):

$$\mathbf{e} = \begin{bmatrix} 1 & 0 & \cdots & 0 \\ 0 & 1 & \cdots & 0 \\ \vdots & \vdots & \ddots & \vdots \\ 0 & 0 & \cdots & 1 \end{bmatrix}$$

This portfolio has some interesting properties. Taken together, it can be seen as a riskless asset paying exactly one unit of purchasing power in any possible state. Taken individually, each security can be seen as a bet on one particular state of nature.

Market completeness

If there is an Arrow security for each state, we can produce any arbitrary payoff structure from the appropriate mix of Arrow securities. But the existence of a complete set of Arrow securities is not even necessary. What we need is a set of securities from which we can build an Arrow security for each state. This is the case, for example, if there are as many securities as states and the payoff vectors of the securities are linearly independent.[29] We then say that markets are complete. A more intuitive definition is the following (from Lengwiler, 2004):

> Markets are complete when individuals can trade assets in such a way as to affect the payoff in one specific state without affecting the payoffs of other states.

In reality, markets may deviate from the ideal of a complete markets' setting. Most importantly, markets may be *incomplete*. For example, normally there is no individual security nor any combination of securities that pays one dollar in the state in which a couple are married by a particular date. However, one exception will be mentioned in Chapter 13 A.

Arbitrage

Markets may have *redundant* securities, creating opportunities for riskless arbitrage if securities are mispriced (see Box 7.1). The presence of riskless arbitrage opportunities violates the most basic requirement of market efficiency. In fact, such opportunities are rare, although a number of redundant securities exist (e.g. most options). Note also that redundancy and incompleteness do not contradict each other: Markets can have redundant securities and still be incomplete.

Box 7.1 Riskless and risky arbitrage

Arbitrage is commonly defined as the exploitation of a mispriced security (Brunnermeier 2001, p. 190). Mispricing means that the price of an asset differs from its equilibrium price. There are two notions of arbitrage: Riskless and risky.

Riskless arbitrage (the only true arbitrage in a theoretical sense) is possible when an asset is mispriced *relative* to other assets. Assume that asset A can be duplicated by a bundle B of other assets, such that both generate the same future payoffs. If there is a price difference between the two assets, an investor can buy the cheaper asset and sell the other asset short. This adds up to a portfolio with a safe, positive net return.

Riskless arbitrage requires the existence of some redundant assets. The standard example are options which are often redundant, as they can be duplicated by a portfolio consisting of (long or short) positions in the underlying asset and some riskless asset like government debt. In fact, the idea of option valuation behind the famous *Black–Scholes formula* rests on perfect dynamic duplication of the option.

Risky arbitrage, by contrast, aims at exploiting some *absolute*, rather than relative mispricing. An example of an absolutely mispriced asset would be a bond whose yield at current market prices lies below the yield of an otherwise identical risk-free government bond. Selling the first bond short and buying the second yields a safe positive net return if both positions are held *until maturity*. If, however, the investor needs to liquidate the position early, the terminal value is uncertain. Although the relative price between the bonds is fundamentally wrong, exploiting it is not free of risk.

Incomplete markets lead to a number of complications (Lengwiler, 2004). In particular, prices in equilibrium are not unique. More precisely, existing securities do not have unique prices, and missing securities are not priced at all. Incomplete markets thus may not be able to reflect all available information.[30]

Alice runs for president of the local bird watchers' club, while Bob hopes for a promotion to become head of the crazy customer complaints' department. Let us indicate the four different outcomes (capitals standing for success, lower case letters for failure) by AB, Ab, aB, ab. Both Alice and Bob can hardly await the outcome. They also know that other people, like the members of the bird watchers' club or colleagues at Bob's office, have better information. Asking them would be the straightforward strategy; experience tells us, however, that people tend to give answers that please the inquirer rather than express their true beliefs.

Alice and Bob try something else: They define securities and open markets for these securities, hoping that observed prices will tell them the odds of success.

The necessary securities

What securities do they have to define? As usual, (at least) one for each state. However, be careful as things become tricky! The possible states are AB, Ab, aB, and ab, not—as one might think naïvely—A, a, B, and b.[31] Only a set of Arrow securities for states AB, Ab, aB, and ab (or any other set of assets from which this set can be constructed) constitutes a complete market. Securities for A, a, B, b do not. In addition, only the complete market, but not the incomplete market allows information to be represented in prices.

The purpose of this chapter is to examine how heterogeneous beliefs are aggregated. But assume for a moment that all observers share the same beliefs. They all agree on the probabilities of the different outcomes, $\Pr(\omega) = \pi_\omega$.[32] These probabilities are given in Figure 7.1.

Equilibrium prices

If everybody is risk-neutral, the prices of securities in equilibrium reflect the probabilities of the underlying states. As all securities pay \$1 in the respective state, prices p must satisfy $p_\omega = \pi_\omega$, or, more explicitly:

$$\mathbf{p} = [p_{AB}, p_{Ab}, p_{aB}, p_{ab}] = \begin{bmatrix} 1 & 0 & 0 & 0 \\ 0 & 1 & 0 & 0 \\ 0 & 0 & 1 & 0 \\ 0 & 0 & 0 & 1 \end{bmatrix} \begin{bmatrix} \pi_{AB} \\ \pi_{Ab} \\ \pi_{aB} \\ \pi_{ab} \end{bmatrix},$$

The probabilities in Figure 7.1 imply an equilibrium price vector of $p_{AB} = 1/2$, $p_{Ab} = 1/4, p_{aB} = 1/8, p_{ab} = 1/8$. Observing these prices, Alice and Bob can tell that the odds that both will succeed are $1/2$, while the odds that neither will succeed are $1/8$. Alice alone is successful in $1/4$ of the cases, Bob alone in only $1/8$.[33]

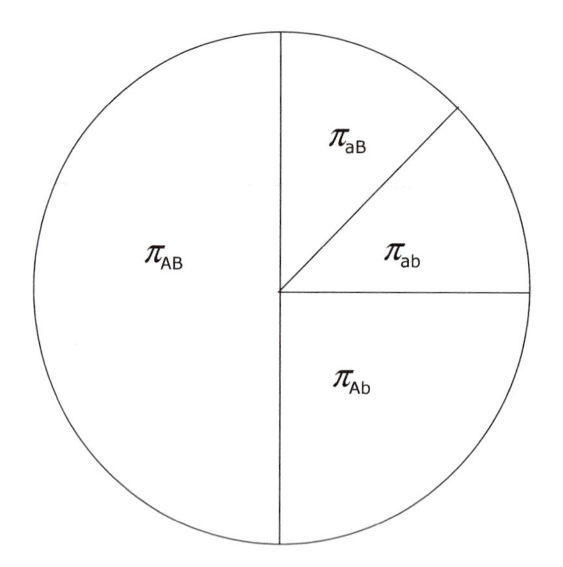

Figure 7.1 Alice and Bob both hope for success (denoted by capital letters A and B, respectively). They may also fail (small letters, a and b). Each possible state, i.e. both successful (AB) or failing (ab) as well as only one succeeding (aB or Ab), occurs with a certain probability π. Since one state will be realized for sure, the probabilities sum up to one.

When prices tell . . .

Both Alice and Bob may be primarily interested in the simple question: "Will I make it?", rather than whether the other will be successful. This is no problem; observed prices tell that Bob has a chance of $\pi_B = \pi_{AB} + \pi_{aB} = 5/8$ to be promoted. There is a problem, however, if Alice and Bob set up markets naïvely and define securities according to the outcomes they are interested in, rather than according to the states in the sense of the elements of Ω. While aB, for instance, is a state of nature, B is not. B is the union of two states AB and aB, i.e. $B = AB \cup aB$. A market with four securities, A, B, a, b, is not complete. It does not allow individuals to trade assets in such a way as to affect the payoff in one specific state, without affecting the payoffs of other states. Readers may verify that there is no combination of the four assets that pays $1 in ab, and nothing in the other states.

. . . and when they do not.

Such an incomplete market does not allow information to be mapped on to prices. The price vector $\mathbf{p} = [p_A, p_B, p_a, p_b]$, does not indicate the probability markets attribute to, say, state ab. The market is not only incomplete with respect to tradeable payoffs, it is also informationally incomplete. An individual who knows that neither Alice nor Bob will make it (ab) cannot bet exactly on this state. The market (commodity) structure is too coarse compared to the information structure.

7 C b Aggregating heterogeneous information

In the previous section we only looked at a market's ability to reflect information in principle, that is, at things like market completeness. We tacitly assumed that if contingent prices exist and are unique, then they correctly reflect individuals' information. This was not very critical as individuals did not have conflicting information. In this section we try to show how heterogeneous and potentially conflicting information is aggregated. When we say that we try to show, this is meant literally. Models of price formation in markets with private information are extremely demanding. We will work with strongly or, rather, excessively simplified models. Dissatisfied readers are invited to consult the classical paper by Grossman (1976) or the discussion in Brunnermeier (2001).

We continue with the example of Alice and Bob hoping for election and promotion respectively.

Alice and Bob introduce four (perfectly divisible) Arrow securities, each paying \$1 in one of the four possible states AB, Ab, aB, and ab. They sell these securities in equal quantities to Victor, Wilma, Xenia and Yves, before these four individuals learn about probabilities. We call this round zero of market transactions (= the issue of the securities), which takes place at prices $p = p_{AB} = p_{Ab} = p_{aB} = p_{ab} = 0.25$.

After having acquired the securities, Victor, Wilma, Xenia and Yves obtain the following information about the probability of the potential outcomes:

- Victor knows that $\pi_A = 3/4$.
- Wilma knows that $\pi_B = 5/8$.
- Xenia knows that $\pi_{AB} = 1/2$.
- Yves knows that $\pi_{ab} = 1/8$.

What will happen?

We assume that each of the four individuals knows that (s)he is the only one with some information on the particular probability. They also know that someone else knows the other probabilities. Despite having a "monopoly" on the knowledge of one probability, the four behave competitively in the market, i.e. they bid prices corresponding to their individually known probabilities.

Victor's turn

Once individuals have obtained their private information, issue prices cease to be equilibrium prices. Victor, for example, finds that the outcome A is undervalued in the market. He is willing to buy the contract written on *AB* and the one on *Ab*. He has no preference on either of these two contracts, so let us assume he buys equal quantities. His competitive bid is 3/8 for each contract as $2 \cdot (3/8) = (3/4)$, which is Victor's privately known probability $\pi_A = \pi_{AB} + \pi_{Ab}$. For consistency's sake (probabilities add up to one), the prices of the other two contracts (ab, aB) must fall. Plausibly, Victor will not only buy the undervalued contracts, but also sell the overvalued contracts, thus depressing their prices to 1/8. The resulting prices after the first market round are also represented in Figure 7.2

Wilma's turn

Assume that it is Wilma's turn now. Wilma thinks that contracts paying in the event that the outcome includes *B* are undervalued, and those on *b* are overvalued. She thus buys *AB* and *aB*. Again, she tries not to influence the relative price relation of 1/3 between *AB* and *aB*, since she knows that someone else knows the respective probabilities, and that knowledge is likely to be reflected in market prices. She takes an agnostic view to best avoid the risk of trading against an insider. For *AB* and *aB* she bids 15/32 and 5/32, respectively. This leaves their relative price unchanged, but adds up to 5/8, Wilma's belief of π_B. The second market round ends with prices that are depicted in Figure 7.2

Xenia and Yves

Both Victor and Wilma would not find profitable trading opportunities, as prices correspond to their private knowledge. Xenia and Yves, however, do. To speed up procedures we will treat them simultaneously. Both discover that the contract

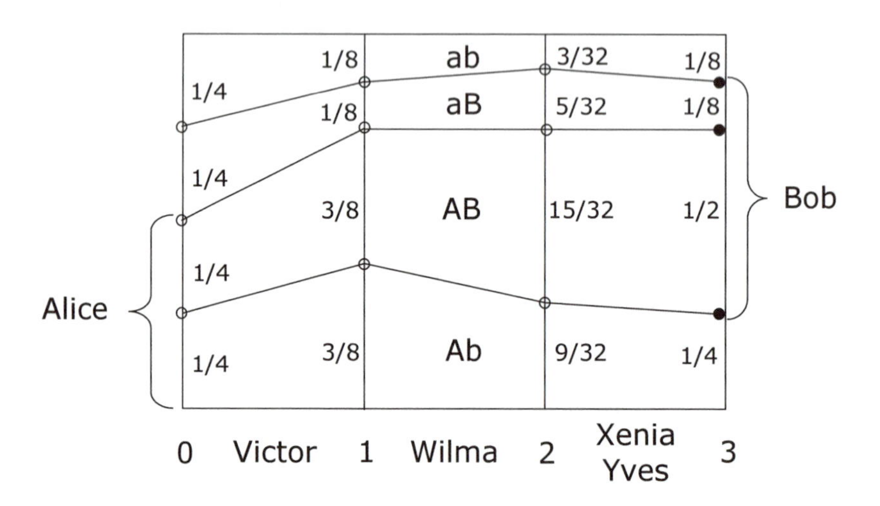

Figure 7.2 Prices for securities *Ab*, *AB*, *aB* and *ab*, before and after each trading round express the probabilities that Alice will be elected and Bob will be promoted.

on which they have some information (directly or indirectly), *AB*, is undervalued by 1/32. They thus bid them up by that amount. According to our assumptions they should also sell a share of each of the three complementary contracts. For simplicity's sake, however, we assume that they do not stray into each other's home turf. Xenia thus sells *aB* and *Ab*, but not *ab*; Yves sells the same two contracts, but not *AB*. They thus bid down *aB* and *Ab* by 1/32 each. This leads to prices consistent with the private information held by all four participants. After the third round, the market is thus in equilibrium, as can easily be verified from Figure 7.2.

Market equilibrium

The market has integrated all information originally held in dispersed and private form. Alice and Bob can now deduce their chances of being elected/promoted from the market prices. They do not need to know which of the market participants knew what, to deduce that the odds for Alice are 3/4, while those for Bob are 5/8. The market tells them all they can possibly know. It cannot remove any remaining uncertainty, since informed individuals themselves have imperfect information. But the market can provide Alice and Bob with all the information that is available. Note also that (as long as transactions are anonymous) the "informants" can keep their privacy; nobody could complain about Xenia thinking that with a 50 per cent chance at least one of the two will not make it.

Admittedly this is only an example designed to help the reader's intuition. While the example illustrates the market's ability to put together different pieces of information, it also brushes over a number of important conceptual issues. To make the example work, we had to make a few informational assumptions:

- Individuals' information is complementary, but not conflicting;
- Each individual knows that he/she is the only one with information on one particular probability;
- Each individual knows that the others are in a similar position;
- Each individual knows his/her information is not incorporated in the market price as long as the price does not correspond to the probability belief.

To these we had to add a fairly restrictive behavioral assumption:

- Individuals behave competitively, even though they know they have a monopoly on their particular piece of information.

The informational assumptions may not be very problematic *per se*. Yet, they are definitely restrictive. A real understanding of how markets work requires more relaxed assumptions. For example, one might want to see how the market aggregates some conflicting prior beliefs, like Wilma thinking that Bob's chances of being promoted are 5/8, and Walter thinking that Bob will make it with a probability of only 3/8. It is tempting to assume the two would "compromise" at a price of 1/2, but—within the logic of the above example—there is no reason why they should. What we need in such a case is a model that shows what happens if an individual initially believes Bob's chances are 5/8, but sees that some other market participant(s) stubbornly bid(s) down the price on the respective contract(s) to 3/8 (and vice versa). In other words, we need a model that describes revisions of individual beliefs on the basis of observed market prices.

The literature on market microstructure (see O'Hara, 1995) has developed several such models. These highlight many interesting aspects of information aggregation in markets. In particular they show that the market's ability to aggregate information, once one examines the issues closely, is not as obvious as one might believe on the basis of everyday experience.

We cannot go deeper into models of market microstructure. We will briefly discuss some of the paradoxes encountered. The main complication arising in market models—the need to know what others know—will be discussed in Chapter 8. For the rest we will trust that the basic intuition illustrated by the above example is not entirely misleading and that the market indeed is a powerful (if not failsafe) information processor. For budding Nathan Rothschilds, though, we strongly recommend some deeper study of market microstructure models using for example O'Hara (1995).

7 C c Market efficiency

Allocative efficiency

Economists' main concept to assess outcomes of market or other transactions is *efficiency*. Traditionally, the term efficiency applies to *allocative* efficiency. An allocation is efficient when the scarce resources of an economy are optimally distributed among individuals. "Optimal" in this context means a so-called Pareto optimum (after Vilfredo Pareto, 1848–1923): An allocation is Pareto optimal if there is no other which makes at least one person better off without making another person worse off. In other words, no Pareto improvement is possible in such a situation.[34]

When individuals have imperfect information, their utility levels depend on what they know or believe. Under these circumstances, the Pareto optimality criterion can (or has to) be applied at different informational stages:

- ***Ex ante* efficiency** refers to utility levels expected *before* individuals receive their private information.
- **Interim efficiency** refers to utility levels expected once individuals have received their private information.
- ***Ex post* efficiency** refers to utility levels expected when all information has been released (i.e. when everybody knows the true state of the world), and everybody has the same (common) knowledge.

Ex ante efficiency implies interim efficiency which, in turn, implies *ex post* efficiency. Conversely, if an allocation is *not ex post* efficient, it is *not* interim efficient;

if an allocation is *not* interim efficient, it is *not ex ante* efficient (Brunnermeier, 2001, Ch. 1).

Note that allocative efficiency may require imperfect information. Informational efficiency (to be discussed below) in the sense of full revelation of information through prices may prevent risk-sharing (Brunnermeier 2001, p. 26). This is called the insurance destruction effect or Hirshleifer Effect, and will be discussed in Section 13 D.

Operational efficiency

Markets, especially financial markets, are said to be operationally (or internally) efficient if they work at the lowest possible cost. The ideal would be frictionless markets, free of transaction costs, trading restrictions, taxes, etc. The operational efficiency of markets is also important for another interpretation of efficiency, so-called informational efficiency, i.e. the ability of prices to reflect information (to be discussed in the following section).

Most economists would agree that operational efficiency helps information to become built into prices. If individuals cannot buy or sell quantities consistent with their information, information is not revealed. One example are short sale restrictions. These may prevent negative information from becoming absorbed in prices and become an important factor in the dynamics of asset price bubbles.[35]

Informational efficiency in the theory sense . . .

Markets are informationally efficient if prices are a sufficient statistic for all the information in an economy (Brunnermeier, 2001).[36] Observing prices thus would lead to the same equilibrium outcome as if all information were shared among all market participants.

Informational efficiency does not mean that we can work back from observed prices and infer all information that went into prices. Sometimes, reverse engineering is possible, sometimes not. In the former case prices are said to be fully revealing. In the latter case they are partially revealing.[37]

In our example of Alice running for club presidency and Bob hoping for promotion, prices were not fully revealing. They did express the information Alice and Bob were primarily interested in (their individual chances), but they did not reveal the precise pieces of information that had gone into them.

. . . and in the sense of empirical research

In empirical finance literature, informational efficiency is defined somewhat differently as the absence of profit opportunities. The *Efficient Market Hypothesis* rests on three assumptions (Fama, 1970): (i) No transaction costs, (ii) information is available without cost, and (iii) rational expectations (agreement on the same model generating future returns).

There are three levels of efficiency commonly distinguished in the literature:

- **Strong form**: Prices reflect all public and private information in the sense that it does not leave any profitable trading opportunity, even for "insiders".
- **Semi-strong form**: Prices reflect all public information, thus leaving profit opportunities to parties with private information.
- **Weak form**: Prices fully reflect the history of past prices. There is thus no profitable trading opportunity based on past prices, or as the saying goes "markets have no memory". Yet, traders may make profits from fundamental research and private information.

We do not join the debate about how efficient markets really are. An introduction to the debate is Shleifer (2000). For a discussion of the different concepts, see (Brunnermeier, 2001).

7 C d Some paradoxes

The Information Paradox (Grossman–Stiglitz)

The *Efficient Market Hypothesis* has an important implication, the so-called *Information Paradox.* Assume that markets are informationally efficient. Prices thus reflect all available information. In such an efficient market no one has an incentive to gather information since the return from information gathering is zero, as all information is already reflected in prices. So nobody gathers information and, since nobody has any information to gain from, nobody trades. Yet, if no-one trades, markets are not efficient. This is a paradox: The efficient market hypothesis only holds if no one believes it; if everybody believes it, it does not hold.

This paradox was put forward by Grossman and Stiglitz (1980). In a similar version, individuals acquire new information for free, but face a trading cost. Again, nobody trades if markets are thought to be efficient, but they are not efficient if nobody trades (Brunnermeier, 2001).

Already quite some time before the discovery of the paradox, Von Hayek (1945) argued that costly information acquisition is a precondition of competitive markets. An investor willing to bear these costs will be compensated by sufficiently high profit opportunities (i.e. inefficiencies). Based on this idea, the paradox fortunately can be solved in a dynamic setting. A tiny (time) advantage is sufficient to provide an incentive to acquire information, which puts the economy back to an almost efficient market setting (as illustrated in the above example).

Individuals from Alice to Zeno acquire pieces of information sequentially. Every morning of every day, prices are efficient with respect to all past information. After the arrival of information, one individual has private knowledge not yet reflected in the price. Markets during the day thus are still efficient in the weak sense (past prices have no predictive content) as well as in the semi-strong sense (all public information is included), but not in the strong sense. This permits the daily "insider" to trade profitably and makes the market incorporate news into prices.

One needs to explain, though, why uninformed individuals trade with the daily insider. A sufficient assumption is that the non-informed traders (i) do not know who is an insider and (ii) have some sufficiently strong trading motive, like a liquidity shortage or some excess liquidity. If the benefit from trading exceeds the expected loss from meeting an insider, trade will take place. It is helpful to further assume that the insider can trade anonymously; this reduces the chance for uninformed traders that the actual counterparty has superior knowledge. Trade thus takes place with a relatively weak individual trading motive. It has also been said that insiders have knowledge of public interest, while non-insiders only have knowledge of private interest (such as their liquidity needs). Traders who only have information of private interest are often called "noise-traders".

A dynamic view allows to reconcile traders' belief that prices are (almost) informationally efficient with the existence of a trading motive. The smaller the individual bits of information that arrive over time, the closer prices remain at efficient levels. In a (dynamic) equilibrium the marginal cost of acquiring information is equal to the marginal profit from trading on information (i.e. the marginal value of information). It thus depends on the cost of collecting information how close observed market prices are to "true" prices.

No trade theorems

Intuition may suggest that trade on financial markets comes from differences in opinion. An individual with a novel piece of information forms beliefs that differ from those of other market participants. While the insider may want to trade, non-informed traders should be suspicious. As explained above, they would risk trading with a better informed person only if they had a strong private motive, like an urgent need for liquidity. As it takes two to tango, so to speak, an information asymmetry should prevent rather than stimulate trade. In the absence of differences in opinion, however, there is no reason for trade either. The enormous trading volumes observed on financial markets thus are quite a puzzle if one takes the stated reservations seriously. One way to explain them would be that privately valuable information (like individual liquidity or diversification motives) is relatively important compared to individual pieces of publicly relevant information. As a result, small crumbs of news could easily dissolve in a sea of noisy ignorance (Brunnermeier, 2001).

Agreeing to disagree

A radical hypothesis says: Among rational individuals, differences of opinion cannot exist. In other words: "We cannot agree to disagree".

If Alice believes the chances of rain tomorrow are 25 per cent, while Bob is sure they are 75 per cent, they might trade one against the other. In a kind of tug-of-war they would pull the price of a security paying one dollar in case of rain in their respective directions, until, if they are not hopelessly stubborn (a possibility Alice would not exclude with respect to Bob), one of them would ask "what does the other know that I don't know and why does it make him (her) so confident?".

If individuals take each other serious, some communication or just a little bit of trade might suffice to align their expectations. Differences of opinion alone therefore are an uncertain basis for explaining trade volume. We will discuss these issues more explicitly in Chapter 8.

The endogeneity paradox

Finally, a paradox described, e.g., by Krugman (1991) arises when individuals want to make practical use of the information they have inferred from prices. By "use" we mean that individuals condition their actions on observed prices. If, in turn, actions affect the fundamentals that are reflected in prices, the "right" price *ex post* differs from the price observed *ex ante*. Yet, rational traders would anticipate that prices will affect action and that action affects the equilibrium price. With rational traders the price *ex ante* thus already incorporates the future action—but then it cannot reflect the very information that would *lead* to the action. The action does not take place. Yet, if the action does not take place, the price remains meaningful. This is reminiscent of the Grossman–Stiglitz Paradox discussed above.

For an illustration, think of the above example with Alice running for presidency.

Market prices tell Alice, she will be elected president of the bird-watchers club with a 75 per cent chance. A 25 per cent probability of not being elected may not satisfy Alice. Like politicians not satisfied with the polls she may decide to intensify her campaign. Let us assume that Alice gives a tea party where money is collected for the Hedge Fund—a fund devoted to the preservation of hedges (a nesting environment for birds). Alice thus manages to increase her chances to 90 per cent. That is fine for Alice—or so it seems. In fact, rational investors should have foreseen that Alice, seeing the market giving her "only" a 75 per cent chance, will try to push her chances higher. If so, they would not have bet their money on a 75 per cent chance, but on a 90 per cent chance. Yet, seeing the market forecasting victory with 90 per cent probability, Alice might not have given the Hedge Fund party!

For an agent who would like to act on market information, the market as an information aggregator in the sense of Von Hayek may actually be *too* efficient. The action conditional on the observed price is already incorporated into the price.

Prices thus are helpful, except when you need them most, namely as guides to action. To *homo œconomicus*, the information contained in prices is what water was for the mythological Tantalus: Reaching up to his lips, the water would immediately recede each time Tantalus tried to take a mouthful to quench his thirst. We will illustrate this problem with two Applications below.

7 D Application: Terrorism futures and prediction markets

A bold project . . .

In July 2003 word leaked out that an internal think tank of the US Department of Defense (the Pentagon) planned the introduction of "futures markets applied to prediction" (FutureMAP). The market was designed as a futures exchange where traders could speculate on indices of economic and political stability in a number of countries (particularly in the Middle East), conflict indicators, and specific events like a biological attack on Israel, a North Korean missile strike, or a regime change in Syria. Trading in some first contracts was scheduled to start 1 October 2003; other contracts would be introduced later.

. . . with a short life

On 29 July, a few days after its public disclosure and two days after a press conference, the plan was abolished. Politicians from all camps felt that wagering on the fate of foreign leaders and the likelihood of terrorist attacks was morally repugnant and grotesque. In addition, the prospect that terrorists could even financially benefit from their dreadful acts was considered absurd. So the proposal, dubbed "terrorism futures" by critics, was abandoned. The responsible agency, DARPA (Defense Advanced Research Projects Agency), came under heavy criticism, and its head, John Poindexter, soon had to resign. It comes as a particular irony that Poindexter could follow the predicted odds of being ousted: Contracts paying US$100 in case of his resignation before the end of August 2003 were traded in an offshore betting exchange (for details see Wolfers and Zitzewitz, 2004, and the literature cited therein). Before mid-August, he asked to resign as of 29 August.

"A good idea with a bad press"?

Was the DARPA project so bad after all? DARPA officials (inspired by some academic economists) argued in favor of the political analysis market that "markets like this are extremely efficient, effective and timely aggregators of dispersed and even hidden information," a formulation of the *Efficient Market Hypothesis* that would soon be ridiculed as "DARPA-speak" by critics. Information aggregation had become an issue for defense and intelligence agencies after the terrorist attacks of 11 September 2001. Evidence collected in the aftermath suggested that a great deal of information on the terrorists' plans had actually been around, but that nobody had actually managed to aggregate it. Hence the planned recourse to markets.

Amidst the political and moral uproar, the voices of the few economists who joined the debate were relatively feeble. The economist Hal Varian defended the Pentagon scheme in an article "A Good Idea With a Bad Press"[38] with an argument to be discussed in Problem 7.4. Others pointed out that bets on some political events had been available on the internet for quite some time.

The history of betting markets

In fact, the project could quote a long history of betting markets as predictors of political events. Rhode and Strumpf (2004) show that between 1884 and 1940 betting markets predicted the outcome of presidential elections remarkably well. Betting odds had more predictive power than any other available information. Betting markets were also quite large but later lost some attraction with the advent of public polls and with the legalization of betting on horse races. More recently, the Iowa Electronic Market (run by the University of Iowa) started listing bets on presidential elections and other events. Several betting exchanges list events from sports and politics. Indices for the success of movies and movie stars, predictions of awards, and a related set of derivative contracts are traded (in virtual currency) at the Hollywood Stock Exchange (for details see Wolfers and Zitzewitz, 2004).

Economists have analyzed the predictive power of betting markets (Rhode and Strumpf, 2004; Wolfers and Zitzewitz, 2004). They find mixed evidence on market efficiency. But the analysis of betting markets should not distract from the predictive power of "ordinary" financial markets. The oil price is a predictor of political stability in the Middle East, while the gold price is a predictor of inflation. A striking example of the market's detective abilities is the stock market reaction to the 1986 Challenger crash (Maloney and Mulherin, 2003). While the event was widely observed, it took several months for an esteemed panel to determine which of the mechanical components failed during the launch. By contrast, securities' trading in the four main shuttle contractors seemingly singled out the firm that manufactured the faulty component very quickly.

Some firms have even introduced internal prediction markets in order to forecast printer sales (Hewlett-Packard) or the timeliness of software projects (Siemens). In both cases market predictions were more accurate than firms' traditional internal tools. Wolfers and Zitzewitz (2004) thus see some scope for the future use of prediction markets for firms' planning purposes or as sources of political information. They point out that the predictive power of markets derives from three sources: (i) The incentive for truthfully revealing private information (see also Chapter 15), (ii) incentives for research and information discovery, and (iii) the algorithm for aggregating opinions.

Of course, the design of such markets is a challenging task, the details of which we cannot fully explore here. But first of all, contracts need to be clearly specified. This is not a trivial task. A contract traded on the Iowa markets was written on a seemingly clear criterion: The number of seats won by each party in the 1994 US Senate elections. However, the day after the election one senator switched sides (before all votes were counted).

Wolfers and Zitzewitz (2004) also point out that "to the extent that the valuable information generated by trade in these markets is not fully internalized into the profits earned by these firms [the exchanges], prediction markets are underprovided." This is an example of the underprovision of information as a public good, as discussed in Chapter 6.

Would it work?

The DARPA project, to be sure, was mainly rejected for political and moral reasons. Given the evidence on the predictive power of betting markets, there was hardly much of an *empirical* basis for rejecting the project (although the advocates of the project may have somewhat underestimated the difficulty of valuing small probability events). However, there may be a *theoretical* problem with this kind of market, namely the potential *endogeneity* of prices. In 1904, Andrew Carnegie coming back from vacation, stated: "From what I see of the betting, . . . I do not think that Mr. Roosevelt will need my vote. I am sure of his election . . ." (quoted from Rhode and Strumpf, 2004, p. 132). Yet, if all observers had thought like Carnegie, Theodore Roosevelt may never have become president!

It is hardly realistic to assume that a defense agency wants a forecast for the mere fun of it. The most productive use of a prediction of a terrorist event would be to prevent it from happening. Indeed, when canceling the FutureMAP project, DARPA said it would have explored "the power of future markets to predict *and thereby prevent* [our emphasis] terrorist attacks" (DARPA news release, 29 July 2003).

Yet, what happens if a forecast implied in market prices can be expected to trigger some preventive action? If prevention is successful, the event does not happen. If the event does not happen, markets do not forecast it. But if markets do not forecast it, the Pentagon cannot react and prevent it. So it may or may not happen. Here we may have one more informational paradox. We will look deeper into it within the following Application on the use of market prices by bank supervisors, because there is no doubt that bank supervisors want information to enable them to take action!

7 E Application: Should bank supervisors look at market prices?

7 E a The charm of market prices

Prudential supervision of banks faces two difficult tasks: Information and enforcement. A supervisory authority first needs *information* about the solvency of a bank. Second, it needs to *enforce* measures required to restitute the solvency of troubled banks.

Information

The information part is itself difficult. Modern banks are financially, legally, and organizationally complex entities. Accounting data, the traditional basis of solvency assessment, are alone of little value for risk measurement and management. Supervisors thus collect a variety of data through on-site inspections, etc. And increasingly, they rely on banks' own sophisticated internal reporting systems.

Market price data promise to help the supervisor in their tasks. Markets aggregate available information into prices. If markets are strong-form efficient, prices even include confidential supervisory information. In this case, the supervisors would waste resources by collecting information themselves, as market prices already tell everything that there is to know. Even if markets are not strong-form efficient, they may provide supervisors with valuable complementary information.

Enforcement

While the diagnosis part is difficult, prescribing the cure can be unpleasant. First, it is often not obvious what intervention is optimal. Second, it takes a great deal of courage for any outsider, even for a supervisor vested with a legal obligation, to tell a bank what to do. A supervisory authority thus may be tempted to postpone intervention, hoping the bank may recover by itself. This is known as supervisory forbearance.

Market prices can also help with enforcement. Tying supervisory action in advance to a market indicator reaching some trigger level provides the supervisor with a commitment strategy. Banks know the supervisor will have to intervene; they cannot speculate on leniency in the event of trouble and have to get their act together on time. At the same time, the supervisor becomes more accountable to the public: As market prices are observable, everybody would detect if bank supervisors postpone some due intervention.

An example

In 2002, the two big Swiss banks, UBS and Credit Suisse Group (CSG), like most internationally active banks, felt the joint impact of negative events like weak stock prices in the aftermath of the internet bubble and debt problems in several countries (Russia, Argentina). These factors led to a temporary increase in the yield spreads on the two banks' subordinated debt. As Figure 7.3 shows, the hike in spreads was much more pronounced in the case of CSG. In fact, unlike UBS, CSG was hit by some bank-specific problems. In particular, the CS Group suffered losses at one of its important insurance subsidiaries as well as at the Group's investment banking arm, Credit Suisse First Boston (CSFB). As Figure 7.3 shows, the market assessed the default probability for CSG at around 2.3 per cent at its peak.

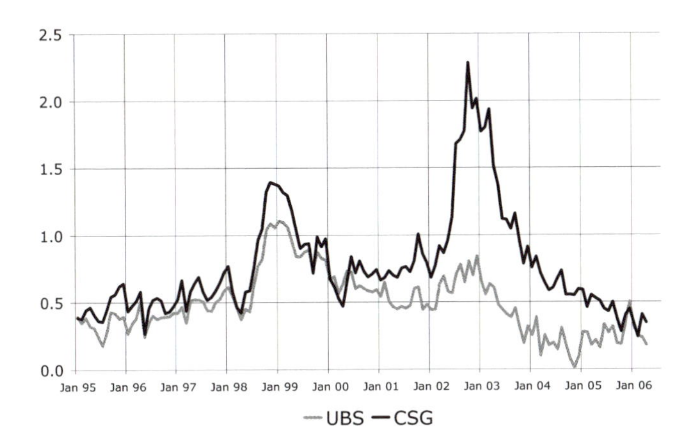

Figure 7.3 Yield spreads over government bonds on subordinated debt of Swiss big banks (UBS and CSG) show different spikes in the 2002 crisis.

Policy proposals

Several authors have proposed that market data be used as a source of supervisory information or even as triggers for action. One particular advantage of market prices is that some people have put their money behind them. Therefore "market participants have an incentive to look through reported accounting figures to the real financial condition of a bank and to price a bank's securities based on their best estimates of the distribution of the securities future cash flows" (Evanoff and Wall 2001, p. 1).

The most likely candidates for solvency indicators are share and bond prices. Shares are traded regularly in relatively liquid markets. The drawback of share prices is that they reflect the whole distribution of future cash flows, including very favorable developments. Supervisors, however, are more concerned with adverse developments. Information about the lower tails of cash flow distributions is most likely to be found in the prices of subordinated bank debt. In the resolution of a bank, subordinated debt is only repaid if all other due liabilities have been honored. The yield spread on subordinated debt—the difference between its yield and the yield of a riskless (e.g., treasury) bond of similar maturity—is often considered to be the ideal proxy for a bank's insolvency risk and the main candidate as a trigger for supervisory actions. (For a more detailed treatment of subordinated debt and a cautious view on the risk-measurement quality of its yield, see the respective Application in Chapter 14 and the literature cited therein.)

An intense debate on the pros and cons of using market information for supervisory purposes has developed among academics, central banks and supervisors. A broad US study (BoG/DoT, 2000) examines the suitability of subordinated debt prices as supervisory indicators.

In line with that study, most supervisors take a cautious view. While they recognize the potential information content in market prices, they hesitate to make any automatic links between market data and supervisory actions. "I do not need a spring gun!" as a Swiss supervisor put it. The view of the supervisory profession is probably well represented by a well-balanced statement by Alan Greenspan, longtime Chairman of the Treasury:

> Significant changes in a banking organization's debt spreads, in absolute terms or compared with peer banks, can prompt more intensive monitoring of the institution.
>
> (Greenspan, 2001)

7 E b The supervisor's dilemma

The main problem in the use of market data for enforcement purposes is the endogeneity paradox: Observed prices include all information, including expected supervisory actions. A supervisor therefore cannot use the same price both as a basis for intervention and as a source of information (about the world *without* the expected intervention). Above we illustrated this dilemma with the example of Alice organizing the Hedge Fund tea party. Here we will present it in a bank supervisory setting. See Birchler and Facchinetti (2006) for more details.

Think of two banks, both of which are rated as "troubled" by the supervisory authority. The supervisor conducts an on-site examination at each of the two banks. For Bank A the outcome of the on-site examination is surprisingly good, for Bank B it is surprisingly poor. In the market, the subordinated debt spread of one bank falls, while the other rises. Which bank is which?

The answer seems obvious: Of course, the spread of Bank B, the one with a surprisingly poor outcome, must increase. Unfortunately, this may be wrong. It would be correct if the supervisor took the same stance with both banks. Yet, the supervisor is likely to take a much tougher stance with Bank B. The latter, though initially in a worse condition than Bank A, will be in a relatively better shape *after* the supervisory intervention. DeYoung, Flannery et al. (2001, p. 902), from where this example is taken, conclude that "the anticipated regulatory response frequently dominates the information's implications about current bank conditions."

How should a supervisor read the changes in the two banks' debt spreads? A naïve supervisor might conclude that the on-site examiners, pointing at Bank B, did a bad job: Markets after all seemed more concerned about Bank A than about Bank B. A sophisticated supervisor, by contrast, would understand that bond spreads already include expected supervisory intervention. Observed spreads thus do not tell the supervisory authority where they *should* intervene, but where they are *expected to* intervene.

As the example shows, prices are endogenous to expected intervention. Once intervention is also endogenous to prices, things become tricky, indeed. The following example illustrates the difficulty of reading information from prices if prices incorporate expected action.

A subordinated debt spread policy

Assume that banks have to issue subordinated debt. The supervisory agency declares (or is obliged by law) to act, whenever the yield spread (the difference in yield to an equivalent government bond) a bank pays on its subordinated debt reaches 3 per cent. A spread of 3 per cent over treasury would suggest a 3 per cent p.a. chance of failure. The intervention of the supervisor against a bank hitting this ceiling could consist of either (i) mandatory recapitalization or (ii) closure and liquidation.

In the case of (i), rational investors would anticipate that whenever a bank's subordinated yield spread hits the 3 per cent ceiling, recapitalization would follow. After recapitalization, the bank would be quite safe again. No investor would thus sell subordinated debt at a discount corresponding to a yield spread of 3 per cent. Thus, the spread would never hit the threshold, and the authority would never intervene. Here we have the paradox again. This is reminiscent of the exchange rate target zone literature: A perfectly credible announcement by a central bank to keep the exchange rate within a target zone never requires intervention, as explained in Krugman (1991).

In the case of (ii), rational investors would anticipate that whenever the subordinated yield spread reaches the trigger rate of 3 per cent, the bank would be closed and liquidated. Liquidation often destroys value; subordinated debt holders thus would most likely take a loss corresponding to more than the 3 per cent yield spread. Anticipating this, nobody would buy subordinated debt with a spread of, say, 2.99 per cent. In fact, the spread would directly jump from a value below 3 per cent to a value above it.

Are market prices useless for policy?

As both examples show, the behavior of market prices depends very much on the anticipated supervisory reaction. Therefore, they are hard to decode for supervisors. It may seem that market prices are quite useless for supervisory policy. This is not necessarily true. Under relatively strong assumptions, market prices remain fully revealing despite rationally anticipated supervisory intervention (Birchler and Facchinetti, 2006). The conditions under which the game between supervisors and market participants has (i) no, (ii) one, or (iii) more than one equilibrium remain an area for future research.

Such research (like Bond, Goldstein and Prescott, 2006) seems all the more important, given the endogeneity problem which arises in different contexts (presented here in a supervisory setting). Yet, central banks may also use market prices as a basis for decisions. Important information for a central bank is the level of consumer prices. Some economists suggest that a central bank should also look at asset prices in order to prick bubbles early. Yet, if a central bank does, both consumer prices and asset prices become endogenous to central bank decisions. In the case of consumer prices, this may not be as severe as in the case of prices for daily traded assets which are very susceptible to the endogeneity effect.

7 F Conclusions and further reading

In this chapter we have tried to give an intuitive insight into the market's ability to extract private information from individuals and to aggregate or integrate them into prices. Ideally, the market works so well that market prices are a sufficient statistic for the information behind them. In this case observation of prices leads to the same equilibrium as knowledge of the underlying information itself.

Some securities and the related markets do not exist, because the underlying event is not in people's perceived state space. In plain English: because nobody even thinks of the event as a possibility. Herein lies a trap for those who believe that prices are good predictors of future events. Indeed, prices may well reflect the probabilities of those events that we think may happen at all. Yet, prices contain no information regarding what events are possible in the first place. Prices only exist where a market exists; a market exists where a bet has been defined; and a bet can be defined if someone has thought of a possible event. Prices are like answers: They first require a question to be asked.

In reality, markets are incomplete. They are more incomplete, the further we try to look into the future. While the future will undoubtedly surprise us (or those of us who will live long enough to see it), human imagination nevertheless tries to cast a ray of light into the dark of the next few decades.[39]

We also tried to show that although prices tend to aggregate information, this information may not be ready for use. As long as some players try to predict future developments (like terrorist attacks or banking crises) with the intention of preventing those events, prices become endogenous to preventive efforts and may cease to reflect the original unconditional information. The attempt to use prices as a basis for action may destroy their very meaning.

For further reading we recommend the following texts: Brunnermeier (2001) and Lengwiler (2004) explain all the theoretical concepts like markets, efficiency, information revelation through prices, etc. As mentioned above, the evidence on the informational efficiency of financial markets is summarized in Shleifer (2000, Ch. 1). An overview of the relatively new but already expanding field of market microstructure is O'Hara (1995).

The present chapter was written around some inherent difficulty: Prices do not only reflect information about fundamentals (like the probability of rain tomorrow), but also information about beliefs. We tried to explain the aggregation power of markets without explicitly making this difference; some lack of clarity here and there may be the price we (or the reader) paid. It is very timely therefore to devote the following chapter to this important issue.

Checklist: Concepts introduced in this chapter

- Contingent claims
- Arrow securities
- Complete markets
- Riskless arbitrage

- Equilibrium prices
- Aggregation of information through trading
- Informational efficiency (in a theoretical sense)
- Market efficiency (strong, semi-strong, weak)
- The Information (or Grossman–Stiglitz) paradox
- No-trade theorems
- The endogeneity dilemma

7 G Problem sets: Two heads know more than one

For solutions see: www.alicebob.info.

Problem 7.1 Aggregating heterogenous information
Semi-finals in the soccer world championship. Switzerland is still in the game and only three other teams are left. You are very keen on finding out the chances wether Switzerland makes it into the final game (event S) or not (event s). One of the other teams is Brazil who may make it to the final (event B) or not (event b). You have no information on the probability that your team makes it nor any information about the information structure of other individuals, like experts, coaches and players, as they won't tell you. To find out the probability you set up securities so that the prices tell you the chance of making it into the finals.

(a) What kind of securities would you set up (*Hint:* what are the relevant states of nature?)
(b) Marius knows that in the final Switzerland will play against Brazil with a probability of $\pi_{SB} = \frac{1}{2}$. The Brazilian coach knows that their chances of making it into the final are $\pi_B = \frac{3}{4}$. The Swiss coach knows they will make it into the final with a chance of $\pi_S = \frac{1}{2}$. Calculate the equilibrium prices of the different securities.
(c) What assumptions are necessary for prices correctly reflecting the probabilities? Do you think these assumptions hold in this setup?
(d) What would be the problem in setting up the securities in such a way that they pay contingent on the states S, s, B and b?

Problem 7.2 Airline safety
Could we infer the safety of individual airlines from a thorough analysis of ticket prices?

Problem 7.3 Target zones
Assume the exchange rate for the Swiss franc is at 1.20 CHF/USD. Now, the Swiss National Bank announces that it will keep the exchange rate in a band between 1.00 and 1.30 CHF/USD. Do you expect any changes in:

(a) the level of the exchange rate?
(b) the volatility of the exchange rate?

Compare your solution with Krugman (1991).

Problem 7.4 The intra-agency market

In an article "A Good Idea With a Bad Press" in the 31 July 2003 edition of *The New York Times*, the economist Hal Varian defended the Pentagon scheme discussed in Section 7 D above:

> If such a market were put in place, should the Secret Service monitor it?... And the most important question of all: If you were a potential target, wouldn't you want the best possible forecasts of possible attempts on your life? I would. I would also prefer that such forecasts not be quite so public. A private market in the probability of specific assassination attempts – say within an expert community like the C.I.A. – could make sense, assuming, of course, that it provided useful information.

(a) What are the pros and cons of such an *agency-internal* market?
(b) What incentives for agency staff might such a market create?

Problem 7.5 Market data based supervision

Assume that the bank supervisory authority closes down a bank whose subordinated debt has a market yield exceeding the treasury bond rate for the same maturity by three per cent. What would be the impact on:

(a) the behavior of the bank's subordinated debt yield?
(b) on the frequency of supervisory intervention?
(c) on the risk taking attitude of the bank's management?

8 Knowing facts or reading thoughts?

8 A Introduction

The Greek philosopher Socrates (469–399 BC) coined the phrase "True knowledge exists in knowing that you know nothing". Not quite as old is the joke about the man who believed he was a mouse: After some years in the mental asylum, one day he finally managed to convince himself that he was not a mouse. On the day of his release, the director bids farewell, and the man walks towards the gate. Less than five minutes later, he is back, trembling. "What's wrong, Sir?", the director asks. "There is a c-c-cat w-w-waiting on the street." "But you know that you are not a mouse, Sir, right?" "Well, yes, sure I know. But how do I know the cat knows?"

What do the philosophical paradox and the joke have in common? Both play on the distinction between two levels of knowledge—knowledge about facts versus knowledge about knowledge. In one case, knowledge of facts (we know nothing about the world) is confronted with our knowledge about *our own* knowledge (we know that we do not know anything for sure). In the other case, knowledge of facts (I am a man) is confronted with knowledge about what *someone else* (the cat) knows.

The distinction between factual or fundamental knowledge and knowledge about knowledge is also a crucial distinction in economics. As an illustration, take the Nathan Rothschild legend from Chapter 1. Rothschild knew that England had won at Waterloo. Was this a valuable piece of information? No, not *per se*. What made it valuable was the meta-information it came with. First, Rothschild knew that *nobody else* knew who had won. However, he also knew that the other traders at the London stock exchange knew that *he* knew who had won; at least they would quickly learn once he bought or sold conspicuous quantities. Therefore, he had to sell in order to mislead them and to buy through a straw man. Clever as the scheme looks, it should nevertheless have failed! Rational traders should have anticipated that Rothschild would try to mislead them. They should have *known* that he knew that they knew about his information advantage. A clever trader might have *bought* on seeing Rothschild sell! Traders thus missed the unique chance to buy Rothschild's portfolio at fire-sale prices. This may be the reason why the story is only a legend; the true Nathan Rothschild may have been too clever to fall into a self-made trap.

The example makes clear that knowledge about facts (who won the battle) can become quite unimportant compared to knowledge about knowledge. This is why we devote a full chapter to the distinction between so-called meta-information or higher-order information and fundamental information. Or, to put it simply, reading thoughts versus knowing facts. Examples of the impact of higher-order information (or higher-order uncertainty) abound. Thinking about what others think is important in both economic and everyday life. In the financial markets higher-order beliefs may trigger (or prick) asset bubbles or create systemic risk. In the social sphere, higher-order information helps to explain fairly different phenomena like conformism, the use of mediators in conflict situations, or why we apologize.

As usual, we shall present the main ideas in a non-technical way. In the theory section we start with knowledge about one's own knowledge (remember Socrates). We continue with knowledge about others' knowledge (remember the man and the cat). The key concept in this chapter, and one of the most important concepts in information economics and game theory, is *common knowledge*. The idea of common knowledge explains something we would count as one of the top ten phenomena in economics: An announcement (for example, by a central bank) can have a dramatic impact, even if it does not reveal anything new. The present chapter also lays the foundations for the two subsequent chapters, Chapter 9 on coordination and and Chapter 11 on the macroeconomics of information.

8 B Main ideas: Fundamental versus strategic uncertainty

The main idea of this chapter, as we mentioned earlier, is that there is information about facts, and there is information about information. In this respect information fundamentally differs from any other economic good: Information can refer to itself. An apple cannot be about another apple. But some information can be about some other information.

Information about information, so-called "higher-order" information, is not just the stuff of philosophical paradox or jokes about men and cats. One of the most important findings in modern economics is that knowing facts is not the only thing that matters, but that reading thoughts is often more important. "It is now clear that much of any economic importance largely depends upon what people know. But the issues run deeper than this first level—it is important to keep track not only of what people know, but what they know about what others know, and perhaps what they know about what others know about what others know, and so on" (Samuelson 2004, p. 368).

The higher orders of knowledge are important in situations of *strategic uncertainty*. While fundamental uncertainty refers to some unknown play by nature (the non-strategic player; see Chapter 4), strategic uncertainty is the effect of unknown play by other players. Strategic uncertainty prevails whenever (i) our payoff depends on others' play and (ii) we are uncertain of their strategies. Rolling dice or throwing a boomerang is a game against nature and, as such, is subject to fundamental uncertainty. Playing poker is a game among humans (though nature deals good and bad hands), and is thus also subject to strategic uncertainty.

Bluffing, a typical technique in poker, is a reaction to (and a source of) strategic uncertainty. Obviously, knowing what others know and what they know about what we know is important in these kinds of games.

Higher-order information is particularly important in financial markets. Stock prices are what they are, not what they ought to be. Prices do not always reflect the "true" fundamental value of the shares (the sum of the discounted values of future dividends). For this reason, it is important not only to know what a share is worth, fundamentally speaking, but what the market will believe it is worth at the expected time of selling. Obstinately betting on fundamental values (assuming they are known), may lead to huge transitory losses. As practitioners say, "it is better to be right for the wrong reason than to be wrong for the right reason" (Paulos, 2003). Yet, in order to be right, one should know what others think about what others think. John Maynard Keynes formulated this insight in his immortal beauty contest analogy (see Box 8.1).

Box 8.1 The stock market as a beauty contest

John Maynard Keynes (1893–1946) wrote his *General Theory of Money, Interest, and Employment* during the years of the Great Depression and its immediate aftermath. One event that had marked the beginning of the economic crisis was the stock market crash of 1929. Keynes was concerned that financial markets could get the better of the real economy. Financial assets often were not priced according to their fundamental values, but rather according to beliefs. Keynes introduced the idea of higher-order beliefs by analogy to a newspaper beauty contest (Keynes 1936, pp. 156ff.):

> Professional investment may be likened to those newspaper competitions in which the competitors have to pick out the six prettiest faces from a hundred photographs, the prize being awarded to the competitor whose choice most nearly corresponds to the average preferences of the competitors as a whole; so that each competitor has to pick, not those faces which he himself finds prettiest, but those which he thinks likeliest to catch the fancy of the other competitors, all of whom are looking at the problem from the same point of view. It is not a case of choosing those which, to the best of one's judgement, are really the prettiest, nor even those which average opinion genuinely thinks the prettiest. We have reached the third degree where we devote our intelligences to anticipating what average opinion expects the average opinion to be. And there are some, I believe, who practise the fourth, fifth and higher degrees . . .
>
> It is the long term investor, he who most promotes the public interest, who will in practice come in for most criticism, wherever investment funds are managed by committees or boards or banks. For it is in the essence of his behaviour that he should be eccentric, unconventional

and rash in the eyes of average opinion. If he is successful, that will only confirm the general belief in his rashness; and if in the short run he is unsuccessful, which is very likely, he will not receive much mercy. Worldly wisdom teaches that it is better for reputation to fail conventionally than to succeed unconventionally.

The beauty contest analogy has become famous and has been quoted countless times. Keynes' insight inspired many authors, such as Shiller (2000), as well as some work by Morris and Shin (2002) and Allen, Morris and Shin (2006) we will discuss below.

On common knowledge, a key concept in this chapter, the Stanford Encyclopedia of Philosophy has the following example:[40]

> A waiter serving dinner slips and spills gravy on a guest's white silk evening gown. The guest glares at the waiter, and the waiter declares "I'm sorry. It was my fault." Why did the waiter say that he was at fault? He knew that he was at fault, and he knew from the guest's angry expression that she knew he was at fault. However, the sorry waiter wanted assurance that the guest knew that he knew he was at fault. By saying openly that he was at fault, the waiter knew that the guest knew what he wanted her to know, namely, that he knew he was at fault.

With the public declaration, the waiter's fault becomes "common knowledge" between him and the guest. Common knowledge is the highest level in an infinitely tall building. There is information about facts, information about information about facts, information about information about information about facts, and so on—an infinite chain leading in the limit to common knowledge. A public statement, like the waiter's, establishes common knowledge. There is no remaining doubt about the facts, nor about who knows what.

The difference between common knowledge and public knowledge may appear slight. However, it is not. In many contexts, whether economic or social, it makes a great difference whether a fact is "only" known to everyone or is common knowledge. Even tenth-order knowledge, for example, is not common knowledge. The incidence of common knowledge can have a dramatic impact in many situations—from stock markets to marital harmony.

8 C Theory: Higher-order information

8 C a Knowing what you know

What do we know about what we know? Even the philosopher who claimed that we are ignorant assumed we could know about our ignorance. We have introduced the state-space model of information (Chapter 3) as a tool to describe what an individual knows. But we have yet to introduce the axioms on which the state-space model rests. These are called the "axioms of knowledge".

The axioms of knowledge

There are five axioms of knowledge (see Samuelson, 2004). These describe what we know and what we know about what we know:

1. *Awareness*
   We know the possible states of the world.
2. *Omniscience*
   We know all implications of what we know.
3. *Knowledge*
   We can only know things that are true (events that have happened).
4. *Transparency*
   We cannot know something without knowing that we know it.
5. *Wisdom*
   We know what we do not know.

Together, these five axioms constitute the state-space model of information.

Some implications

The axiom of *awareness* excludes the existence of states of which we are not aware. When Alice throws a standard dice, it is unfeasible that she is not aware of the possibility that the result, the number shown on the top surface, could be 1. At the same time, she is not aware that the result could be a 7; therefore this result is impossible. Unlikely events may happen, but a real surprise is not possible. Applied to real life, this is a very restrictive assumption.

The axiom of *omniscience* states that if we know something, then we also know all its implications. Knowing that both event E and event F have happened is equivalent to knowing that the event "intersection of E and F" ($E \cap F$) has occurred. If Alice knows the number on her dice is in the upper half and odd, she knows it is a 5 (and vice versa).

The axiom of *knowledge* establishes that we do not err; we cannot know that a dice shows an odd number when in fact it shows an even number. The axioms of *transparency* and of *wisdom* state that we know our information structures (we know which states lie within the same subset and which do not). When Alice knows that she can distinguish whether a dice shows an odd or an even number, she knows that she can distinguish between a 1 and a 2, but not between a 2 and a 4. These axioms have an infinite number of layers: If Alice knows that she knows the result, she also knows that she knows that she knows, etc.

The five axioms of knowledge describe what an individual knows about the world and about his or her own knowledge. Yet, most economic transactions involve more than one person.[41] Therefore, it is very important to know what *others* know. This is where higher-order beliefs come in.

8 C b Knowing what others know

Ground level zero

Information has many different levels. We count the levels like Europeans count the floors of a house: The ground floor, information about facts or "fundamental information", is zero-order information. Levels above the ground floor represent meta-information, starting at first-order information (I know what you know). We feel this way of counting best highlights the difference between knowing facts and knowing knowledge. Be aware, however, that a large part of the literature use American terminology and counts the ground floor as the first floor.

The higher levels

The different orders of information are represented in Table 8.1 (following Ayres and Nalebuff, 1997).

The double helix of knowledge

Alice's and Bob's information is in fact interrelated. If Alice knows that Bob knows ($1a$), it follows that Bob knows ($0b$).[42] Symmetrically, if Bob knows that Alice knows ($1b$), Alice knows ($0a$). There are two chains of knowledge: For example, $2b$ implies $1a$ which implies $0b$. Conversely, $2a$ implies $1b$ which implies $0a$. This is why Ayres and Nalebuff (1997) compare the higher orders of knowledge to a double helix. This double helix is of infinite length. The limiting order, *common knowledge*, implies all lower orders.

8 C c The magic of common knowledge

Definition

An event is *public knowledge* among a group of individuals, if everyone in the group knows it. In Table 8.1, $0a$ and $0b$ together represent public knowledge.

Table 8.1 The different orders of information

Order	*Alice's knowledge* a	*Bob's knowledge* b
0	A knows	B knows
1	A knows that B knows	B knows that A knows
2	A knows that B knows that A knows	B knows that A knows that B knows
...	... and so on ...	... and so on ...
∞	"common knowledge"	

The event is *mutual knowledge* if everyone knows that everyone else knows. Mutual knowledge implies public knowledge. The event is *common knowledge* if everyone knows that everyone knows that everyone knows, and so on *ad infinitum*. Common knowledge implies mutual knowledge.

The idea of common knowledge, first defined by the philosopher of language David Lewis (1941–2001), was introduced to economics by the Nobel laureates Robert Aumann and Thomas Schelling. Common knowledge is a key concept in game theory, and thus in many economic contexts.

The emperor's new clothes

In Hans Christian Andersen's fable "The emperor's new clothes" (Sunder, 2002), two swindlers promise the emperor a superb new cloth. The cloth will be made of silk and gold threads so fine that they are only visible to intelligent people. To stupid people, however, they are invisible.

When the emperor is invited for a first show of the garment, he can only see a loom but no cloth. Yet, for fear of looking stupid, he does not dare say anything. Instead he asks his courtiers who cannot see any cloth either but who, for the same fear of looking stupid, break out in praise of the magnificent garment. When the emperor finally walks in a procession to show off his new clothes to the public, everyone applauds their beauty. Only a small child says: "But he doesn't have anything on!" Now, suddenly, everyone agrees that the emperor is naked.

The child spoke out something that everyone already knew. Why, then, is it possible that people changed their minds so quickly? According to Sunder (2002), the child's remark was so powerful because it changed higher-order beliefs. Several observers, including the emperor, could not see any clothes, but feared that they were in fact stupid. Others, who saw no clothes either, knew they were not stupid; but they feared that they might be branded stupid by others. Others may also have reasoned: I know the emperor is naked, and I know the others know; yet, as the others may not know whether all others know, they will still pretend to see the clothes; therefore I had better pretend to see clothes, too.

The child's remark turned the situation around. It became clear that the inability to see the clothes was not linked to stupidity (the common presumption being that the child was innocent, but not stupid). Not only did this become clear to everyone; it also became common knowledge. Sunder (2002) concludes: "The higher the order of beliefs on which the applause was based, the easier it would have been for the child's remark to change their minds. As a house of cards grows taller, the easier it becomes to topple it."

The lever of common knowledge

A chairman of the Federal Reserve Board is hardly as naïve as the child in the fable. Yet, as the example of former chairman Alan Greenspan's famous "irrational exuberance" (see Box 8.2) speech illustrates, an innocent remark can have severe consequences if it happens to move the lever of common knowledge.

Box 8.2 "Irrational Exuberance"

On 5 December 1996, Alan Greenspan, then chairman of the Federal Reserve Board, during a rather low-key dinner speech (Greenspan, 1996) asked in passing: "But how do we know when irrational exuberance has unduly escalated asset values . . .?" The following day stock markets around the world lost about 4 per cent. What had happened? Alan Greenspan's question sounds no less innocent than the child's remark about the emperor. It is not even a statement, just a question. In particular, the chairman made no forecast; to the contrary, he claimed that even the Federal Reserve was unable to predict stock prices. However, Greenspan made it clear that one could believe that the stock market was overvalued without necessarily being stupid. Given the worldwide interest in Greenspan's speeches, the inequality of "bearishness" and stupidity became common knowledge. A portfolio manager thus could recommend selling without threatening her reputation. A plunge in prices was possible even if nobody was fundamentally pessimistic. The mere knowledge that others knew that others knew means that pessimists would now speak up and recommend selling which might have triggered some sales. This is not even unlikely; indeed most shares picked up soon afterwards. But, sudden common knowledge that selling was not a sign of stupidity made selling rational in the short run.

"Irrational exuberance" would shake markets once more. The Greenspan speech did not prevent the biggest bull market in US history, known as the "new economy bubble", "internet bubble" or, as it happened in the late 1990s, the "millennium bubble". Believing that the "new economy" was an emperor without clothes, Robert Shiller, a Yale economist who was known almost exclusively within the still relatively new field of "behavioral finance" at that time, wrote a book (Shiller, 2000) in which he offered a grim outlook for stock investors. For the title of his book Shiller borrowed Greenspan's expression "Irrational Exuberance". Princeton University Press sent out the first review copies of the work in early March 2000, shortly before the NASDAQ composite index reached a record level of 5078 on 24 March. The subsequent first slide in stock prices led Princeton Press to bring forward the official publication date from 7 May to 8 April. After the publication, share prices fell further; two years later the NASDAQ had fallen by more than 70 per cent, back to the level where it stood at the time of Greenspan's speech. While markets plummeted, sales figures for "Irrational Exuberance" rose, catapulting the book into the top 20 on the New York Times bestsellers' list for hardcover non-fiction.

We certainly do not suggest that Robert Shiller "sunk the NASDAQ". We would not even claim that anybody believed in stock market forecasts of a university professor. But the concept of common knowledge makes clear that his book may nevertheless have been a factor. It did provide academic backing to people who secretly entertained private doubts about the level of tech stock. Even if, deep down, no trader would give a cent for academic advice, the book (and its prominent reviews in publications like *Business Week*, the *Financial Times*, the *New York Times*, the *New Yorker*, and *Barron's*) made one thing clear about common knowledge, namely that being bearish in March 2000 was not proof of low intelligence. This might have been sufficient to tumble a house of cards that had grown as tall as new economy stock prices.

Common versus pooled knowledge

When, exactly, is an event common knowledge? Trying to strike a balance between readability and precision, we will argue by use of an example. For more rigorous treatments we refer to Brunnermeier (2001) and Samuelson (2004).

Assume Bob and Alice have the following information structures on the outcome of rolling a dice: Alice's partition is $\{(1),(2,3),(4,5),(6)\}$, Bob's is $\{(1),(2,3,4),(5,6)\}$. What may be common knowledge among them?

The two information structures are represented in Figure 8.1. They are themselves common knowledge between Alice and Bob (otherwise they would not be correctly drawn, see Samuelson (2004)). Common knowledge thus cannot be defined by concepts entirely outside the notion of common knowledge itself.

Look at event $\{1\}$. If the dice shows 1, both know. And, knowing each other's information structure, both know the other knows the result is 1. If event $\{1\}$ occurs, it is therefore common knowledge. Now, compare event $\{6\}$. If the dice shows a 6, Alice knows. But she also knows that Bob cannot distinguish 5 and 6. Thus the event $\{6\}$ is not common knowledge. But could event $\{5,6\}$ be? Assume the true outcome is a 6. Alice knows; this implies that she also knows event $\{5,6\}$ occurred. Bob also knows event $\{5,6\}$ occurred. Event $\{5,6\}$ thus is public knowledgebetween the two. Alice also knows that Bob knows that event $\{5,6\}$ occurred (first-order knowledge). Yet, Bob does not know whether Alice knows. Given his information, the event may be $\{5\}$ which Alice cannot distinguish from $\{4\}$. Event $\{5,6\}$ is therefore not common knowledge. According to the same logic, nor are events $\{4,5,6\}$ or $\{3,4,5,6\}$. Yet, event $\{2,3,4,5,6\}$ is common knowledge, if one of the respective numbers is thrown.

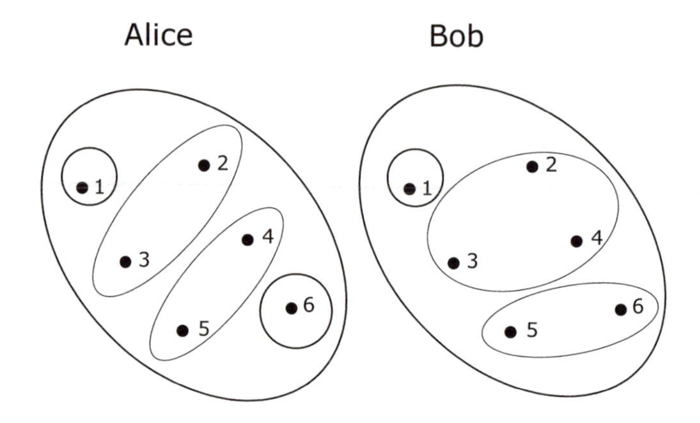

Figure 8.1 In the event {1} or in the event {2,3,4,5,6} Alice's information is known also by Bob, and vice versa. In addition, both know this and know that the other knows, and so on. The two events are common knowledge.

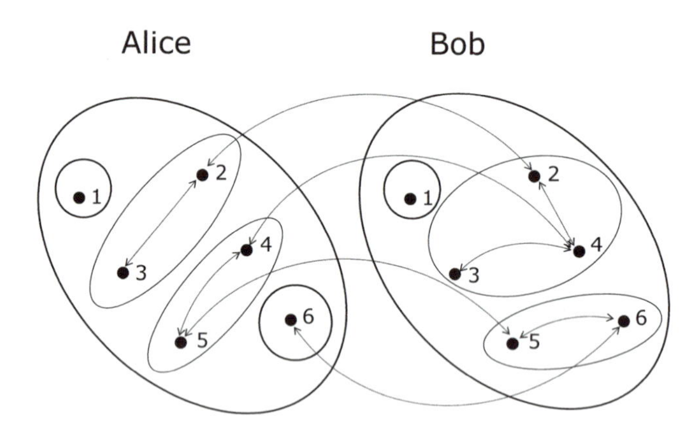

Figure 8.2 An information structure characterized by subsets of the state space may be seen as a "group of islands" with "cities". A traveller is allowed to either move to another city on the island or to fly to another island in another archipelago. This is called "island hopping". The partition consisting of subsets (groups of cities) that cannot be connected by island hopping describes common knowledge.

The "island hopping" analogy

There are only two events in Figure 8.1 that become common knowledge when they happen: {1} and {2,3,4,5,6}. As Figure 8.2 illustrates, common knowledge can be illustrated by the idea of "island hopping". Look at information structures as groups of islands (the subsets of the state space) and at individual states of nature as cities on those islands. Assume a traveller starts in one city (the true state of nature) on Alice's archipelago. From there, he is allowed two kinds of moves: (i) walking to another city on the same island, or (ii) jumping from any city to the corresponding city on Bob's archipelago, and so on. After any move, the same rules apply with respect to the new location. Areas of common knowledge are represented by the area a traveller can touch, but not leave while observing the above rules.

The join . . .

In the example in Figure 8.2, a traveller starting in {1} can never get to any other city; conversely, a traveller starting in {2,3,4,5,6} can get to any other city in this subset. The information structure:

$$P_c = \{(1), (2, 3, 4, 5, 6)\}$$

represents the areas that are common knowledge, whenever they contain the true state of nature. Common knowledge is represented by a coarser information structure than individual knowledge. Therefore, P_c is also called the finest common coarsening or the *meet* of Alice's and Bob's information structures.

. . . and the meet

The common knowledge structure P_c tells us what Alice and Bob commonly know without any communication. It does not represent what they might know from *pooling* their individual knowledge. If Alice and Bob can in fact talk to each other (or mutually observe actions, or trade with each other), they can use the full joint potential of their individual knowledge. The information structure resulting from joining individual information is found by intersecting events (islands) across the two individual information sets represented in Figure 8.2. Under the joint information only two states, {2} and {3}, cannot be distinguished, as they lie on a common island on both archipelagos. Therefore, the pooled or joint knowledge of Alice and Bob reads:

$$P_j = \{(1), (2, 3), (4), (5), (6)\}$$

P_j is called the *join* of the two information structures.

The bridge between individual knowledge, pooled knowledge (the join of information structures) and finally common knowledge (the meet of individual information structures), is communication. Communication can take different forms: speaking, taking observable actions, or trading (if trades are observable). A combination of trading and the development of common knowledge leads to a phenomenon that looks like "subterranean information processing" (Paulos, 2003), the subject of our next example.

8 C d Subterranean information processing

Information may travel underground. We will illustrate this effect of subterranean information processing with an example adapted from Paulos (2003). Events stretch over three periods. During the initial period, some information on a company becomes available. Yet, the share price does not react. During the second period, no new information emerges, and the share price remains unchanged. During the third period, there is again no new information, but the share price suddenly moves strongly.

Hedgeville Hopper, a small local airline, is in financial trouble. If it does not receive financial support by 1 April, it will have to file for bankruptcy. There are two solutions: (i) obtaining credit from the local bank, Hedgeville Mutual, or (ii) selling to Locust Air, the fast-growing regional carrier. Unfortunately for the troubled airline, these are mutually exclusive: Mutual will not lend if there are negotiations with aggressive Locust; Locust will not buy if there is a chance that Hedgeville Hopper will fall deeper into debt with the bank.

Alice and Bob both have some information about the likely outcome. From his rear window, Bob can observe whether Hedgeville Hopper officials go to see the bankers at Mutual's head office. He would not know, though, whether the bank has accepted a deal or not. Alice would know when a deal with Locust was struck (as part of her present part-time job she is responsible for chilling champagne at Locust headquarters). She would not know if a deal was proposed or rejected. These information structures are illustrated in Figure 8.3. The information structures (who knows what), are common knowledge, but the actual information is not.

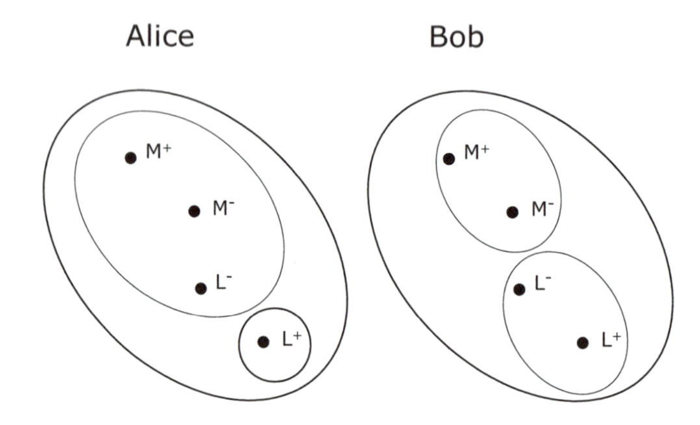

Figure 8.3 Alice knows whether a deal has been struck with Locust (L^+); Bob knows whether Hedgeville Hoppers sought a deal with Mutual or (if not) with Locust (M or L).

Alice and Bob decide to use their information in the stock market. Each has some resource, for example a pension fund account, against which they can borrow a considerable sum. They agree on a simple strategy: If there was at least a fifty per cent chance that the company would get a deal from either Mutual or Locust, they would *buy* a certain amount of Hedgeville Hopper shares. If the odds of a deal seemed less than fifty per cent, they would *sell* the same quantity short. Unfortunately, on 30 March, they have an argument, and they will not talk to each other for the next few days. But they at least trust each other with regard to the agreed stock market strategy.

On 1 April, Bob observes that a management team from Hedgeville Hopper in fact visits Mutual headquarters. This means Hopper decided to go with Mutual. As Bob thinks an acceptance of a deal (M^+) is just as likely as a rejection (M^-), he buys Hopper shares. Alice has not been asked to put the champagne on ice at Locust headquarters (she cannot know that no talks are actually being held with Locust). Locust may have rejected or may not even have received a proposal. Alice must conclude that a positive deal has only a one-in-three chance (one outcome among M^+, M^-, L^-). She thus sells. As Bob buys what Alice sells, the share price does not move.

On 2 April, there is no further news. Yet, via the unchanged share price, the parties know about each other's trading. Alice knows Bob has bought (otherwise the share price would have fallen); Bob knows that Alice has sold (otherwise the share price would have gone up). Bob's buying does not tell Alice anything new: She knows that from Bob's perspective the odds of a deal must have been fifty per cent anyway, irrespective of whether Mutual was contacted or not. Nor does Alice's trading reveal anything new to Bob: He already knew that there could not be a successful deal with Locust, since the offer went to Mutual in the first place. Therefore, neither Alice nor Bob has a single piece of new fundamental information. Yet, information sets have changed: The fact that no deal has been struck with Locust has mutated from public knowledge to common knowledge. Alice now knows that Bob knows she knows, and so on. The new information structures are represented in Figure 8.4.

The new information structures lead to the following trading decisions. Bob, knowing that Locust is out and having no further information about the deal with Mutual, still must think that a deal with the bank has a fifty-fifty chance. Therefore, he continues to buy. Alice, having no new information (except that she knows that Bob knows that no deal was signed with Locust), continues to sell. Again, the Hedgeville Hopper share price does not move.

On 3 April, still no further news emerges. Yet, miraculously, the Hedgeville Hopper share price jumps. What has happened? From the trading patterns of 2 April it has become common knowledge between Alice and Bob that Hedgeville Hopper did not propose Locust a deal. Otherwise Bob would have *sold* on 2 April. (If the proposal had gone to Locust, Bob would have observed no contact with Mutual; the fact that Alice sold on 1 April would have told him Locust rejected an offer.) Alice then knows that the offer went to Mutual. In the absence of any specific information, she thinks the odds of a deal are fifty-fifty. Therefore she

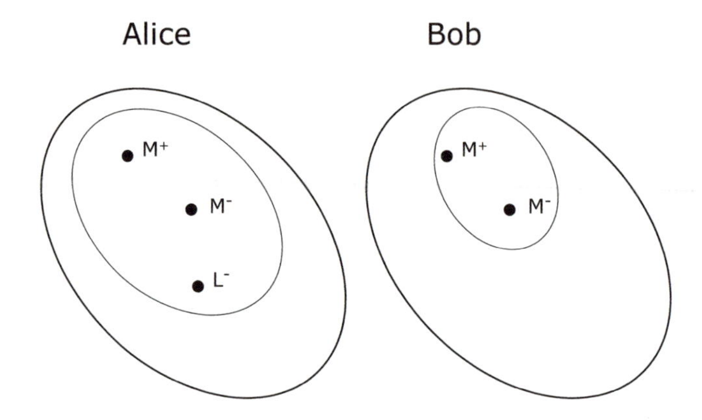

Figure 8.4 It has become common knowledge that no deal has been struck with Locust (L^+).

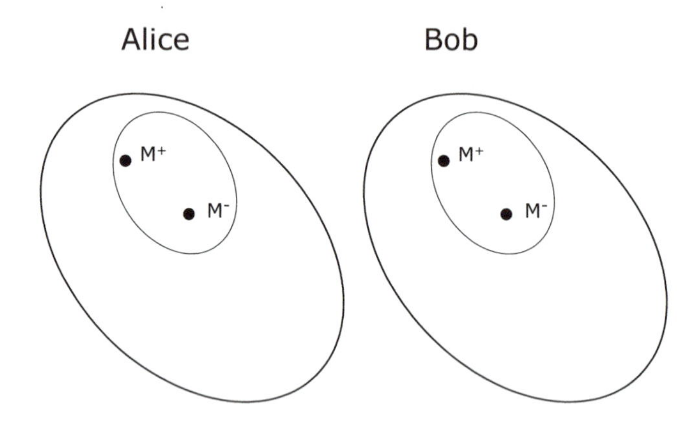

Figure 8.5 It is common knowledge that Hedgeville Hoppers sought a deal with Mutual, but not whether it was accepted or rejected (M^+ or L^+).

buys. So does Bob; having no new information he also believes in a fifty per cent chance of a deal with Mutual. As both buy, the share price rises. Their information structures are given in Figure 8.5.

What looks surprising is that the stock price rises on 3 April, although nothing has happened and no new fundamental news emerged since 1 April. The only thing that has happened is that the market processed the information available to the two investors. It converted private information into public information and, ultimately into common knowledge. Technically speaking, the market enlarged the scope of common knowledge from the so-called meet of the two initial information structures to their join. Without a single word being spoken, the market achieved what credible direct communication among the parties could have achieved.

The example illustrates that financial markets are capable of what Paulos (2003) calls "subterranean information processing". Private knowledge becomes common knowledge, when individuals can observe (or infer from) each others' actions. As mentioned in Chapter 3, a market transaction can be a linguistic signal in the same way as the spoken word. Even the absence of a transaction may convey information. As we have seen above, in markets people put their money where their beliefs are. The money standing behind market transactions makes these very credible signals. This makes the market a powerful instrument of communication. Common knowledge is a key part of this instrument.

8 C e Can we agree to disagree?

Box 8.3 We cannot agree to disagree

Human beings disagree all the time. They have different views on politics, different tastes and different expectations as to tomorrow's weather. Yet, in 1976, Robert Aumann in a paper entitled *Agreeing to Disagree* presented a startling result: Rational individuals cannot agree to disagree. More explicitly, let's take two people who are interested in discovering the truth. Let them share some common prior beliefs. Give each of them some diverging private information, and let them talk to each other. In the beginning, they will have different views, given their diverging private information. Yet, they should not end up in disagreement. In brief, if posterior beliefs are common knowledge, they cannot rationally differ (Samuelson 2004, p. 376; Binmore 1992, Ch. 10.7, pp. 472ff.).

The intuition is: If two individuals take each other seriously, they should revise their beliefs upon hearing the other's beliefs. It is not even necessary to know why the other has a different view—as long as people trust each other to be honest and to use information in a rational way.

Nor is it necessary that individuals are able to state their posterior views explicitly. Trading on their information at publicly observable prices is generally sufficient to establish common posteriors. The negative way to put this is: When individuals entertain no uncertainty about each other's information structures and priors, they cannot expect to gain from any mutual trade (Samuelson 2004, p. 375).

Aumann's finding that we cannot agree to disagree (Box 8.3) may seem to spare us a great many unproductive arguments. But this is far from the truth. It applies to objective truths, not to tastes. Tastes remain an area of unlimited scope for disagreement, even among rational people. Disagreement about facts may also remain, as long as individuals do not take each other's views seriously.

The argument Alice and Bob had a few days ago went like this. They were peacefully playing dice. Evening fell and the light started to fade. It was Alice's turn. In order to win she needed a 4. Unfortunately, the dice rolled on the floor. "I guess you lost," Bob said; "In the dark, I can only say that it is neither a 1 nor a 5 or 6. The probability of a 4 is only one in three!" "No!" Alice exclaimed, "I can recognize points on the four corners. As we agree that it does not look like a 6, I can claim with at least 50 per cent certainty that I won!" Bob insists: "I say 33 per cent." "And I say 50!" "Why do you think you know better?!" "Why do I think I always know better, you mean?" "I didn't say that, Alice." "But you meant it!" "Why would you not check that dice, rather than reading what I may mean or not?" "Check it yourself!" Alice says, walking off hastily.

Events should not actually have taken that turn. Even if both were too angry or too proud to check the dice, they could have done better. True, they had different views. In fact, their information structures were those represented in Figure 8.1. The problem is that Alice and Bob did not take each other's beliefs seriously. Otherwise Bob might have argued: "I can understand why you think you are sure that the dice shows a 4 or a 5; However, I'm sure that it is either a 2, 3 or 4. The only conclusion that tallies with the information that we both have is that you rolled a 4. You've won. Congratulations!"

Alternatively, Bob could have said: "I do not believe your observation. Therefore, I bet $20 against $10 that the dice does not show a 4." "Oh, I bet $100 against $10, it is a 4 or a 5!", Alice might have replied. The odds offered by Alice would signal that she firmly believes her observation. This may have led Bob (who still believes that 5 can be excluded) to counter her offer: "In this case, I bet $100 against $10 it *is* a 4." "In this case, I do not even bet a single dollar against a 5!," Alice would have rationally replied. Again, both would have agreed on the dice showing a 4.

To conclude: If direct communication does not work, a market (like betting) should be able to align posterior beliefs (as shown in the above section on "subterranean information processing"). In fact, the alignment of beliefs through mutual offers prevents a trade (a bet) from taking place. This illustrates the link between the disability of rational individuals to disagree and the no-trade theorems discussed in Chapter 7.

8 C f The bias towards public information

Bob, not seeing Alice presently, has made an appointment to meet with some of his old friends. Their appointment is at 5 p.m. in their local village. The problem is that neither Bob nor his friends can tell the actual time precisely. They all have old wristwatches which are known to be inaccurate. In addition, they can all see the clock on the church tower, but being two hundred years old it is notoriously unreliable, too. Bob would like to arrive with the others, rather than be objectively on time. He equally dislikes waiting and being waited for. He knows his friends share the same preferences.

On the afternoon of the appointment, the wristwatch shows 3 p.m. and the church clock 4 p.m. What time should Bob think it is, and when should he go to the meeting point?

The answer to the first question is based on statistics. Assume that the time displayed on both timepieces is normally distributed around the true time, but that the wristwatch is twice as precise as the clock on the church tower. The best estimate for expected time is the precision-weighted mean of the two individual times (see Chapter 11, and (Myatt, Shin and Wallace, 2002)), the precision of a continuous signal being the reciprocal of its variance. If the wristwatch shows 3 p.m. and the church—which is half as accurate—shows 4 p.m., then 3:20 p.m. would be the expected true time.

This leads to the second question: When should Bob go? Given his best estimate of the present time of 3.20 p.m. and the meeting being at 5 p.m., he might decide that he has 1 hour 40 minutes left. Yet, Bob is not likely to base his decision on the expected true time. Recall that he wants to arrive simultaneously with the others. Consequently, the arrival of the others becomes important. Unfortunately, these share the same problem. Everyone would like to arrive at a time everyone thinks, everyone else thinks (and so on) that everyone will arrive on.

This means that everyone will finally arrive according to "church time". The church is publicly visible, and therefore it is common knowledge. The time on individual wristwatches, by contrast, is private knowledge. Even though the public signal is less reliable (less precise) than the private signals, individuals will use it as a basis for their decisions—for the simple reason that everyone knows that everyone knows it, and so on.

This is more or less the essence of the argument in Morris and Shin (2002), who show that "excessive" reliance on public information can be perfectly rational. Excessive means that we often follow what we perceive as public opinion, although we think we "know better". On the financial markets, observed prices can deviate from prices that reflect average fundamental beliefs. During such asset bubbles, people mistrust or ignore their private judgement in the presence of public information.

An example of public information are announcements by the central bank. Central bank communication therefore is a delicate art (see Chapter 11), particularly in the era of mass media. Above we cited AlanGreenspan's remark on "irrational exuberance", which had a remarkable influence on prices, even though it was void of any fundamental information. The "hub" through which such information becomes public is the press and other media. Shiller, in his *Irrational Exuberance* (2000), points out that the history of bubbles (in shares, tulip bulbs etc.) actually started around the same time as the history of newspapers. He believes that the media played an important role in the internet hype of the second half of the 1990s (see also Paulos, 2003).

There is also evidence that the decline in market values of internet stocks in spring 2000 by 45 per cent was not precipitated by new fundamental information such as disclosures on web-traffic statistics, earnings, or earnings forecasts, but rather by changes in investors' perceptions of already available information before. Keating, Lys and Magee (2003) tried to explain share price movements of internet stocks by fundamental factors. They find that after the decline set in, investors suddenly became concerned about firms' cash availability and future capital-raising opportunities due to qualified audit opinions and media attention. Adjusted cash burn (the fraction of its cash a firm would lose per year) became statistically significant, even though this was roughly known prior to summer 2000. Investors, it seems, grew anxious about firms' chances to continue as a going concern, once they began to believe that others might believe others . . . were becoming anxious.

8 D Application: Keynes in the lab

8 D a A social intelligence indicator

We have quoted Keynes' famous dictum that stock market investment is about "anticipating what average opinion expects the average opinion to be". Experimental economists have tried to test whether people really do. However, they also wanted to know how far people think—whether "there are some . . . who practise the fourth, fifth and higher degrees . . . ", as Keynes put it. As a test, they developed the numbers guessing game (see Box 8.4).

Box 8.4 The numbers guessing game (or "k-game")

The number-guessing or "k-game" has very simple rules: Each participant picks a number from 0 to 100. A prize is awarded to the player whose number is closest to 2/3 of the average.

What number should a participant pick?

One might reason as follows: The average between 0 and 100 is 50; two-thirds of it is 33; therefore, this seems the right number to pick. However, if everyone thinks like that and picks 33, the winning number would be 22. But what if everyone anticipates that number 22 will be picked, or alternatively the optimal response of 15?

This logic can be carried arbitrarily far. Obviously, the game has one (and only one) Nash equilibrium at 0. This is what a rational player would pick, under the assumption that rationality is common knowledge.

From the numbers that subjects actually pick in a controlled experiment, one can infer how many steps these subjects think ahead. The number of steps, often called k, has given the game its name of "the k-game". We may call k a subject's "social intelligence indicator". It measures two facets of intelligence, though: How far the subject thinks ahead, and how far the subject thinks others think ahead. A person who does not think ahead at all, has $k = 0$. A rational person who perceives the others as having $k = 0$ has $k = 1$, exhibiting a fairly autistic form of rationality. A person, by contrast, who thinks that the others are clever and manages to be at least as clever would have a high k.

8 D b Experimental evidence

In the experiment, the Nash solution 0 is not what participants in a k-game typically pick. Numbers between 20 and 30 are much more frequent. Camerer, Ho and Chong (2004) review 24 published experiments based on the "k-game". None of these experiments has a winning number below 15. Game theorists, self-selected readers of a newspaper, portfolio managers or Caltech students exhibited somewhat deeper thinking than CEOs, 80-year-olds, or the Caltech board. People also tend to think a bit harder when the stakes are higher. But only if the game is repeated several times with the same group, will the selected numbers eventually edge closer to 0.

There are two possible reasons why people would pick numbers bigger than 0 in the k-game: (i) They do not see far enough due to irrationality; (ii) they understand that picking 0 is the equilibrium, but they do not think that others will pick 0, as these may not be fully rational themselves or may doubt other players' rationality. It is possible that everyone may *know* that 0 is the equilibrium; yet as long as this equilibrium is not *common knowledge*, it is unlikely to be played. The point is that a player wins by thinking *one* step ahead of the other players, but not by thinking many steps ahead. The equilibrium choice of 0 is not a best response to non-equilibrium play.

Experimental evidence suggests that most people only think very few steps ahead. Camerer, Ho and Chong (2004) try to formalize these findings. They make two assumptions:

- A "step k player" believes that opponents are distributed over steps 0 through $k - 1$. This entails a certain degree of overconfidence.

Step 0 players are assumed to behave randomly. In our example, they would pick any of the possible values between 0 and 100 at random, their average guess being 50. Step 1 players anticipate this behavior and choose 33.3 ($= 2/3 * 50$).[43]

- Players' types k have frequencies given by a Poisson distribution (see Box 8.5).
  Step 2 players in our example would choose 2/3 of the Poisson-weighted average of step 0 and step 1 players.

To illustrate the second assumption, Figure 8.6 depicts the Poisson distribution for four different values of τ. The figure shows the probability of different values of k as a function of four different values for τ. Take, for example, the very left bar of the first group. This bar shows that with $\tau = 1$ the probability of $k = 0$ is approximately 0.37 ($=1/e$). With low τ, the probability mass for k is to the left: with higher τ it moves to the right. With $\tau = 3$ the median for k is 3.

Two values for τ (1.61 and 3.7) are taken from Camerer, Ho and Chong (2004). These authors distill empirical Poisson τ's from the two dozen experiments they review. They find an estimate (median) for τ of 1.61. As Figure 8.6 shows, the average player performs 1 or 2 rounds of computation and less than half of the players performs more than one round. The fourth value for τ used in the figure is 3.7, the median value Camerer, Ho and Chong (2004) find for game theorists.

Box 8.5 The Poisson distribution

Under a Poisson distribution, the discrete variable k (taking integer values: $k \in \{0, 1, 2, \ldots, \}$) has the probability distribution

$$f(k; \tau) = \frac{e^{-\tau}\tau^k}{k!}.$$

A Poisson distribution is described by the single parameter τ, which is its mean and variance (as well as the third moment).

8 D c Why do game theorists fail to play better?

As Figure 8.6 shows, more than half of game theorists do more than two rounds of computation. This is more than "ordinary" citizens do, but still not impressive. Are game theorists less sophisticated than the assumed players in their models? Or is it plausible that game theorists are sophisticated, but do not believe their colleagues are? We do not know. Anyway, there is no such being as "the" game theorist. Just like any other group who participated in theexperiment, game theorists are characterized by considerable heterogeneity (as illustrated by the corresponding

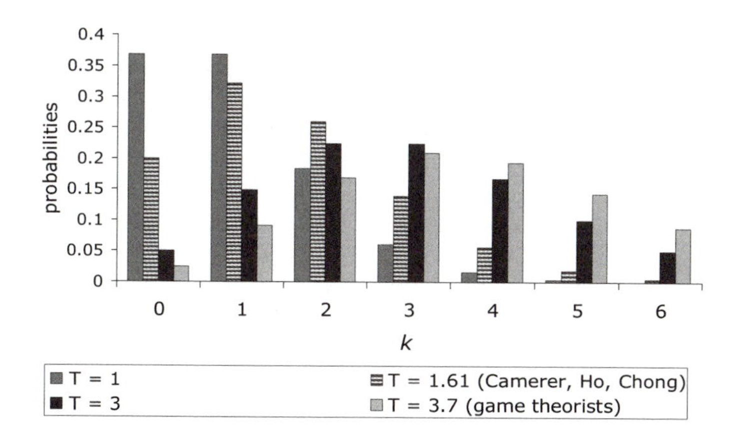

Figure 8.6 Empirical Poisson distribution derived from experiments reported in Camerer, Ho and Chong (2004). The vertical bars, for different values of τ, indicate the frequency of players looking k rounds ahead.

Poisson distribution included in Figure 8.6). Heterogeneity may be both the cause and the effect of the difficulty of picking the right number.

8 E Application: Conformism and learning from debate

8 E a Conformism and systemic risk

Above we used the example of Bob and his friends who were more concerned about arriving simultaneously with all of the others than about being objectively on time. They exhibited a certain degree of "conformism". Conformism is a well-known psychological phenomenon (and often a useful individual survival strategy). It is not only known from practical life, but also documented in experimental literature. People tend to declare a color as "green" after hearing a few other people call it "green", even if in reality it is blue.

It takes two ingredients for a conformist movement to dominate a group: First, group members must exhibit some taste for conformism, like Bob's and his friends' desire to arrive at a meeting at the same time rather than at the right time. Second, a publicly observable signal must provide individuals with a benchmark with which to conform. This can be acentral banker's statement, the example of a fashion leader, the announcement of some trend (like "the new economy" of the 1990s) in the media. Such a public signal, assisted by the lever of common knowledge, often activates latent conformism and triggers parallel behavior.

There is an illuminating scene in the Oscar-winning 1989 Peter Weir movie "Dead Poets' Society". A teacher asks pupils to walk in a circle. When he starts clapping his hands, pupils instinctively and unconsciously synchronize their steps. They "obey" a public signal, although nobody asked them to do so.

This reads like a parable on the role of public information in financial markets. Whenever the press claps its hands—declaring the birth of a "new economy" or the like—watch out! There are several mechanisms in financial markets with a tendency to synchronize individuals. Using dynamic hedging techniques, assessing portfolio managers against "the market" as a benchmark, requiring pension funds to sell shares when prices fall, punishing analysts for dissident views—these all are basically conformist attitudes. Once an external event, like a public signal, synchronizes a few players, others tend to follow. The result is systemic failure, the financial system being unable to insure everyone against the same shock at the same time.

Danielsson and Shin (2003) illustrate systemic risk or, to use their term, endogenous risk, by an analogy from the physical world. On 10 June 2000, the 325-meter-long Millennium Bridge across the Thames was opened. Unfortunately, the innovative structure built without tall supporting columns almost collapsed during the opening. Some unidentified event—a gust of wind, perhaps—induced the thousands of people who were on the bridge to synchronize their steps. The small left-right-left lateral movements of these individuals accumulated to such a degree that the bridge began to wobble violently. In the event, the bridge had to be evacuated.

8 E b Who should speak first?

Every organization faces a dilemma in trying to make the best of its members' information. If the top boss speaks first in a meeting, it becomes difficult for people lower down in the hierarchy to voice dissent. Some important information is thus lost. If, conversely, people on the lowest level first present their opinions, superiors on the next level may find it cautious to confirm their views; finally the top executive is confronted with overwhelming unanimity on the lower levels. Like theemperor and his new clothes, the top executive may not admit dissent for fear of looking stupid or stubborn. Again, some important information is lost during the decision process.

Both the highest or the lowest ranks speaking first may lead to a loss of information and to suboptimal decisions. This dilemma is discussed in Ottaviano and Sørensen (2001). Although it is presented in the frame of a hierarchy, the tendency to conformism does not require a hierarchical structure. It is sufficient that people dislike voicing opinions that conflict with what someone else has said before. This is not the imitation we found in an informational cascade. Here, individuals know what they believe, but they hesitate to admit it. In a cascade, individuals lack some information and try to infer it by observing others' decisions (see Chapter 10).

Would the danger of a conformist bias mean that groups decide worse than independent individuals? Three economists at the Bank of England (Lombardelli, Proudman and Talbot, 2005) conducted an interesting experiment. The task was setting interest rates. Economically literate students had to move interest rates with the goal of stabilizing a notional economy subject to some external random shocks. Interest rate decisions were taken either by individuals or by groups, that is, through voting. Experiments showed that committees did better than all but the very best individuals. The experiments were conducted with and without the possibility of pre-voting discussion among members. Surprisingly, discussion within groups did not improve decisions. In discussions, the benefits of information exchange may be compensated by the negative impact of some dominant (but not necessarily competent) "alpha" members.

8 F Application: Betrayals and mediation

In social relationships—friendship, marriage, parenthood, teaching, teamwork, etc.—dealing with information is often simultaneously important and tricky. First, social relations typically are of a repeating nature. What is said—or not said—today may become important tomorrow or even years from now (an example is given in Box 8.6). Second, values like mutual respect, self-esteem, honor, reputation in social relations play an important role, as Ayres and Nalebuff (1997) point out.

What advice would an economist give to a husband who finds that his wife is having an affair? Assume that the husband still loves her and would want to stay married, if only she would stop cheating on him. What could he do?

One option is to confront her and ask her to stop. However, he might think twice. Confronting her changes the information set. True, the wife already knows that she is cheating on him, and would therefore not learn any fundamental (zero-order) news. But she would learn that he knows (first-order information). Apart from the shame of being caught, she might fear he could exploit her sense of guilt. In addition, by confronting her, he would learn that she knows that he knows (second-order information). This would let him see her future behavior as giving in (if she stops) or as obstinacy (if she continues). Worse, she would also know (third-order information). This again would place him in a dilemma between playing tough, with the risk of losing her entirely, and playing soft, thus seemingly accepting her extra-marital affair. The higher orders of information make things quite tricky indeed.

Fortunately, the husband may have a better option. He could remain silent and communicate to his wife through a third person like a common friend or a professional mediator. This would permit him to exchange the fundamental information without creating higher-order information. The mediator could ask the wife: "If your husband would find out, would you be ready to finish your affair in order to save your marriage?" This would not tell the wife that her husband knows, nor would he know whether she knows. The exchange of fundamental information would not invoke considerations of prestige, guilt, trust, and the like. Without making it sound like a threat, the mediator could also communicate to the wife that, "according to her personal impression", the husband might want to stay married, if she quits her affair.

A further option would be to remain silent and tolerate the relationship without losing face. Eventually, the wife may figure out that her husband suspects something. But as long as their knowledge is only mutual knowledge (on an arbitrary high level) but not common knowledge, both could still act "as if" nothing was wrong. This helps to postpone a solution—maybe forever. How poisonous this strategy can be for a relationship is illustrated by Harold Pinter's play *Betrayal* (see Box 8.6).

A final possibility (though not strictly an option) is that one day the husband comes home and finds his wife in the arms of her lover. This immediately establishes common knowledge. As in the case of the clumsy waiter, common knowledge makes any discussion about the facts redundant. At the same time it forces the involved parties to find some settlement. In the case of the waiter, the restaurant would pick up the dry cleaner's bill. In the case of the unfaithful wife, a settlement might be somewhat more painful.

Box 8.6 *Betrayal*

Betrayal is a play by Harold Pinter (the 2005 Nobel laureate for literature). The three main characters are Emma, her husband Robert, and Jerry, Robert's best friend. The backbone of the plot is Emma's year-long love affair with Jerry. Emma betrays her husband with his best friend, Jerry; Jerry betrays Robert, his best friend, by never telling him of his relation to Emma. But Emma also betrays her lover Jerry by keeping quiet about the fact that she has told Robert about her affair long ago. Robert in turn betrays Emma, and doubly so: Learning of her affair he answers that he feels more attached to Jerry than to her, anyway. And later, Emma learns that Robert had affairs with several other women over the years. But not enough: Robert also betrays Jerry by concealing that he knows about the affair. Finally, there is an allusion suggesting that Jerry's wife may have had a secret life as well.

The different levels of knowledge drive the conversations. When Jerry and Robert have their regular lunches, Robert does not reveal that he knows. Defeated in bed, he plays his informational superiority in conversation. When Robert invites Jerry for a game of squash, Jerry feels compelled to invent ever more lame excuses to avoid or postpone the game.

Later, when both know, they have the following conversation:

ROBERT: I thought you knew.
JERRY: Knew what?
ROBERT: That I knew. That I've known for years. I thought you knew that.
JERRY: You thought I knew?
ROBERT: She said you didn't. But I didn't believe that.
. . .
JERRY: But you betrayed her for years, didn't you?
ROBERT: Oh yes.

JERRY: And she never knew about it. Did she?
ROBERT: Didn't she? *Pause*
JERRY: I didn't.
ROBERT: No, you didn't know very much about anything, really, did you? *Pause*

So, the play is not only driven by what its characters do, but rather by what they know and what they know about what the others know. It is a striking illustration of the role of higher-order information and beliefs in human relationships, particularly in deteriorating relationships perhaps.

8 G Conclusions and further reading

In many situations of economic and social life, it is not only important what people know about the facts, but also what they know about what others know. We have presented several examples for the importance of higher-order knowledge, ranging from stock markets to marital problems.

In particular, we have tried to illustrate the importance of common knowledge. Common knowledge provides already known information with a lever that can turn a situation from bubble to bust. Public signals normally are common knowledge and thus are amplified by this leverage effect. In financial markets, an important type of public signal are central bank announcements. Greenspan's 'Irrational Exuberance' speech is just one example. Central banks are usually very careful in what they say, and how they say it. It has also increasingly become clear that informational aspects may play an important role in choosing optimal monetary strategies. Inflation targeting, for example, is not only a formula to describe how a central bank should react (by a change in interest rates, say) to changes in economic fundamentals. It is also a way of influencing expectation, i.e. of convincing markets that the central bank is able and willing to prevent any surge of inflation in the future (see Chapter 11 D).

Higher-order beliefs play an important role throughout this book. They stand behind the theorem that "we cannot agree to disagree" and the no-trade paradoxes mentioned in the previous chapter. The role of higher-order beliefs may even challenge the dominance of fundamental valuation models in finance (Sunder, 2002). Higher-order beliefs are further important in coordination problems (Chapter 9) and in a macro-economic context (Chapter 11). In the presence of information asymmetries, higher-order information can lead to market failure, another no-trade effect (Chapter 13). Finally, the importance of higher-order beliefs may have left its imprints on our genes. In Chapter 4 we asked how it comes that humans often try to bargain with nature. Here, we think, is the right place for an answer (Box 8.7).

Box 8.7 "Touch wood". Why we bargain with nature.

Humans are social animals, just like chimpanzees (our closest kin), and other primates. In society, an individual's genetic fitness depends on social success, much more than on success in the individual fight against nature. The biggest challenge to Robinson Crusoe was probably not the fight against a hostile environment. Rather, "it was the advent of Man Friday on the scene which really made things difficult for Crusoe. If Monday, Tuesday, Wednesday and Thursday had turned up as well, then Crusoe would have needed to keep his wits about him" (Humphrey 1976, p. 305). This is why evolution gave humans a mind for outwitting, bargaining, or manipulating—even when dealing with nature.

Like Robinson Crusoe, prevailing among a group of cannibals may require somewhat different wits than successfully negotiating a business contract. Yet, there are two common key abilities: Empathy and mind reading (Singer and Fehr, 2005). Empathy has been defined as "the recognition and understanding of the states of mind, including beliefs, desires and particularly emotions of others without injecting your own" (Binmore 1992, p. 212) or as the ability of understanding—and outwitting—each other (Humphrey, 1976). Empathy is not sympathy; we do not have to like the cannibal, but we have to understand how he works—the better, the more hostile he is. The idea that empathy is the root of a human sense for fairness and morality, as developed in Adam Smith's (1723–1790) *Theory of Moral Sentiments* (1759), is confirmed by recent findings in neuroscience which point to "the intriguing notion that to recognize an emotion in others, one needs to be able to experience it oneself." (Camerer, Loewenstein and Prelec 2005, p. 22).

Mind reading and outwitting each other means guessing each other's emotions, goals, possible strategies and actions, and—last but not least—knowledge or beliefs. These are sophisticated tasks, and guessing and processing beliefs, particularly higher-order beliefs, may be the most demanding. In particular, they require good brains. Indeed, "many neuroscientists believe there is a specialized 'mentalizing' (or 'theory of the mind') module, which controls a person's inferences about what other people believe, feel, or might do" (Camerer et al. 2005, p. 34). The human brain—at first sight an evolutionary luxury—may be "the result of an arms race within our species, aimed at building bigger and better internal computing machines for the purpose of outwitting each other" (Binmore 1992, p. 212). For such reasons, some genomic researchers believe that social science, rather than biology, is likely to explain why humans have a relatively powerful brain.

So, why do we bargain with nature? It seems that understanding and outwitting each other is so important for evolutionary success that humans ended up bargaining even with deaf and blind nature. Evolution has made us somewhat contradictory creatures. We are smart enough to solve higher-order belief games, like translating a Fed chairman's remarks into share valuation. At the same time, we stick to fairly hopeless rituals to "outsmart" nature, like putting on the left shoe before the right. Intelligence and superstition, though contradictory at first sight, may be evolutionary twins. Their common origin is the "arms race in terms of outwitting each other", including nature. And the race may not be over yet: Recent genomic research has found evidence that the human brain is still evolving (Dorus, Vallender et al., 2004). Touch wood.

For further literature on this subject, we recommend the works cited in this chapter. An easy-to-read and rather intuitive introduction is Paulos (2003, Ch. 8). An inspiring set of examples are proposed by Sunder (2002) and Koessler (2000). More *al dente* treatments are Samuelson (2004), Ayres and Nalebuff (1997) or Binmore (1992, Ch. 10). These texts also provide links to the relevant original papers by Lewis, Aumann and Schelling. Finally, the role of higher-order beliefs in financial markets, including the role of public signals as well as systemic or exogenous risk, is discussed in Morris and Shin (2002), Allen, Morris and Shin (2006), and in Danielsson and Shin (2003).

Checklist: Concepts introduced in this chapter

- The axioms of knowledge
- Private, public, and mutual knowledge
- Common knowledge
- The join and the meet of two information structures
- Subterranean information processing
- Agreeing to disagree
- The bias towards public information
- Keynes' beauty contest
- Conformism
- The Poisson distribution

8 H Problem sets: The art of outguessing others

For solutions see: www.alicebob.info.

Problem 8.1 Insider information

You are witness to a conversation from which you conclude that an important new drug developed by company A has successfully passed the tests and will be cleared for sale by the health authority soon. You are sure that the information is true. Do you know enough to bet your wealth on the company? (Hint: what questions would you try to answer before buying?)

Problem 8.2 Island hopping

Alice and Bob have different information about the outcome of rolling a dice. Alice can distinguish all outcomes except 4 and 5. Bob can not tell 2 from 1, nor 4 from 3. What set of numbers is common knowledge between Alice and Bob if the true outcome is:

(a) 1?
(b) 4?
(c) 6?

Problem 8.3 The two commanders

A famous example (Sunder, 2002) in game theory is the following: Two commanders win a battle if they attack simultaneously, but lose it if only one attacks. They can send messengers back and forth to inform each other when they are ready to attack. The messengers might be cut by the enemy, however. They both agree that each would immediately attack as soon as they were sure that both would in fact attack, but no sooner. After how many successful messenger runs do they attack?

Problem 8.4 Last chance

Three prisoners with a life sentence are standing in a line in front of a judge. Each wears either a black or a white hat. None can see his own hat's color, but only the hats in front of him. The judge promises he will set free the first of the three who can tell the color of his own hat (and can explain why). The judge even gives them a piece of information: He tells them that at least one hat is white. First, there is silence. But after about a minute the front man says: My hat is white. Is he right, and how may he know? (Please answer using the terminology developed in the present chapter.)

Problem 8.5 Individuals versus groups

"Groups are more intelligent than individuals." Discuss.

Problem 8.6 Miss Switzerland

Miss Switzerland is chosen by a jury of eight experts from 16 final candidates. Each expert has one vote; a ninth vote is cast by the public via telephone on the election night. Assume (contrary to reality) that public voters who pick the future winner share an award of one million Swiss francs. For whom should you vote (as a participant in public voting), if you are to maximize your expected revenue?

9 Coordination problems

9 A Introduction

Jim and Della are a loving but poor couple in "The Gift of the Magi" by O. Henry.[44] Christmas is approaching and neither can afford a gift even remotely appropriate for their beloved. Yet, both have a possession in which they take great pride. Jim has inherited a gold watch from his father, which he had inherited from his father; Della has wonderful hair, "a cascade of brown water". And so, love triumphs over poverty. On Christmas Eve, Jim and Della surprise each other: Jim proudly presents Della with a set of combs for her hair. Della, in turn, bought a gold chain for Jim's watch. Only, to buy the combs, Jim had to pawn his watch, while Della cut and sold her hair to buy the chain.

This is an example of what economists prosaically call a "coordination failure". Coordination problems arise because individuals often have an interest in coordinating their actions, but find it difficult or impossible to do so. In particular, coordination may fail when partners cannot communicate or, like Jim and Della, choose not to communicate.

Coordination problems occur in many contexts—economic, financial, political and social. Examples include conventions (like left- or right-hand traffic), the choice of a common language (English as an international language; or everybody using "green" to denote the color green), technical standards (see Box 9.1), corruption, political insurgence, currency attacks and bank runs.

Coordination problems arise from a two-way externality. The action of one agent influences another agent's payoffs and vice versa. Actions are referred to as "strategic complements" (in the case of a positive externality) or "substitutes" (in the case of a negative externality). Coordination problems are characterized by multiple equilibria.

A famous example of multiple equilibria in the field of finance are bank runs. We will discuss bank runs and the well known model of Diamond and Dybvig (1983) in the Application section. We also offer a graphical illustration and a numerical version of the model, two things we could not find in the literature.

Box 9.1 HD-DVD versus Blu-ray

The 1990s witnessed the triumph of the CD as a medium for the storage of electronic data and particularly of music. For large data volumes such as are required for movies or video games, the CD's capacity was too small, though. Eventually, the CD gave way to its successor, the more potent DVD. Yet, soon the demand for sharper images, richer sound and more interactive features led to the development of technologies superior to the DVD. In 2006, two new devices hit the market: the Blu-ray Disc (BD) and HD-DVD. Both technologies use a blue-violet laser to read information on the disc. Due to the shorter wavelength of blue compared to red light, the new discs can store more information than DVDs or CDs, which use red lasers.

While both HD-DVD and Blu-ray play the old DVDs, the technologies are not compatible with each other. Producers of content (movies) as well as producers of hardware (computers, home electronics) had to decide which of the formats to use. Consumers face the same choice. Obviously a coordination of producers and consumers on one of the standards would be an equilibrium. But Hollywood and manufacturers have split their support of the new formats. The Blu-ray camp unites Hewlett-Packard, Dell, Apple Computers, Hitachi, Panasonic, Samsung, Philips, Sharp, Sony, 20th Century Fox and Walt Disney. HD-DVD supporters are Microsoft, Toshiba, NEC, Sanyo and Universal Pictures. Some firms, like Paramount, Warner Bros. and Thomson, have announced that they will support both standards.

The competition between HD-DVD and Blu-ray is a typical example of a standards or format war. Consumers and producers of content (movies) as well as of devices (computers, discs) would benefit from a generally accepted common standard. Yet, given that both available formats are backed by strong alliances and a great deal of money is at stake, the chances of imminent coordination are rather slim.

9 B Main ideas: Red or white?

Alice and Bob, having made up after their recent argument (Chapter 8), plan to have dinner together. Alice will cook, Bob will bring the wine. In Chapter 4 Bob was very keen to bring the right wine (white with fish, red with meat). He still is. But, Alice, too, is eager to make the right choice of menu (fish with white wine, meat with red wine).

Table 9.1 Symmetric equilibria: Cell entries show the pairs of payoffs (Alice, Bob) from the four possible combination of actions

		Bob	
		white	red
Alice	fish	1, 1	0, 0
	meat	0, 0	1, 1

Alice and Bob are now in a symmetrical situation: (i) both have identical preferences—which wine "fits" with what food; (ii) these preferences are common knowledge. Both would like to have matching items on the dinner table. However, neither of them has full control over their consumption bundle. One buys the food, the other the wine. The problem is that both have to decide before they know the other's decision, just like J. J. Rousseau's hunters.

Alice's and Bob's actions are complementary. Complementary actions, or "strategic complements" mean that two players can mutually increase each other's utility by making the right decisions.

Alice and Bob are playing the game in Table 9.1. This game has two equilibria: fish+white and meat+red. In both outcomes, neither player will regret his or her decision.[45] The two other outcomes are not equilibria.

Box 9.2 Deer or hare?

The French philosopher Jean-Jacques Rousseau (1712–1778) sketched an early example of a coordination game in his *Discourse on the Origin and Foundations of Inequality among Men* (1754):

> In this manner, men may have insensibly acquired some gross ideas of mutual undertakings, and of the advantages of fulfilling them: that is, just so far as their present and apparent interest was concerned: for they were perfect strangers to foresight, and were so far from troubling themselves about the distant future, that they hardly thought of the morrow. If a deer was to be taken, every one saw that, in order to succeed, he must abide faithfully by his post: but if a hare happened to come within the reach of any one of them, it is not to be doubted that he pursued it without scruple, and, having seized his prey, cared very little, if by so doing he caused his companions to miss theirs. (Translation: G. D. H. Cole)

Game theorists (see, e.g., Hargreaves and Varoufakis 1995, p. 214) have read Rousseau as describing a situation represented by the following payoff matrix:

		Hunter 2	
		deer	hare
Hunter 1	deer	2	1
		2	0
	hare	0	1
		1	1

This game has two equilibria (in pure strategies). A superior ("good") equilibrium (deer, deer) and an inferior ("bad") equilibrium (hare, hare). Rousseau describes primitive man as not hesitating to follow the hare. But how should a perfectly rational being surrounded by other rational beings decide?

One might argue that a rational player who believes that the other players are rational as well should play "deer". Yet, being rational and thinking that others are rational as well is not quite sufficient for successfully playing "deer". The rationality of all players must be *common knowledge* (as defined in Chapter 8). Once more, beliefs about beliefs become important. If a sufficient number of players are not sure whether a sufficient number think that a sufficient number think ... that others are rational, too many will go for hares, and the deer may escape.

Strategic complements and substitutes

Multiple equilibria have their root in a positive externality. One agent's action (to cook fish, say) has a higher payoff if the other agent chooses a particular action (to bring white wine). Agents' actions are called strategic complements. A nice example of complementary actions is two people rowing a boat. Rowing only works if both scullers pull their oars with equal force. All pairs of equal effort are equilibria.

The opposite of strategic complements are strategic substitutes. One agent's action creates a negative externality for the other agent. This is typically the case in a zero-sum game. If there is one cake to share, increasing one agent's share reduces the maximum share available to another agent. Such games tend to have only one equilibrium.

The difference between strategic substitutes and complements is illustrated in Figure 9.1. The figure shows agents' reaction functions (best responses to the other agent's hypothetical actions). The functions are drawn for a continuous action

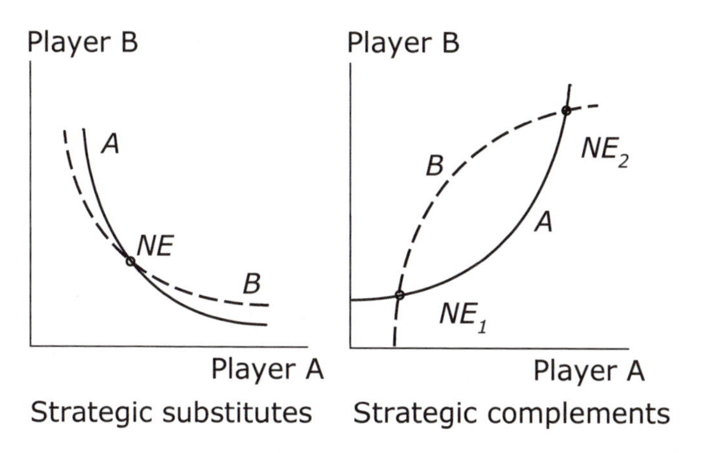

Figure 9.1 Reaction functions have a negative slope and a unique intersection (equilibrium) in the case of strategic substitutes (left part), but positive slope and multiple intersections (equilibria) in the case of strategic complements (right part).

space (not just a discrete choice like white or red wine). An example for strategic substitutes (left part) are the quantities of apples two local farmers offer on the market. An increase in one farmer's quantity depresses the price for both farmers. A higher quantity is a public "bad". An example of strategic complements are the individual efforts of members of a research team: The greater one member's effort, the higher the chance of success for the whole team. A greater effort is a public good (compare Section 6 C c). While, with symmetrical players, strategic substitutability leads to unique equilibria, strategic complementarity is associated with multiple equilibria.

Multiple equilibria in financial markets

Coordination is a key issue in many social and economic contexts. The following are particularly important:

- In a *currency attack*, speculators have better chances for success if they attack simultaneously (see, e.g., Myatt, Shin and Wallace, 2002). Everybody attacking and nobody attacking are both equilibria. Attacking in insufficient numbers or with insufficient funds can be very costly, a lesson speculators against the Hong Kong Dollar in 1997–98 learned the hard way (Box 9.3).
- Another example are *bank runs*. In one equilibrium all depositors leave their money in the bank and the bank is fine. In the other all withdraw, and the bank fails (Section 9 D).
- In many areas of economic activity participants have a strong interest in common *technical standards*. Often there are several candidates for common standards (remember HD-DVD and Blu-ray, Box 9.1) and therefore

multiple equilibria. Another example are credit cards. A credit card is only useful if it is widely accepted by sellers and it is only accepted if it is widely used by buyers. Credit cards are examples of so-called two-sided markets (see Box 9.4).

Box 9.3 Defending the Hong Kong Dollar

On 1 July 1997, the United Kingdom handed over sovereignty of Hong Kong to the People's Republic of China. The following day, the Thai baht collapsed under a wave of heavy speculation. The fall of the baht marked the beginning of a currency and banking crisis which subsequently hit several South-East Asian economies.

The Hong Kong Dollar, which was pegged to the US-Dollar, came under speculative pressure in October 1997. The Hang Seng stock index dipped by almost 25 per cent in a few days. Yet, the Hong Kong Monetary Authority (HKMA) promised to defend the currency. First, it warned banks against imprudent borrowing in (i.e. speculating against) the Hong Kong Dollar. On 23 October, the HKMA issued a circular to remind banks that for last resort liquidity support under the Liquidity Assistance Facility (LAF) penalty rates might be imposed on repeated borrowers. Second, the HKMA stood by the peg of the Hong Kong Dollar. However, this meant raising short-term (overnight) interest rates to arbitrarily high levels. On several occasions these hit three digit figures, and at one point in August 1998, they exceeded 500 per cent (per annum).

The high interest rate was necessary to keep the currency peg; yet, at the same time it further increased the downward pressure on the stock market. This inspired some speculators to adopt an indirect strategy. They took short positions in the Hang Seng and then sold Hong Kong Dollars. They hoped the automatic high interest effect of selling the currency would give them a lever to send the Hang Seng further down and thus cash in on their short positions.

Yet, in mid-August 1998, HKMA and the Government declared war on speculators and opened a second front. The Government started buying component shares of the Hang Seng Index. Acquiring shares for more than 100 billion Hong Kong Dollar (about US$13 billion), it temporarily became the largest shareholder of some of the companies.

In September 1998, the front of speculators crumbled. The currency peg between the Hong Kong Dollar and the US-Dollar at 7.8:1 withstood unscathed. The Government later divested itself the shares with a profit of some 30 billion Hong Kong Dollar (about US$4 billion). Speculators ended up with heavy losses.

[Source: HKMA Annual Reports 1997, 1998, 1999; http://www.info.gov.hk/hkma/eng/public/index.htm]

Box 9.4 Two-sided markets

In February 1950, Frank McNamara paid for his meal in a New York restaurant using a new device: his recently invented Diners Card (*The Economist*, 10 December 2005, p. 84). Inventing the card (then of cardboard) was one achievement; perhaps the even bigger achievement was to solve the "chicken-and-egg problem". On the one hand, restaurants would not accept the card unless a sufficient number of consumers were ready to use it; on the other hand, consumers were only interested in a card accepted by a large number of restaurants.

This is an interesting type of coordination problem. Consumers as well as restaurants are interested in coordination in order to avoid duplication of systems and wallets crammed with different cards. There are many instances of such two-sided markets (Rochet and Tirole, 2004). A newspaper is more attractive for advertisers if it is read by many people, and it may sell better if it carries advertisements. A computer system like Windows is more attractive if there are many applications; the programming of applications is more rewarding if there are large numbers of potential customers using the appropriate systems.

In most cases, direct communication between both sides of the market is not really feasible, at least not at a reasonable cost. Yet, the owner of the system—who is most interested in the coordination both sides of the market—can try to act as a bridge. In most cases, system owners subsidize one side of the market, while making a profit on the other. Diverse business models exist and operate profitably (Rochet and Tirole, 2004).

Hope for coordination: The idea of a focal point

Situations with multiple equilibria confront players with two related challenges. One is to end up in an equilibrium, rather than in a disequilibrium. The other is to end up in a "good" equilibrium rather than in a "bad" equilibrium, if equilibria differ with regard to players' utility level.

Often the game itself does not give players much of a clue on what they might successfully coordinate. Thomas Schelling has put forward the idea of a "focal point" (Hargreaves and Varoufakis 1995, p. 205). Sometimes one among several equilibria has some salient features. People who have an appointment in New York but forgot the place as well as the time of day are most likely to turn up at Penn Central Station at noon. Or, they may draw on past experience to guide them.

Alice and Bob perhaps had white wine and fish last time and meat with red wine the time before. This would suggest it is the turn of meat and red wine again. If it is common knowledge between them that both remember, they have a fair chance of coordinating.

When the reasoning behind a focal point is too sophisticated, the focal point is not common knowledge, and parties may fail to anticipate each other's thoughts correctly.

The chances of coordination increase when some communication among the parties is possible. In the basic case of Table 9.1, a one-way announcement by Alice that she plans to cook fish might be sufficient to ensure coordination. However, even with communication, coordination is not failsafe.

9 C Theory: Coordination and multiple equilibria

9 C a Multiple equilibria

Modeling strategic complementarity

Strategic complementarity means that the utility of agent i increases in a given action by other agents. There are two simple ways to model others' action: (i) a constant number of agents can vary the level of some action (effort in a soccer team), (ii) a variable number of agents can participate in some fixed effort (going to the same bar).

Complementarity is not a fixed property of the relation of two actions. Rather, the degree of complementarity may *vary* with the action level, or with the number of participating persons. A good illustration is going to a bar. Being the only person in a bar is not much fun. People who visit the same bar (or learn the same language, for instance) thus produce some positive externality for each other. A bar may become too crowded though, so complementarity is limited in this case.

A team effort

Here we will use an example with a fixed number of agents and a flexible (continuous) action space. The example very much resembles the model of the production of a public good (information) we used in Chapter 6. Much of what follows draws on the overview article by Vives (2005).

Individuals cooperate in a team effort. Each team member chooses an individual effort level in isolation. The benefit for team member i, q_i depends on her own efforts e_i as well as on the efforts of all others e_{-i}. We assume that the effort of team member i becomes more productive the higher the average of other members' efforts, i.e.:

$$q_i = e_i \bar{e}_{-i}.$$

The cost of effort, c, is an increasing convex function of e:

$$c_i(e) = \theta_i \frac{1}{\alpha} e_i^{\alpha},$$

where θ measures the idiosyncratic component of effort cost, while $\alpha > 1$ is a measure of the convexity of the effort cost function.

Individual utility is defined as benefit minus cost, i.e.:

$$U_i = e_i \bar{e}_{-i} - \theta_i \frac{1}{\alpha} e_i^{\alpha}.$$

Efforts are obviously complementary. Each individual contributes to the productivity of all other individuals. There are two differences to the setup used to model the production of a public good in Section 6 C c: (i) There, the benefit to an individual was equal to the sum of all efforts; here it is equal to an individual's own effort multiplied with the average effort made by others; (ii) the cost function used here is more general; setting $\theta = 1$ and $\alpha = 2$ yields the cost function we used above in Section 6 C c.

Equilibria

It is easy to see that this game has multiple equilibria. Let us use the simple case with 2 homogeneous players ($\theta_i = \theta$) and $\alpha = 3$.[46] The utility function of player 1 reads:

$$U_1 = e_1 e_2 - \theta \tfrac{1}{3} e_1^3.$$

Optimal effort is defined by the first-order condition

$$\frac{\partial U_1}{\partial e_1} = e_2 - \theta e_1^2 = 0.$$

The two players' optimal reactions to the opponent's decision thus are:

$$e_1^*(e_2) = \sqrt{\frac{e_2}{\theta}},$$

$$e_2^*(e_1) = \sqrt{\frac{e_1}{\theta}}.$$

These reaction functions are drawn in Figure 9.2. For homogeneous players they are symmetrical with respect to the 45°-line. Their intersection points define the two Nash equilibria of the game:

$$e_1^* = e_2^* = 0,$$

$$e_1^* = e_2^* = \frac{1}{\theta}.$$

Efficiency of outcomes

The equilibrium with the low effort $e = 0$ is Pareto inferior to the equilibrium with the higher effort $e = 1/\theta$. Yet, even the equilibrium with the higher effort falls short of the social optimum. This is because players do not take into account the

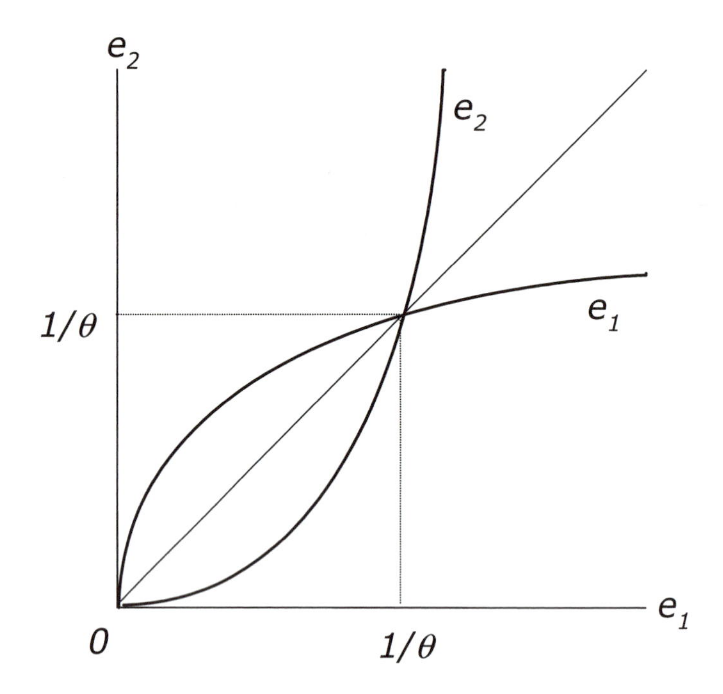

Figure 9.2 The two Nash equilibria are given by the intersections of players' reaction functions (at $e = 0$ and at $e = 1/\theta$).

positive externality of their own effort on the opponent's utility. A social planner would maximize aggregate utility:

$$U = 2e^2 - 2\theta \tfrac{1}{3} e^3.$$

The first-order condition yields a socially optimal effort

$$e^{**} = \frac{2}{\theta} = 2e^*.$$

The game thus has a *social optimum*, $e^{**} = 2/\theta$, which is not an equilibrium, and two equilibria (neither of which is a social optimum): a (relatively) superior equilibrium, $e^* = 1/\theta$, and (iii) an inferior equilibrium with $e^* = 0$. This illustrates all three aspects of coordination problems. First, players may fail to reach an equilibrium. Second, they may coordinate on an inferior equilibrium. Third, they may reach the superior equilibrium, but still miss the social optimum.

9 C b The problem of equilibrium selection

Our model explains why there may be multiple equilibria. But it says nothing about which equilibrium will result (if any). The payoffs of the game offer only limited guidance to the players.

Table 9.2 Dual equilibria (C, C) and (D, D)

		Player 2 C	Player 2 D
Player 1	C	$2/3, 2/3$	$-1/3, 0$
	D	$0, -1/3$	$0, 0$

The team effort simplified

Assume for simplicity's sake that in the above example $\theta = 1$. The two equilibria then are $e_1^* = e_2^* = 0$ and $e_1^* = e_2^* = 1$. As we are not interested in non-equilibrium strategies, we assume that players have a discrete action space: $e_{1,2} \in \{0, 1\}$. We will interpret action $e = 1$ as "cooperate" (C) and action $e = 0$ as "defect" (D). In a bank run the respective actions are stay (C) and run (D); in a currency attack they are attack (C) and not attack (D). In the present example, individual payoffs are $U(C, C) = 2/3$, $U(D, D) = 0$, and $U(C, D) = U(D, C) = -1/3$ as represented in Table 9.2.

What is rational?

Both players prefer equilibrium (C,C) to equilibrium (D,D). However, the payoff matrix does not help them to coordinate on (C,C). It is rational for one player to play (C) if there is reason to assume that the other player will play (C) as well. One may argue that it is "rational" to play (C), as rational opponents would mutually anticipate each other to play (C), and thus each would indeed play (C). However, a player playing (C) runs the risk of ending up with utility $-1/3$. By contrast, playing (D) ensures a utility of zero, whatever the other player's response is. Playing (D) thus is the "safe" strategy. If one player expects that the other will play safe with a probability of at least $2/3$, the best choice is in fact (D).

From a theory point of view multiple equilibria are unsatisfactory, as there is no explanation predicting which equilibrium (if any) will in fact be chosen. Several attempts have been made in the literature to solve this problem. The basic ideas are to add some heterogeneity of players or to perturb the game slightly. All these changes, however, modify the nature of the game in a fundamental way. In that sense, they do not solve the problem of equilibrium selection, but merely present a "similar" game, in which the chosen equilibrium is unique.

9 C c Types of coordination games

Chances for coordination depend on the actual payoff structure of a coordination game. Therefore, we look at some variations on the payoff matrix introduced in Table 9.1 (Myatt, Shin and Wallace, 2002). We first analyze these games in the absence of communication, before we examine the impact of communication on the likely outcome.

Alice and Bob both need new computers. The most important decision is the choice of the operating system. They consider Linux[47] or Windows. Both Alice and Bob would like to use the same operation system.

A simple game

Our point of departure is the *simple* (symmetrical) coordination game we used in Table 9.1. For the Linux–Windows analogy, the payoff matrix is given in Table 9.3.

There are four possible outcomes which we denote by {Lin, Lin}, {Lin, Win}, {Win, Lin}, {Win, Win}. We assume that using identical systems has a payoff of unity for the individual users. The payoff from using different systems is normalized to zero. Payoffs are common knowledge. Yet, individual decisions are not.

Both players have to decide in isolation. They face a simultaneous game with two (pure strategy) Nash equilibria: {Lin, Lin} and {Win, Win}, i.e. outcomes in which none of the players regret their decision.[48]

A focal game

Our next example is what we could call a *focal* coordination game. Unlike in the simple coordination game, both players have a common preference for one of the two systems. They still want to coordinate, but preferably on Linux, say. The payoffs are given in Table 9.4. Their coordination problem now consists of two tasks. They would like to have matching systems, yet they would also like to coordinate on the preferred equilibrium {Lin, Lin}. The game now has a focal point in terms of the combination {Lin, Lin}. Both players might reason: "If we can only coordinate by chance, why not aim at the better outcome?" If it is common knowledge among both players that the other is rational enough to think alike, they successfully end up in the superior equilibrium {Lin, Lin}.[49]

Table 9.3 The simple game has symmetric equilibria: (L, L) and (W, W)

		Player 2	
		Lin (L)	Win (W)
Player 1	Lin (L)	1, 1	0, 0
	Win (W)	0, 0	1, 1

Table 9.4 One superior equilibrium, (L, L)

		Player 2	
		Lin (L)	Win (W)
Player 1	Lin (L)	2, 2	0, 0
	Win (W)	0, 0	1, 1

Table 9.5 Conflict of interest

		Player 2	
		Lin (L)	Win (W)
Player 1	Lin (L)	2, 1	0, 0
	Win (W)	0, 0	1, 2

A conflict game

Our third case is a game with a coordination *conflict*. As before, both players would like to coordinate on one system. However, Player 1 prefers {Lin, Lin}, while Player 2 prefers {Win, Win}. This game is represented in Table 9.5.

This game has no focal point for coordination. Both players know that they end up with nothing if they just pick their own favorite. But if both play the other's favorite, they also fail to coordinate. Players thus need to know something about each other if coordination is to be successful. Knowledge about "types" is one possibility. If it is common knowledge that Player 1 is "tough" and that Player 2 is "soft", both should rationally play {Lin, Lin}. History is another source of clues on coordination.

Alice and Bob may have a tradition of deciding conflicts by alternating between leaving an advantage to one or the other. It may be common knowledge whose turn it is this time. For example, technical decisions may be Alice's domain, while Bob specializes in difficult social choices.

A risk-dominance game

Our fourth and last case is a *risk-dominance* game as represented in Table 9.6. The game is similar to the game in Table 9.4. Both players would like to coordinate, preferably on Linux. However, Windows is a relatively safe bet. A player choosing {Win} gets the same payoff whatever the other player does. Playing {Win} is

Table 9.6 Risk dominance of (W, W) over (L, L)

		Player 2	
		Lin (L)	Win (W)
Player 1	Lin (L)	2, 2	0, 1
	Win (W)	1, 0	1, 1

a risk-dominant strategy (Myatt, Shin and Wallace, 2002). This game has two equilibria with focal properties. While {Lin, Lin} is the better of the two equilibria, playing the relatively safer strategies leads to {Win, Win}. As a consequence, coordination is very difficult.

9 C d The role of communication

Communicating what?

A natural reaction to a coordination problem is to ask: "Why don't they talk to each other?" If Alice announces what she is going to play, it is easy for Bob to fit his own play to Alice's. If Bob can respond, coordination would seem almost assured.

One objection is philosophical: Talking, or communication, cannot be a general solution to coordination problems. Communication requires a common language; a common language is a convention that already presupposes some successful prior coordination (Lewis, 1969).

Here, we will assume that a common language exists. We also assume that individuals are able to exchange messages and that they want to do so—unlike Della and Jim for whom it was important to surprise the beloved. We will focus on the obstacles to successful coordination that arise from the *limited credibility* of individual messages.

Such messages can relate to different subjects. There are (i) announcements of intended actions, (ii) statements about payoffs and (iii) contract offers. "I will sue you if you do not stop infringing my patent," is an announcement of intended action (i). "Do you really believe your patent would be upheld in court?" is an invitation to revise expected payoffs (ii). "Why don't we pool our patents?" would be a contract offer (iii). A contract offer links future payoffs to specific actions, thus combining elements of (i) and (ii).

Cheap talk

We will focus on the first of the three above cases: Announcements of intended play. The other two, statements about payoffs and contracting may be just as important in practice. Yet, both imply *changes* to the underlying game. Our focus thus is more narrow. We examine how communication affects the way games with multiple equilibria are actually *played*, not how they can be *transformed* into different games. This limits our attention to pure, non-binding announcements of intended play. Such announcements are often called "cheap talk" (Farrell and Rabin, 1996).

Can cheap talk, in the sense of non-binding announcements, help parties to coordinate? In order to tackle this question, we have to specify the game under consideration. Communication is an exchange of pre-play messages; it is a game preceding the original coordination game. Players now face a two-stage game with (i) a communication stage, followed by (ii) a decision stage.

Like any communication, cheap talk can be one-way or two-way. In the former case, one party announces intended play; then the decision stage is played. In the latter case, both parties first announce intended play.[50] The announcing party first plays the decision game and then announces (correctly or incorrectly) the action chosen. They may do so simultaneously (no party knows the other's announcement when formulating their own announcement) or sequentially (one party announces, the other gets the message and then announces too).

In terms of the above example, players now have a choice of four actions: $\{linLin, linWin, winLin, winWin\}$, with lower-case expressions referring to announcements. Expected payoffs from the communication stage depend on both parties' expected play in the subsequent decision stage. In the different types of coordination games discussed above, communication may lead to different outcomes. We have to examine them separately.

Self-committing messages

Communication only works if participants exhibit a minimum of rationality. Complete fools cannot communicate. One possible definition of rationality is based on credibility. The basic notion is of a *self-committing* message. For one-way communication it would read as follows (Van Rooy, 2003):

> SELF-COMMITTING MESSAGE IN ONE-WAY COMMUNICATION:
> An announcement by Player 1 is self-committing if it is optimal for her to honor the announcement, if Player 2 believes and honors it.

Sequential two-way communication means that the game has two communication stages (announcement/reply) plus the decision stage. For sequential two-way communication, we can generalize the notion of a self-committing message in one-way communication:

> SELF-COMMITTING MESSAGE IN SEQUENTIAL TWO-WAY COMMUNICATION:
>
> 1 An announcement by Player 1 is self-committing if it is optimal for her to honor the announcement, if Player 2 believes it and sends a consistent reply.
> 2 A reply by Player 2 is self-committing if it is optimal for him to honor the announcement made in the reply, if player 1 believes and honors it.

Communication in the simple game

We start with the simple coordination game (Table 9.3). In this game a unilateral announcement has a good chance of leading to coordination.

Whatever Player 1 announces is rational and should be honored, if Player 2 is rational too. There is a problem, however, if neither is convinced about the other's rationality. Coordination may fail if rationality is not common knowledge.[51] In the symmetrical payoffs of the simple game, coordination succeeds if both players think that rationality is common knowledge with 50 per cent probability.

Two-way communication is not necessarily an improvement on one-way communication. If messages are sent simultaneously, they may match or not. The coordination problem is transferred to the communication level. In the so-called simple game, simultaneous two-way communication destroys the benefit of one-way communication. Coordinating simultaneous two-way messages is just about as difficult as coordination in the game itself.

We will therefore focus on the sequential variety of two-way communication. Sequential communication has a big advantage: The second player can confirm the first player's message. In the simple game, it is quite irrational to send a conflicting message; with consistent messages, coordination is likely to succeed.

Communication in the focal game

In the focal game with a Pareto superior and inferior equilibrium (Table 9.4), communication may not even be necessary for successful coordination on the superior equilibrium. If communication is possible, rational players announce the choice of the action that leads to the superior equilibrium.

Communication in the conflict game

Things become tricky in the conflict game, however, where the coordination motive is compounded with a conflict of interest between players (Table 9.5). In one-way communication, Player 1 has a "hard" option (announce "Lin") and a "soft" option (announce "Win"). Neither of the two options is a priori more rational than the other. Announcing to play hard is a good choice if Player 2 is likely to believe it. Otherwise Player 1 may be better off ensuring at least an equilibrium by choosing the soft option. Similarly, in sequential two-way communication, the rational reaction of Player 2 depends on the credibility of player 1's announcement. If Player 1 announces "Lin", Player 2 may believe and give in. Or, Player 2 may play hard and counter with "Win", hoping that Player 1 will be intimidated while taking the risk of coordination failure. The outcome of pre-play communication very much depends on the players' mutual beliefs about each other's types. In other words, sequential two-way communication is likely to end like the game without communication.

Note that Player 1 faces a special problem in sequential two-way communication. Player 2 knows that after replying "Win", whatever Player 1's original announcement, it is rational for Player 1 to give in and play {Win}. Doing so,

Table 9.7 Seducing game

		Player 2	
		Lin (L)	Win (W)
Player 1	Lin (L)	2, 2	0, 1.5
	Win (W)	1.5, 0	1, 1

Player 1 at least assures coordination (with a payoff of 1, rather than 0). Obviously Player 2 will play {Win}, too. If it is common knowledge that Player 1 is rational, her announcement is in fact irrelevant.

Communication in the risk-dominance game

Our final example is a game in which there is a superior equilibrium {Lin, Lin}, but also a riskless action {Win}. Payoffs from the riskless action do not depend on the other player's action (Table 9.6). In one-way communication, both possible announcements by Player 1 are rational. However, if Player 2's rationality is common knowledge, it is optimal to announce "Lin". In sequential two-way communication, "Lin" again is the better announcement. If Player 2 confirms, coordination on the superior equilibrium is likely to succeed. If he counters with "Win", Player 1 can still play "Win", securing a safe return of 1.

Some authors like Aumann and others (see Van Rooy, 2003) claim, however, that the requirements we have set for rationality or credibility are not strict enough. Consider the game in Table 9.7.

The game is almost identical to the risk-dominance game of Table 9.6 or Rousseau's deer hunting game. The only difference is that the payoff of playing {Win} against {Lin} is set to 1.5 instead of 1. Players thus unambiguously prefer the other player to play {Lin}. A message by Player 1 "I will play {Lin}" in these circumstances is slightly less credible. It might just be that Player 1 intends to play the soft option {Win} but attempts to seduce Player 2 to play {Lin}. The closer the payoff of playing {Win} against {Lin} get to 2, the more tempting such a seducing strategy becomes. If the payoff even exceeds 2, the game becomes a so-called "prisoner's dilemma" (see Chapter 10, Table 10.2; Gibbons 1992, Ch. 2.3). In a prisoner's dilemma, the announcement to cooperate (here: play {Lin}) is not credible, as the announcing player has an interest *not* to honor the announcement.

Self-signaling messages

We thus may need a stricter requirement for an announcement to be credible. Such a requirement is that the announcement be *self-signaling* (Van Rooy, 2003):

SELF-SIGNALING MESSAGES:

- An announcement by Player 1 is self-signaling if the sender wants the message to be believed, if, and only if, she is going to honor the message.

- An announcement is credible if it is both self-signaling and self-committing.

Final remark

We hope the above examples may have helped to develop some intuition on the scope and limits of successful communication in coordination games. For simplicity we had to restrict our analysis to games with two actions for each player. A more serious limitation of our approach is its focus on theoretical results. We recommend studying the evidence on communication in *laboratory experiments* (Hargreaves and Varoufakis, 1995).

Box 9.5 Leadership and advertisement

Leadership and advertisement are practical examples of coordination on an equilibrium by sending a message. An important function of *leadership* may be to provide a focal point for the coordination of a firm's employees. A firm often has several different paths to success. While the choice among alternative strategies may be important, it may be even more important that all employees follow the same strategy, whatever it is. A leader who clearly communicates her vision makes it common knowledge where she wants to go. Even if she has no means to ensure compliance (i.e. the announcement of a corporate strategy is indeed cheap talk), employees coordinate if they want the firm to be successful (Koessler, 2000).

Like leadership, *advertisement* can help to coordinate on technical standards or on common networks. Standards or networks are more valuable to one agent if they are also used by other agents (languages, software, bars). If a sufficiently important firm or a consortium of firms publicly commits itself to some standard, it facilitates coordination on that standard to competitors, suppliers and customers. Public advertisement creates common knowledge, while bilateral advertisement, like direct phone marketing, would not (Koessler, 2000). One should thus expect an empirical correlation between the social or network character of a good and the fraction of marketing occurring *publicly*.

9 D Application: Bank runs

Bank runs are among the most devastating economic events. A run on an individual bank forces a bank to curb lending and its borrowers to terminate productive investment. A run on several banks or the whole banking system also destroys deposits, an important part of money in an economy. Yet, the story of the man and the cat very clearly illustrates the deeper logic of a bank run.

Researchers have tried to understand what exactly happens in a bank run. The leading, if not uncontested, hypothesis is that bank runs are an example of coordination failure. If depositors only could agree not to withdraw, they would all be better off. Yet, coordination among a large number of agents is likely to fail, particularly if some depositors try to free-ride on others' patience and withdraw against their collective interest.

The classical model of bank runs by Diamond and Dybvig (1983) is still the point of departure for most research in the field, a number of criticisms notwithstanding. Diamond and Dybvig (1983) try to show that (see Freixas and Rochet (1997, Ch. 7)):

- Banks issuing demand deposits can insure agents who need to consume at different random times.
- The demand deposit contract provides better risk sharing than a securities' market.
- However, the demand deposit contract also has an undesirable equilibrium (bank run) in which all depositors withdraw immediately.

In a bank run, even those depositors withdraw who believe that the bank in the absence of a run would be healthy. If a large number of depositors simultaneously try to withdraw, a bank becomes illiquid and has to get rid of assets through fire sales or by asking around for cash. Both may be costly, and chances are that the fight for liquidity pushes the bank over the edge of insolvency. Therefore, perfectly healthy banks can fail if they fall victim to a run.

The vulnerability to runs is a consequence of the very nature of banks. They issue liquid deposits in order to finance long-term investment. This is the main contribution of banks to the efficiency of the economy. In normal times, the formula works well, as only a fraction of depositors need cash in any single period. Only a run reveals that banking has an Achilles' heel.

9 D a The Diamond–Dybvig model

The model economy has three periods, as illustrated in Figure 9.3: an initial period ($t = 0$) and two consumption periods ($t = 1$ and $t = 2$). There are a large number of individuals. All of these are born equal. They have the same endowment (one unit of consumption good in $t = 0$), the same utility function (they are risk-averse), and the same investment opportunities. However, at the end of the initial period ($t = 0$), when investment decisions have already been made, individuals learn their "type". They can be either early diers who can only consume in the first consumption period ($t = 1$), or late diers who only derive utility from consumption in the second consumption period ($t = 2$). Types are not observable, but the fractions of early diers (k) and of late diers ($1 - k$) in the population are known.[52]

There are two investment technologies. The short-term investment (storage) yields one unit in $t = 1$ per one unit invested in $t = 0$. Storage can also be used from $t = 1$ to $t = 2$. The long-term investment (production) yields $R > 1$ in $t = 2$

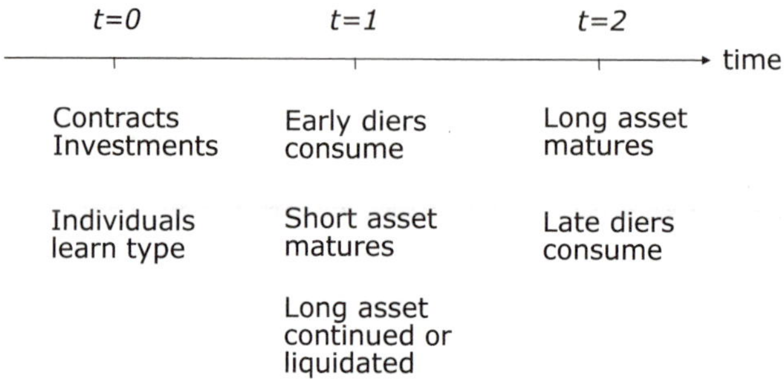

t=0 t=1 t=2 time

Contracts
Investments

Individuals
learn type

Early diers
consume

Short asset
matures

Long asset
continued or
liquidated

Long asset
matures

Late diers
consume

Figure 9.3 At $t = 0$, each individuals has one unit to divide between a short-term asset (storage) and a long-term asset (production). The short-term asset yields a gross payoff of 1 at time $t = 1$. The long-term asset yields $R > 1$ if held until $t = 2$ or $L < 1$ if liquidated early at $t = 1$.

for every unit invested in $t = 0$.[53] The long-term investment can be liquidated in $t = 1$, but it only yields L ($L < 1$) if liquidated.

As a consequence of costly liquidation, individuals find themselves in a dilemma in $t = 0$. Investing in long-term technology promises a relatively high yield R. However, the risk is that the individual learns he is an early dier; liquidating in $t = 1$ yields only L, i.e. less than if he had just kept his funds in storage. Conversely, an individual who only invests short term, and then learns she is a $t = 2$ consumer, will regret her decision. The compromise, investing part of funds short term, the other part long term avoids both extremes, but will always be partly inefficient *ex post*.

The individual dilemma can be mitigated by an appropriate institutional setting. We will compare three settings: individual autarky (as a benchmark), a securities' market, and insurance through a banking arrangement.

9 D b Autarky, markets, insurance and banking

Autarky

Under *autarky*, each individual faces the dilemma described above. In Figure 9.4, one of the two extreme options is represented by point {1,1}: The individual keeps all his funds in storage. The other extreme is point {R,L}, where the individual fully goes long term. On line *a*, the section between the two points indicates the possibilities an individual can reach by splitting funds between the two investment technologies. The individual optimum is represented by point A, where line *a* is a tangent to an indifference curve.[54]

Securities market

A *securities' market* in $t = 1$ improves over the autarky allocation, as it permits an individual to invest a fraction of her funds short term and another long term.

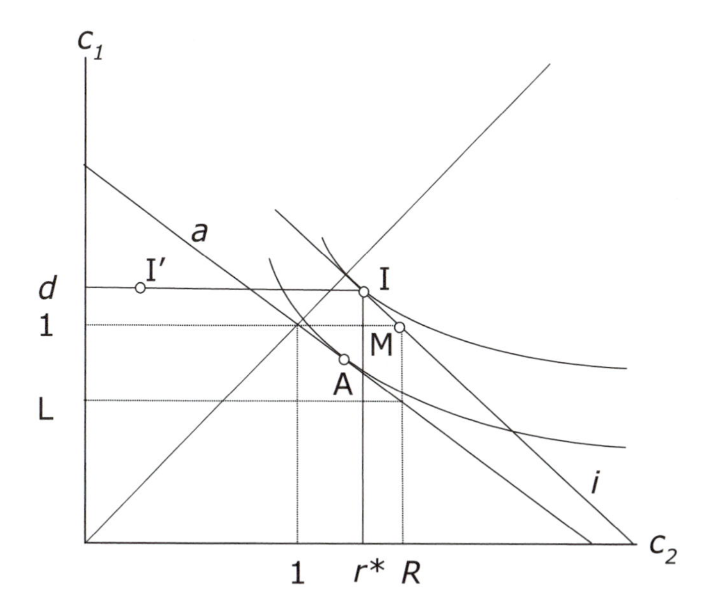

Figure 9.4 Optimal consumption bundles in the Diamond–Dybvig model under autarky (A), a securities market (M), or insurance within a bank (I). If types are not observable, the banking arrangement may lead to inefficient runs (I' instead of I).

After the uncertainty about individuals' type has been resolved, the securities can be traded.

In an equilibrium, the fractions invested in the two types of assets correspond to the fractions of types k, and $(1 - k)$ in the population. From an aggregate point of view, investment is optimal. The securities' markets solves the problem of making income available in $t = 1$ to early diers, and income available in $t = 2$ to late diers. Individuals, once they have learnt their types, trade the share of income they do not need. Early diers sell a security paying in $t = 2$ to late diers, who in turn pay with their $t = 1$ income. The equilibrium price is one unit of $t = 1$ income for R units of $t = 2$ income.

At the end of trading, all k early diers get 1 unit for immediate consumption, and all $1 - k$ late diers will get R units in the next period. The securities market thus permits individuals to reach point M in Figure 9.4. Point M, compared to autarky (point A), represents the same consumption R to a late dier, but one unit of consumption good instead of only L units to an early dier (as no liquidation ever becomes necessary). A market is clearly superior to autarky. The market implements an optimal *ex post* allocation of income. It permits individuals to invest in line with fractions of types in the population. Aggregate investment is efficient in the sense that liquidation is never required.

Yet, efficiency is not synonymous with optimality. The market allocation is not yet the *non plus ultra* as it does nothing to insure individuals against the *ex ante* risk of being either early or late diers.

Insurance

This is where *insurance* comes in. Assume for a moment that types become observable when they are revealed in $t = 1$. In this case individuals could collectively insure. They would agree in $t = 0$ that after revelation of types, late diers will forgive some consumption in favor of early diers in $t = 1$. Early diers will get $d > 1$, but late diers will only get some $r^* < R$. To make such an agreement feasible, investment plans have to be adapted. Individuals need to invest a fraction $d \cdot k > k$ of their endowment in the short-term asset, since they promised $d > 1$ to each of the k future early diers. This leaves only fraction $(1 - d \cdot k)$ for investment in the long-term asset. Feasibility can promise each of the $1 - k$ late diers consumption (in $t = 2$) of r^*, where

$$r^* = \frac{1 - d \cdot k}{1 - k} R.$$

In the optimum, the slope of economy-wide budget constraints equals the marginal rate of substitution.[55] Graphically, the insurance equilibrium is represented by Point I in Figure 9.4. Point I lies on the budget constraint i that goes through M—corresponding to the optimal outcome with a securities' market—and on the highest attainable indifference curve. The insurance equilibrium is an improvement on the pure securities' market equilibrium. It is not only characterized by optimal allocation of available *ex post* income, but also by optimal investment. There is no possible further improvement beyond the consumption bundle $\{d, r^*\}$ associated with the insurance equilibrium.

Banks

Unfortunately types are not observable. Individual consumption or liquidity shocks (being an early or late dier) are private information. As a consequence, the insurance arrangement cannot be implemented via contracting. More precisely, it can only be implemented if individuals have no incentive to lie about their types.[56] Indeed, no early dier would ever claim being a late dier (whatever the payoffs); nor has any late dier an incentive within the insurance arrangement to claim being an early dier, as this would yield d (to store from $t = 1$ to $t = 2$) instead of $r^* > d$. Thus, it seems that the insurance arrangement is implementable by a mechanism under which individuals pool their resources, follow the optimal investment strategy of the insurance arrangement, and self-declare their types after these have been revealed. An individual who has just learnt that he is an early dier would claim d units for immediate consumption; late diers would claim r units in the final period. Diamond and Dybvig call this arrangement a *bank*. The bank issues deposits paying d on demand; individuals who do not withdraw in $t = 1$ get a "liquidation dividend" of r in $t = 2$ when the bank is dissolved. The bank thus is a cooperative with the late diers holding residual claims.

Banks' vulnerability

What might be wrong with this arrangement? At first sight it looks like an economic miracle. It allows individuals to write implicit contracts on non-observable outcomes.[57] Almost, at least. The arrangement has an Achilles' heel, exactly because it rests on the assumption that individuals will be honest about their types. True, rational individuals of both types have an incentive to be honest—but only as long as everybody else is! Once some individuals lie, others have an incentive to lie as well.[58]

Assume that we are in period $t = 1$. The bank's liquid funds $(d \cdot k)$ are exactly sufficient to pay out early diers. The rest is invested in long-term assets for the benefit of late diers. Now, late diers suddenly develop doubts about the rationality of their fellow late diers. What if some of these lose all sense of reason, go to the bank, claim they are early diers and withdraw? The bank's short-term funds can only cover the claims of early diers. In order to honor any demands from late diers the bank has to liquidate part of its long-term investment. Liquidation means a loss. As long as the remaining funds, growing at a gross rate R, will still provide more than d units in $t = 2$, there is no problem. However, if late diers expect that the number of withdrawals is sufficient to push expected $t = 2$ payoff below d, they will in fact withdraw—and quickly! Diamond and Dybvig assume that deposits are subject to a sequential service constraint: Depositors are paid according to a *first-come-first-served* rule. The position of a depositor in the line thus is crucial and running quickly is an equilibrium strategy.

The arithmetics of sequential service

Some arithmetic may clarify the issue. Assume for simplicity's sake that there are no early diers, leaving us with the late diers and enough funds to pay out a return of R to each of them in $t = 2$. Now, late diers start to withdraw, claiming the promised return d for early diers. Each withdrawal triggers liquidation of d/L units of long-term investment. Once a fraction f of the late diers has withdrawn, the bank is left with a residual of $1 - d \cdot f/L$. If kept in the bank, this residual grows at rate R and serves the remaining $(1 - f)$ late diers. Consumption of a late dier, standing in line behind f other late diers, thus is:

$$\hat{r}(f) = \frac{1 - d \cdot f/L}{1 - f} R.$$

This is true as long as the bank still has some funds. Once $f_L = L/d$, the bank has liquidated all its assets and cannot honor any further demands. The remaining $1 - f_L$ depositors receive nothing.

Should I stay or should I go?

What should a late dier who is afraid of withdrawals by other late diers do: stay with the bank or go and withdraw?[59] Of course, she does not want to wait until $\hat{r}(f)$ is zero. Actually, there are two possibilities. If she believes the fraction of withdrawing late diers, f, will be relatively small, such that $\hat{r}(f) \geq d$, she does not withdraw. If, however, she believes that withdrawals will push $\hat{r}(f)$ below d she will also withdraw.

Figure 9.5 presents the situation from the perspective of an individual depositor. The Figure depicts, for $d = 1.1$, the terminal value of a deposit held until maturity as a function of the liquidation value L and the percentage f of late diers who withdraw early. A late dier should withdraw if the payoff at maturity is lower than the face value of a deposit, d. If, for example, the liquidation value of the long-term asset is 90 per cent $L = 0.9$, a late dier should leave her money in the bank if she thinks that at most 54 per cent of other late diers will withdraw. With a liquidation value of 50 per cent, the critical fraction of early withdrawals falls to 17 per cent.

Figure 9.5 shows how an individual who knows the decisions of others should decide. However, a depositor has to decide before she knows what others will do. An equilibrium then is an outcome after which all depositors are happy with their action, once they have learned how other depositors have acted.[60] It is quite obvious from the figure that the outcome "nobody withdraws" is an equilibrium (remember that we are talking of late diers). If nobody does, no individual would want to withdraw. The figure also shows that once a sufficient number of depositors withdraws, everybody would want to withdraw. Therefore, "everybody withdraws" is also an equilibrium. This equilibrium corresponds to a bank run.

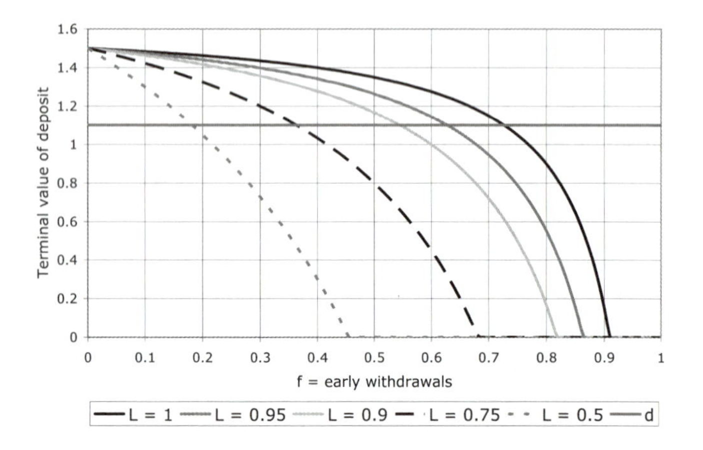

Figure 9.5 An individual's incentive to withdraw a deposit is measured by the difference between the deposit's nominal value (d) (horizontal line) and its terminal value $\hat{r}(f)$ (curved lines). The incentive increases in the fraction of withdrawing depositors and decreases in liquidation value L. Assumptions: $R = \hat{r}(0) = 1.5$, $d = 1.1$.

Table 9.8 Bank run game

		Depositor 2	
		stay	go
Depositor 1	stay	R,R	$\left(2-\frac{d}{L}\right)R,d$
	go	$d,\left(2-\frac{d}{L}\right)R$	L,L

The bank run game therefore has two equilibria.[61] Table 9.8 describes the game in its simplest version: as a game between two depositors (late diers) in a bank without early diers (who may already have withdrawn). There are two strategies: "stay" and "go". Recall that the amount of assets invested long term in the absence of early diers is 2. If one of the late diers goes as well (getting d), the payout for the stayer is $2 - d/L$. The payoffs in the outcome $\{$go, go$\}$, L, is not an effective payoff but an expected payoff. If both depositors go and withdraw, the first in line gets d, the second gets $L(2-d/L) = 2L-d$. In the aggregate they get $2L+d-d$; in expected (but not in realized) terms each therefore gets L.

It is optimal for depositor 1 to stay if depositor 2 stays and vice versa, as $R > d$. Hence $\{$stay, stay$\}$ is an equilibrium. In expected terms, it is optimal for depositor 1 to go to the bank and withdraw if depositor 2 goes and vice versa, if we assume $(2 - d/L)\,R < L$. Hence, for depositors who do not know their place in line, $\{$go, go$\}$ is also an equilibrium. The difference between the two equilibria is that the first is a Pareto optimum, while the second is clearly inferior.[62]

Bank runs against better knowledge

The main message of the Diamond and Dybvig (1983) model is:

> A bank can fall victim to a run, even though depositors know that the bank is solvent!

The vulnerability to runs is the Achilles' heel of banking, but also part of its very nature. Banking (in the sense of Diamond and Dybvig) is an insurance against having to consume early. However, if types are not observable, individuals can always claim they need their funds sooner than they actually do and withdraw money from the bank. The vulnerability to runs is the price for writing implicit contracts on unobservable states of nature (individual j being an early dier, for example).

A bank is immune to runs if it does not promise more than the yield of the storage technology. But then it is not a bank! The instability of banking is an unavoidable consequence of its very function.

9 D c Objections against the DD-model

The Diamond and Dybvig model has been criticized for various reasons (see, for example, Freixas and Rochet (1997, Ch. 7); Brunnermeier (2001, Sec. 6.4)):

- The equilibrium selection is attributed to an exogenous random event called a "sunspot", rather than really explained by the model.
- The banking arrangement is not robust to side deals. In the presence of a bank, a free-rider can invest all his funds in long-term assets. If he is unlucky and learns that he is an early dier he can offer R units of the consumption good in $t = 2$ to a late dier. This is more than the late dier would get from the deposit contract ($R > r^*$). In exchange, the late dier can offer d units of consumption in $t = 1$, which is more than the early dier would get in a securities market.
- The deposit contract may not be optimal. In Figure 9.4 the banking arrangement is represented by two points, point I (no run) and a point I' (run) situated to the left of I. The latter point may yield such a low utility that the securities' market may dominate the banking arrangement. What the optimal contract would be remains an open question.
- The model intermingles a liquidity shock (having to consume early) and a wealth shock (being relatively poor), both implied in being an early dier. Banking improves over a securities' market due to the wealth shock, while in reality banks rather seem to insure against liquidity shocks.
- The intuition of the model (multiple equilibria) is derived from a simultaneous game. The sequential service constraint transforms it to a sequential game, though. The intuition gained from a simultaneous game may not carry over to a sequential game in which, for example, depositors' knowledge about their place in line would have to be modeled.
- In the absence of aggregate uncertainty (about the fraction of early diers in the population, k), bank runs can easily be avoided. A promise by the bank to stop payments in $t = 1$ after paying an amount sufficient to honor all claims of early diers (a so-called suspension of payments, see below) would do. No late dier would then have an incentive to withdraw early.
- The sequential service constraint in Diamond and Dybvig (1983) is unduly restrictive. Instead of paying a fixed amount to withdrawing depositors (as long as it can), the bank may promise depositors amounts decreasing in the number of depositors who have already withdrawn. If appropriately defined, such a repayment rule would give depositors the incentives to declare their types truthfully, as Green and Lin (2003) demonstrate. (The importance of schemes that lead to truth-telling will be discussed in Chapter 15.) At the same time, the bank run equilibrium may disappear, even in the presence of a sequential service constraint.

Despite such objections, the Diamond and Dybvig model has been very influential. Numerous authors have tried to improve upon some aspects of the model and to make it more realistic.[63] One particular improvement is to make the gross return from production, R, and the fraction of early diers, k, stochastic. An uninformed individual observing a queue in front of a bank cannot tell whether others have negative information about R or whether there is just a high fraction of early diers. A queue forming in front of a bank (even if happening by chance) may thus play the role of the sunspot and trigger a run.

If R is stochastic (and has realizations that are smaller than L) bank runs may be efficient. They may force the bank to liquidate before its funds are squandered. The threat of a run is particularly beneficial if banks can influence the distribution of R by exerting an effort or choosing investment risk profiles.

9 D d Policy implications

The Diamond and Dybvig model has also been influential in the political arena. It suggests how several institutions can reduce the danger of bank runs: Suspension of convertibility, deposit insurance, creditor coordination, and a lender of last resort. All these institutions are observed in reality or have played some role historically.

Suspension of convertibility

Suspension of convertibility in the form of "bank holidays" has been used in the financial crisis of the 1930s. In Argentina in 2001, restrictions on deposit withdrawals were introduced under the informal name *corralito* (a child's playpen).

Deposit insurance

Deposit insurance is used in most countries worldwide and has been extensively discussed in the literature (see Freixas and Rochet, 1997). The main problem with deposit insurance is that it may make insured depositors careless regarding the choice of their bank, as we will discuss in Section 16 D. This increases the risk that a bank will run into trouble. Reducing the risk of a run in case a problem exists thus increases the very probability that such problems may arise in the first place.

Creditor coordination

Creditor coordination is the idea behind "soft" bankruptcy provisions like the US "Chapter 11". Creditor coordination was also achieved when the New York Fed brought together the creditors of Long Term Capital Market, an important hedge fund that got into trouble in 1998.[64] Creditor coordination was also at the heart of IMF strategies in the face of the problems of several South-East Asian and Latin American Countries in the 1990s.

Coordination, though in creditors' common interest, can be very difficult. This is particularly true when the number of creditors is large, as in the case of banks with their many depositors. The direct cost of communication with many depositors is only one problem. The other is free-riding (see Chapter 6): When $n-1$ creditors agree to leave their money in a bank, creditor n can withdraw without loss.

Unless a control mechanism can be put in place, creditors thus may pay lip-service a roll-over agreement, but secretly withdraw their funds as soon as they become due.

A lender of last resort

Most countries, in the shape of the national central bank, have a lender of last resort furnishing banks with sufficient liquidity (cash) in case a run builds up (Freixas, Giannini, Hoggarth and Soussa, 1999). After 11 September 2001, for example, the Federal Reserve Board prevented the danger of disruptions in the payment system and financial panic by generously supplying the market with liquidity. In the presence of a lender of last resort a depositor who does not need cash has no reason to run to the bank since the central bank stands ready to provide cash when needed. This role can only be played by central banks, as they have the authority to print money.

Communication is a major challenge in lending of last resort issues. If all depositors are convinced that there is a determined lender of last resort, they have no reason to panic. As a consequence, a credible lender of last resort may never have to intervene. If, by contrast, the lender of last resort is only able or willing to lend limited amounts, depositors have an incentive to run as quickly as possible in order to secure their share in the central bank's scarce resources.

An announcement that a bank will receive liquidity assistance is often sufficient to reassure depositors. Yet, it may also achieve the opposite. Depositors may indeed read an offer of liquidity support to a bank as a concession that the bank may have a solvency problem. The difficulty in distinguishing between liquidity and solvency problems confused investors of Credit Suisse,[65] one of the big Swiss banks, in 1975. On 14 April, the bank had to announce important losses at its Chiasso branch, but did not provide more details over the course of the following days. On 25 April, as rumors flourished, the Swiss National Bank (the central bank) and the other two big banks announced a joint standby facility for Credit Suisse of CHF 3 bn. They hoped to send markets a clear signal, dispelling all doubts about the bank's ability to honor its debt. Yet, they achieved quite the opposite. The announcement (issued without comment) led markets (wrongly) to doubt the bank's solvency. The confusion was compounded when Credit Suisse announced it would not need the credit (for which it had in fact never asked). As a consequence, its share price (which had only lost a few percentage points since 14 April) tumbled by 20 per cent after these announcements.[66]

9 E Conclusions and further reading

Coordination is important in many economic contexts. Yet, typically, coordination often fails. There are two ways in which agents may fail to coordinate. First, parties may choose inconsistent options, thus ending up out of equilibrium. One example is the coexistence of two standards like HD-DVD and Blu-ray. Second, parties may coordinate on an inferior equilibrium. The "QWERTY" system for keyboards (Hargreaves and Varoufakis, 1995, p. 217) is often quoted as an example of coordination on an inefficient equilibrium compared to a general use of superior rivals like, perhaps, "Dvorak".

The game theory background of coordination problems is multiple equilibria. The idea of multiple equilibria provides a relatively simple model to understand disruptions like currency crises or bank runs. At the same time, economists are somewhat dissatisfied with multiple equilibria. Models of multiple equilibria do not answer the key question: Why will one equilibrium be chosen over another? Multiple equilibria are particularly unwelcome in asset pricing. How should one value bank debt, for example, without having a model on whether there will be a run on the bank or not? Attempts have been made to remove the multiplicity of equilibria from coordination games. One idea is to add a bit of unobservable heterogeneity of agents, leading to a so-called "global game" (Myatt, Shin and Wallace, 2002).[67]

Notwithstanding some economists' dissatisfaction, multiple equilibria help to explain many phenomena such as the existence of deposit insurance, a lender of last resort or "soft" bankruptcy chapters designed to facilitate creditor coordination (like Chapter 11 in the US or Sections 11-13 of the Swiss Banking Act).

Coordination games or multiple equilibria are treated in most textbooks on game theory, such as Hargreaves and Varoufakis (1995). A recent review article on coordination games is Vives (2005). Bank run models are discussed in Freixas and Rochet (1997, Ch. 7) and in Brunnermeier (2001, Sec. 6.4). Complementarities arising in macroeconomic contexts are analyzed in Cooper (1999).

9 F Problem sets: "Should I stay or should I go?"

For solutions see: www.alicebob.info.

Problem 9.1 Linux or Windows?

In the Theory section we discussed the problem of Alice and Bob trying to buy computers with the same operating system. We assume that their payoffs are as given in Table 9.4. They cannot communicate.

(a) Which outcomes are (i) equilibria, which ones are (ii) Pareto optima?
(b) How do Alice and Bob decide if they maximize expected income and if each assumes that the other tosses a coin to decide what to buy?
(c) If both randomize their choice, how should they set the odds to install a Linux versus buying a Windows, so that no player regrets the odds chosen (although they may regret the outcome)?

Problem 9.2 A rational echo?

"In sequential two-way communication a reply is rational if it confirms a rational announcement." Comment!

Problem 9.3 Bank runs

In the bank run model of Section 9 D:

(a) Draw a payoff matrix for a bank run in the 2-person case.
(b) Show that there are multiple equilibria.
(c) Why is the existence of multiple equilibria a problem?
(d) Would it help players if they could announce their strategies? Would your answer also hold in the 10,000 player case?

Problem 9.4 Cats and banks

Is there a parallel between the man who was afraid that the cat might not know he was no mouse and a depositor of a bank?

Problem 9.5 The propensity to run

Show in the simple 2-person bank run game of Table 9.8 that the incentive to run first increases and then decreases in $L \in [0, d]$. (Hint: remind that payoffs to depositors cannot be negative.)

Problem 9.6 Bank run versus prisoner's dilemma

What are the differences between a bank run game and a prisoner's dilemma (discussed in Section 10 D)?

Problem 9.7 The remains of the day

Assume that a fraction k of depositors were early diers, and that these have already to withdrawn their funds from the bank. The bank has enough funds to pay each of the remaining late diers a return of r^* in $t = 2$.

(a) Express the amount of assets invested long-term per late dier (i.e. after the withdrawal of the early diers) as a function of r^*.
(b) Now assume that some late diers withdraw as well. Compute the realized return of a late dier as a function of late diers withdrawing prematurely.
(c) Consider the game between 2 late diers. Compute the adjusted matrix of the game. (Hint: Recall that the amount of assets remaining after the withdrawal of the early diers is $2r^*/R$.)

Problem 9.8 Pulling the brakes

As argued in the text, bank runs would not occur if banks could simply reduce or stop payments in the face of excessive withdrawals.

(a) If banks could suspend payments at their discretion, would they often do so?
(b) Why are such suspension arrangements rarely seen in practice?
(c) What would be the optimal way to implement a suspension clause?

10 Learning and cascades

> To every place of entertainment we go with expectation and desire of being pleased; we meet others who are brought by the same motives; no one will be the first to own the disappointment; one face reflects the smile of another, till each believes the rest delighted, and endeavours to catch and transmit the circulating rapture. In time, all are deceived by the cheat to which all contribute. The fiction of happiness is propagated by every tongue, and confirmed by every look, till at last all profess the joy which they do not feel, consent to yield to the general delusion, and, when the voluntary dream is at an end, lament that bliss is of so short a duration.
>
> Samuel Johnson (1709–1784): *Idler* #18 (12 August 1758)

10 A Introduction

In a world of incomplete and dispersed information it is tempting to learn from others. Unfortunately, we cannot directly observe what others know or believe. And if we ask, we may not always get an honest answer. Yet, as the saying goes, "actions speak louder than words". We may best judge other people's beliefs by looking at their actions.

So why not do as they do? A strong tendency towards imitation is not the reserve of children, after all. Farmers introducing new crops, economic forecasters looking at others' forecasts or readers of fashion magazines often follow the example of their peers. Particularly under time pressure, running with the others is often the only way to benefit from their information.

"When you see a Swiss banker jump out of the window, follow! There is bound to be money," the French philosopher Voltaire (1694–1778) said. It may take a bit of courage to follow Voltaire's advice. But what if one, two, or three other people are seen jumping after the banker? At some point, not jumping would require more courage than jumping.

Such behavior, while individually rational, may lead to cascades. In a cascade individuals end up herding. They all run in the same direction, which may be right or wrong. Too often, jumping after a convincing banker has led to a hard landing.

In the present chapter we try to introduce the basic mechanism behind a cascade—how a well-intentioned, and individually rational attempt to learn from observed actions may degenerate into blind folly. Cascades are one example of potentially misguided collective behavior, together with bubbles, runs, fads and other phenomena (see Box 10.1). The chapter also includes an Application section dedicated to learning in repeated games. Within that section we also try to shed some light on the differences between cascading and coordination games (discussed in Chapter 9).

Box 10.1 A glossary of collective behavior

BUBBLE: A persistent deviation of an asset price from its fundamental value (Brunnermeier, 2001). There are rational and irrational bubbles (LeRoy, 2004).

CASCADE: A mechanism that can lead to phenomena like bubbles or herding. Cascades are the result of "observational learning" (learning about other agents' private information by observing their decisions) (Bikhchandani, Hirshleifer and Welch, 1998; Chamley, 2004).

CONFORMISM: (1) Resulting from *preferences*: Agents derive utility from behaving like others; (2) Resulting from payoffs: Agents are punished for deviating behavior (Dasgupta, Prat and Verardo, 2005).

CONTAGION: Financial contagion (for example, among banks) can result from: (1) direct financial linkages (a failing bank not repaying an interbank credit), (2) reputational linkages (confidence in one bank being shattered by the failure of another bank); (3) "pseudo-contagion": common dependence on some event (like the default of a large common debtor) (De Bandt and Hartmann, 2000).

CONVENTION: A generally observed rule like right- or left-hand traffic (Lewis, 1969).

CRASH: The negative instance (often the final stage) of a bubble, sometimes exacerbated by endogenous factors (selling due to loss avoidance or cash constraints). Can be triggered by a sudden switch of beliefs (fundamental or higher-order) in a cascade.

HERDING: Uniform collective behavior due to several potential reasons (agents having identical preferences, endowments, information, etc.; a bubble; a cascade; conformism) (Chamley, 2004).

FAD OR FASHION: Followers (1) adopt a leader's preferences, (2) infer the leader's information from observed action (see **Cascade**), or (3) attempt to signal taste (Chamley, 2004).

RUN: Agents compete for a limited resource to which they have "first-come-first-serve" claims (Chapter 9). There are informed runs (due to bad news) and uninformed runs (in the presence of multiple equilibria). A border case is a run occurring when individuals take the length of an observed queue as a signal of other agents' information (see **Cascade**).

10 B Main ideas: "Always stand at the longest queue."

One Sunday, Alice and Bob go for a hike in the countryside. After a long day they arrive in a small town, starving. The find two restaurants and decide to have dinner immediately. One place looks somewhat better than the other. Yet, once they get closer, they can see that the better-looking place is almost empty, while the other is almost full. At the end they go to the crowded place. They feel vindicated when, half an hour later, a queue builds at the door.

In this example, Alice and Bob behave quite rationally. Under incomplete information it seems natural to learn from what others are doing. Even if other agents are not better informed individually, they may be so in the aggregate. True, Alice and Bob cannot directly observe what others believe, but they can see how the others decided. As decisions are likely to reflect beliefs, ignoring others' decisions would waste valuable information. Given that Alice's and Bob's own belief (the empty restaurant looking better) is not particularly firm, the large number of guests in the other place is overwhelming evidence in favor of that restaurant.

The problem is that everybody may reason like Alice and Bob. If all individuals take the presence of others as a "vote" for the quality of a particular restaurant, a large number of people may go to the same place—potentially against their personal beliefs.

Assume that the first few individuals to arrive believe that Restaurant A is the better place. All subsequent individuals may originally believe Restaurant B is likely to be better. But the presence of some guests in Restaurant A may tilt beliefs in favor of Restaurant B. And with each further guest, the "evidence" becomes stronger. Once a sufficient number of individuals has gone to one place, all others will follow. The decision process ends in a cascade.

In a cascade, true learning stops. The first few individuals can infer their predecessors' beliefs from observed actions (the number of guests). But once the cascade sets in, private beliefs cease to be relevant for decisions. The problem is that a cascade may lead to the right, but also to the wrong direction. Individually, Alice and Bob behave rationally. Yet, for all agents together, the attempt to infer others' private information from their decisions may go awfully wrong.

A cascade is a very special event, triggered by special assumptions. These assumptions are:

- Agents are *uncertain* about the true state of nature (which of two restaurants is better);
- Agents can not observe other agents' *private information*, but only their *actions*;
- Actions are "black-or-white" expressions of beliefs (going to one restaurant or to the other, with nothing in between).

The phenomenon leading to a cascade is *observational learning*. Under observational learning an individual who follows her private signal communicates private information to others. Such an individual creates a *positive informational externality* for subsequent individuals. An individual who follows predecessors' decisions, rather than his private signal, stops to create such a positive externality.

A comparison of externalities also reveals the difference between cascades and successful coordination in a coordination game (Chapter 9). In a coordination game one player's action (cooking meat) has a positive *payoff* externality for another player's action (bringing red wine). In a cascade, one player's action has an *information* externality for other players. The difference can be illustrated by the restaurant choice scenario. If an individual goes to the crowded place because he likes the buzzing atmosphere, he plays a coordination game. If he goes to the crowded place because he takes the number of other guests as a sign of quality, he practices observational learning (a cascade game).

There are many practical phenomena in which cascades, also known as the "bandwagon effect", seem to play a role. It is well documented that bestseller lists for books or download statistics for internet games have an impact on future buyers' behavior. There are even reports of successful attempts by authors to boast about sales figures and, as a consequence, boost sales of their book. Fashion or lifestyle articles are often subject to cascades. A typical cascade was the "Swatch Fever" in the 1980s (see Box 10.2). Cascades also happen in financial markets. Like Swatch collectors, investors during a stock market bubble may be tempted to do what everyone else is doing. Due to the fragility of cascades, bubbles also may suddenly burst when new information arrives or just because of a change in people's beliefs about others' beliefs (Chapter 8).

A cascade may also be a driving force during a panic. Seeing a few people run away is a good indicator of imminent danger. If more follow, running away may seem to be the only choice. This is particularly true when individuals compete for a limited resource, like a seat in a lifeboat or a chance to get out of the door. A financial market example is a bank run (Chapter 9) during which depositors observe the length of the queue in front of the bank. The longer the queue, the greater the chance that others "know something" and the smaller the chance that any money will be left for those who wait any longer.

Box 10.2 The Swatch fever

During the 1970s, the Swiss watch industry, once the pride of the country, got into trouble. Sales of the typical high quality, labor- and skill-intensive Swiss watch suffered from high labor costs and increasing Far Eastern competition. The industry seemed to be in a dead end.

At this point, the vision emerged of a watch that would completely break with Swiss tradition: a cheap fashionable plastic watch produced in a fully integrated process. Such a watch was first produced in 1979, under the name

Delirium, but soon became the *Swatch* (from "second" and "watch"; not, as many people seem to believe, from "Swiss" and "watch"): a relatively cheap and colorful fashion accessory yet still a reliable timepiece. The first 12 models were launched in 1983 at prices around CHF 50 (then around US$35).

The Swatch was an instant success. With each season a new collection followed. In the mid-1980s a collectors' craze swept the nation. Buyers found themselves in the thrall of the delirium. Swatch shops sprang up like mushrooms. Whenever a new model or a collection was launched, long queues built in front of shops. During the peak of the frenzy, and for particularly "hot" models, people would camp overnight on the steps of the shops.

In 1992, the one-hundred-millionth Swatch was produced. In the coming years, sales still went well, but they never reached their peak levels again. The delirium finally was over.

Sources: http://www.swatch.com/, http://www.swatchforum.com/, http://en.wikipedia.org/wiki/Swatch.

10 C Theory: Observational learning

10 C a The basic model of a cascade

Assumptions

The basic model of a cascade was introduced by Bikhchandani, Hirshleifer and Welch (1992). We have chosen to follow this model and its discussion in Bikhchandani, Hirshleifer and Welch (1998), Chamley (2004, Ch. 4) and Brunnermeier (2001, Ch. 5). We will first present the most basic case and then relax some of the assumptions.

At the core of the model there is a simple, discrete decision: the choice between two restaurants located next to each other. One of the restaurants is good, the other bad. We will refer to them as restaurants A and B. Let us assume that A is the good restaurant.

There are n individuals who want to eat. They have the choice between the two restaurants, but no third option, like eating at home. Individuals arrive at the two restaurants in a *given sequence*. They would like to go to the better place, but they cannot tell for sure which one this is.

Individuals have two sources of evidence. First, each individual has an imperfect *private signal* $s_i \in \{a, b\}$ indicating the better of the two restaurants. Second, as individuals decide in a sequence, they can observe their *predecessors' decisions*: Individuals can see how many guests are already sitting in each of the two restaurants. They thus observe the aggregate history of decisions. A stronger assumption would be that individuals see predecessors' decisions in their actual order. The stronger assumption is not necessary for a cascade to occur; it is, however, important for breaking a cascade, as we will show in Section 10 C c.

Individuals go to the place they think is more likely to be the good restaurant. An individual may come to the conclusion that both restaurants are equally likely to be the good one. We assume that in this case the individual follows the private signal.[68]

For the sake of exposition we work with a numerical example. A priori, both places have the same probability $p = 1/2$ of being the good restaurant. The private signal has precision $q = 3/4$.[69]

Social learning

The first individual to decide, called Individual 1, only knows the prior probabilities and his own signal. Individual 2 is in a somewhat better position. In addition to her personal knowledge, she can learn from Individual 1. From observing the action taken by Individual 1 (choosing Restaurant A or B), Individual 2 can try to infer the signal that led to this action. In fact, Individual 2 knows that Individual 1 will always follow his private signal (both restaurants being equally likely to be the good one, a priori). The decision taken by Individual 1 tells Individual 2 about the former's signal. Similarly, Individual 3, seeing in which restaurant(s) the two predecessors are sitting, can try to infer their received signals.

The question is: Can all individuals learn their predecessors' private signals from observed actions? The answer is no. There are in fact two cases:

- As long as individuals follow their private signals, actions truthfully reveal these signals.
- Once an individual decides against his private signal, individuals are later unable to infer their predecessors' private signals correctly.

The chain of events starts in the first of these two states. The action chosen by Individual 1 reveals his private information to Individual 2. We will therefore start with the presumption that individuals initially know the signals of their predecessors. We will then show at which point this presumption is no longer warranted.

Updating probabilities

As long as predecessors' private information can be learnt from their actions, individuals can aggregate their own information with others' information. Each individual combines the publicly known *prior probability* p and the available *signals* (i.e. those inferred from predecessors' decisions plus the individual's own) into *posterior beliefs*. The prior probabilities were assumed to be $p = 1/2$ for each of the two restaurants. Finding posterior probabilities is a matter of using Bayes' Rule (see Chapter 3). We will apply this rule in a moment.

Table 10.1 Individual signal realizations in the cascade example. Although both signal values (a, b) are equally frequent among the first ten individuals, a cascade develops after Individual 8

Individual	1	2	3	4	5	6	7	8	9	10	..
Signal	a	b	a	b	b	a	b	b	a	a	..
α	1	1	2	2	2	3	3	3	4	5	..
β	0	1	1	2	3	3	4	5	5	5	..

Let us assume that for the first ten individuals the private signal takes the values listed in Table 10.1.

We define α as the number of received signals with value a (in favor of Restaurant A), and β as the number of received signals with value b (in favor of Restaurant B). For Individual 5, for instance, $\alpha = 2$ and $\beta = 3$ (including Individual 5's own signal realization, which is b).

The first step

Individual 1 receives signal a. What is the probability that Restaurant A is in fact the better place? Bayes' Rule (which we will apply formally below) suggests the following reasoning: Signal a occurs in two cases: either (i) if A is the good restaurant and the signal is correct, or (ii) if B is the good restaurant and the signal is wrong. This is reminiscent of the probability square we used in Chapter 4: The two above cases correspond to the top left and the bottom right parts of the probability square. Given the prior probabilities $p = 1/2$ and signal precision $q = 3/4$, the two squares represent 3/8 and 1/8 of the unit square. In case (i), Restaurant A is indeed the better place; in case (ii) it is not. The probability that A is the good restaurant, given signal a has been observed, $\Pr(A \mid a)$, is thus 3/4. Due to the symmetry in prior probabilities it happens to be equal to the precision $q = \Pr(a \mid A)$ (see Chapter 3).

A case for Bayes' Rule

Formally speaking, Individual 1 has observed signal history $(\alpha, \beta) = (1, 0)$. For this history Bayes' Rule yields the conditional probability of Restaurant A being the good place (with values for the case $\Pr(A) = \Pr(B) = \Pr(0,0) = p = 1/2$ and $q = 3/4$) on the bottom line:

$$\begin{aligned}
\Pr(A \mid (1,0)) &= \frac{\Pr((1,0) \mid A) \cdot \Pr(A)}{\Pr(1,0)} \\
&= \frac{\Pr((1,0) \mid A) \cdot \Pr(A)}{\Pr(0,0) \cdot \Pr((1,0) \mid A) + (1 - \Pr(0,0)) \cdot \Pr((1,0) \mid B)} \\
&= \frac{\frac{3}{4} \cdot \frac{1}{2}}{\frac{1}{2} \cdot \frac{3}{4} + \frac{1}{2} \cdot \frac{1}{4}} = \frac{3}{4}.
\end{aligned}$$

The first line in the above expression is Bayes' Rule (see Section 3.1), formulated for event A and one signal realization a. On the second line the probability that one signal a occurs is broken down into the probabilities that a occurs in states A and B, weighted by the unconditional (prior) probabilities of these states $\Pr(0, 0)$ and $(1 - \Pr(0, 0))$. The bottom line refers to our numerical example.

Starting the sequence of events

Individual 1, believing the odds are 3/4 in favor of Restaurant A, goes to Restaurant A. This decision is the first in a sequence illustrated in Figure 10.1. Events start at the left corner, where the prior probability of A being the good restaurant is indicated as 1/2. This is the belief held by Individual 1 before he receives the private signal. Then Individual 1 receives private signal a. As we have shown, this updates the probability of Restaurant A being the better place to 3/4. Individual 1 holds a posterior belief corresponding to point 1 in Figure 10.1. This point is in the "camp" of Restaurant A (in the upper half of the figure, Restaurant A is more likely than Restaurant B to be good). Individual 1 consequently goes to Restaurant A. The arrow towards Point 1 represents both Individual 1's signal and his decision. Individual 1's signal thus becomes public knowledge:[70] Other individuals know that rational Individual 1 follows the private signal; from observing the decision, they can tell Individual 1's signal.

The next step

Individual 2 receives signal b. Finding one guest in Restaurant A (and none in Restaurant B), she can infer that this person, Individual 1, had signal a. Individual 2

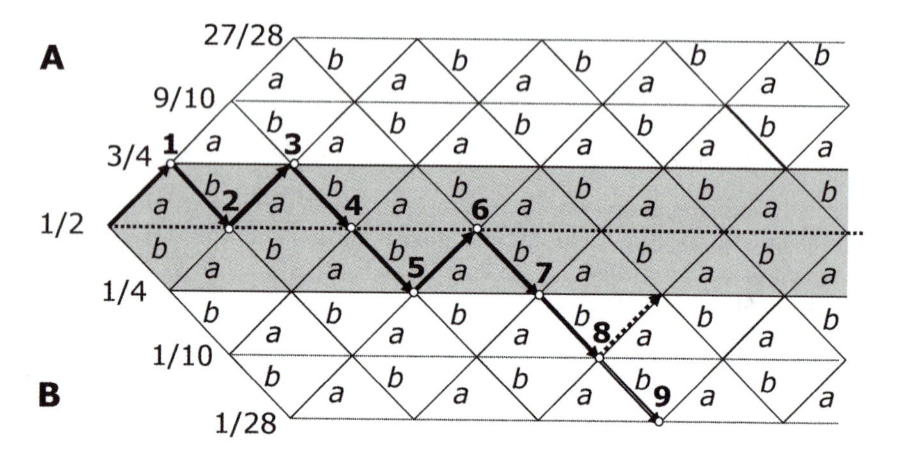

Figure 10.1 Probabilities of A being the good restaurant and the sequence of individual decisions. Individuals observe private signals $s \in \{a, b\}$ as well as the history of decisions. Once the path of signals leaves the shaded area of true learning, further signals are ignored and a cascade occurs.

thus has a record of signals (including her own) of $(1, 1)$. Intuition would suggest that the two opposite signals, a and b, cancel each other out, so that Individual 2's posterior probability of A being the good restaurant would be $1/2$. This is indeed correct, as we will show in slightly more general terms.

Updating in general terms

The general Bayesian updating formula for the probability of A being the good restaurant after a history of signals (α, β) is:

$$\begin{aligned}\Pr(A \mid (\alpha,\beta)) &= \frac{\Pr((\alpha,\beta) \mid A) \cdot \Pr(A)}{\Pr(\alpha,\beta)} \\ &= \frac{\Pr((\alpha,\beta) \mid A) \cdot \Pr(A)}{\Pr(A) \cdot \Pr((\alpha,\beta) \mid A) + \Pr(B) \cdot \Pr((\alpha,\beta) \mid B)} \qquad (10.1)\end{aligned}$$

where $\Pr(A)$ is shorthand for the prior probability of $\Pr(A \mid (0,0))$ and $\Pr(B) = 1 - \Pr(A)$.

Under the simplified assumption that the prior probability $p = \Pr(A \mid (0,0) = \Pr(B \mid (0,0)) = 1/2$, (10.1) can be simplified to:

$$\Pr(A \mid (\alpha,\beta)) = \frac{q^{\alpha} \cdot (1-q)^{\beta}}{q^{\alpha} \cdot (1-q)^{\beta} + q^{\beta} \cdot (1-q)^{\alpha}}.$$

where q, as mentioned above, stands for signal precision ($q = \Pr(a \mid A)$). The expression is further simplified, with different results depending on whether there are more signals a or b (whether α or β is larger):

- If $\alpha > \beta$:

$$\Pr(A \mid (\alpha,\beta)) = \frac{q^{(\alpha-\beta)}}{q^{(\alpha-\beta)} + (1-q)^{(\alpha-\beta)}}; \qquad (10.2)$$

- If $\alpha < \beta$:

$$\Pr(A \mid (\alpha,\beta)) = \frac{(1-q)^{(\beta-\alpha)}}{(1-q)^{(\beta-\alpha)} + q^{(\beta-\alpha)}}; \qquad (10.3)$$

- If $\alpha = \beta$:

$$\Pr(A \mid (\alpha,\beta)) = \frac{1}{2}. \qquad (10.4)$$

Individual 2 has a signal record $(\alpha, \beta) = (1, 1)$. Inserting these values into (10.1) or using (10.4) yields the posterior probability of Restaurant A being the good one $\Pr(A \mid (1, 1)) = 1/2$. We leave it as an exercise for the reader to verify this step by step.

Iso-probability lines

From equations (10.2)–(10.4) it also follows that only the *difference* in the number of signals counts, not their *absolute values*. It is irrelevant for posterior beliefs whether there are no guests in Restaurant A and two in Restaurant B or 1,000 guests in Restaurant A and 1,002 in Restaurant B.

This allows us to draw horizontal lines in Figure 10.1, each representing a uniform probability for all nodes situated on that line. Above the central line for $\alpha - \beta = 0$, giving Restaurant A a probability of being the good place of $1/2$, there is the line for $\alpha - \beta = 1$ (probability in favor of Restaurant A of $3/4$), followed by lines for $\alpha - \beta = 2$ $(9/10)$ and $\alpha - \beta = 3$ $(27/28)$. The same logic leads to the lines and probabilities in the lower half of the figure where Restaurant B is more likely to be the better place.

The sequence of optimal decisions

Individual 2 believes that both restaurants are equally likely to be the better place. In Figure 10.1 she finds herself on the central axis with $\Pr(\alpha, \beta) = 1/2$. Being undecided, Individual 2 by assumption, follows her private signal. She goes to Restaurant B. Again, the solid arrow in Figure 10.1 pointing downward to point 2, indicates both the private signal and the related action. Individual 2's signal also becomes public knowledge to all subsequent individuals.

Individual 3 can infer both predecessors' decisions from observed actions. Individual 3 finds one guest in each restaurant, as illustrated by point 2 in Figure 10.1. Having the private signal *a*, Individual 3 thus has a signal record $(\alpha, \beta) = (2, 1)$. From (10.2) it follows that the posterior probability in favor of Restaurant A is $3/4$, just as it was for Individual 1. Individual 3 thus holds a posterior belief represented by point 3. Consequently he goes to Restaurant A.

Individual 4 is in the same position as Individual 2. Her signal record of $(2, 2)$ yields the same posterior probability as a record of $(1, 1)$ or $(0, 0)$. Individual 4, being undecided (at point 5), follows her signal and goes to Restaurant B. This leaves *Individual 5* in the same position as Individuals 1 and 3. And so the path of decisions (the sequence of solid arrows) meanders on and on—until it is the turn of Individual 8 to choose.

The limits of learning

Until now, all individuals were able to infer the number of predecessors' signals from their actions. Individuals 1 to 7 had no incentive to act against their private information. Yet, this may change . . .

Individual 8 starts with the beliefs represented by Point 7: Three people went to Restaurant A, and four to Restaurant B. This means that prior to Individual 8 receiving a signal, Restaurant B is the better place with probability $3/4$ $(= \Pr(B \mid (3, 4))$, see (10.3)).

If Individual 8 is now to obtain signal a and go to Restaurant A, everything would start at square one, i.e. at the central horizontal line indicating equal odds for both restaurants. Yet, as it happens, Individual 8 draws private signal b. This updates the probability of Restaurant A being the better place ($= \Pr(A \mid (3,5))$) down to $1/10$ (see (10.3)). It is rational for Individual 8 to go to Restaurant B, as it has a $9/10$ chance of being better.

The cascade takes off

Individual 8 is the last individual who follows her private signal. Without doing anything differently from her predecessors, Individual 8 leaves her followers with a different situation. After the move of Individual 8, the history of decisions will never again be identical to the history of private signals.

As a consequence, none of Individual 8's successors will be able to infer *all* past private signals from the observed history of decisions. Individuals know that up to (and including) Individual 8, everybody followed their *private* signal, but that all remaining individuals will follow the majority of their predecessors. In concrete terms: Everybody will go to Restaurant B. This can be seen in Figure 10.1. Individual 8 leaves her successor with a signal history $(3,5)$ summarized by point 8. At this point Restaurant B is the better place with a probability of $9/10$ (two more individuals have received signal a than have received b).

For *Individual 9*, the private signal becomes irrelevant. Irrespective of his private signal the odds remain in favor of Restaurant B. With signal b, the odds in favor of Restaurant B (A) would rise to $27/28$ ($1/28$). Yet, even after a, the signal effectively drawn by Individual 9 (Table 10.1), the odds remain in favor of Restaurant B. Individual 9 thus disregards the private signal (indicated by the dashed arrow pointing upwards) and goes to Restaurant B (as indicated by the solid arrow pointing downwards). The decision by Individual 9 cannot reveal the individual's private signal. From Individual 9 on, the signal history is no longer public knowledge.

Individual 10 knows that she cannot learn anything from the decision of Individual 9. The best guess of her prior odds is therefore those represented in Figure 10.1 by point 8 (as it is known that the first eight individuals followed their signals). She is in exactly the same position as Individual 9: Whatever her private signal, it is best to go to Restaurant B. So do all further successors, observing more and more guests in Restaurant B. Starting with the (rational) decision by Individual 9 to go to Restaurant B, a cascade develops during which all further individuals herd.

In Figure 10.1 there is no way back, once the path of events leaves the shaded area. Individuals in fact follow a simple decision rule. Let δ be the (positive) difference between the guests in the two restaurants (the majority minus the minority). If $\delta \leq 1$, i.e. δ is either 0 or 1, individuals follow their own signal. These individuals start with beliefs located within the shaded corridor. If $\delta \geq 2$, individuals follow their predecessor (or, equivalently, go to the place with the higher number of guests). This is the case as soon as an individual holds beliefs outside of the shaded corridor.

10 C b Informational inefficiency

The inevitability of a cascade

Our simple example (based on Bikhchandani, Hirshleifer and Welch (1998); Chamley (2004)) shows that a cascade starts, as soon as the private information of an individual is much "weaker" than the accumulated information incorporated in past actions. "Weaker" here means that it takes an excess of two past signals in one direction over signals in the opposite direction to make the next and all following private signals meaningless.

The example also makes clear that a cascade occurs with probability one if the number of individuals goes to infinity (if we follow the shaded tunnel in Figure 10.1 from its starting point, we observe that it becomes arbitrarily narrow in perspective as it gets longer). In our example with $p = 1/2$ and $q = 3/4$, a cascade starts as soon as one signal realization has been drawn two times more than the other ($\mid \alpha - \beta \mid = 2$). The probability of this event goes to unity when the number of individuals, n, becomes larger.

The failure of social learning

A cascade is an example of a spectacular failure of social learning. Before a cascade starts (in our example, up to the decision of Individual 8), individuals act according to their private information. Their actions communicate their private information to all successors. Once a cascade starts (with Individual 9), actions cease to convey any private information on later individuals. Public information stops accumulating. Individual 10 cannot conclude that Individual 9 had signal b (which is in fact not the case), as Individual 9 would have gone to Restaurant B even with signal a. Nor does any further information become available to later individuals.

The probabilities of a good or a bad cascade

When only *actions* are observable, rational decisions eventually end in a cascade. Of course, this could be a "good" cascade (everybody going to Restaurant A), but a bad cascade (everybody going to Restaurant B) is a possibility, too. In our example, the probability of a good (bad) cascade is easy to compute. Starting from the initial point, i.e. the left corner in Figure 10.1, it takes two consecutive identical signals to arrive immediately at a cascade. After the first two individuals, a cascade thus starts with probability $q^2 = 9/16$ (good cascade) and probability $(1-q)^2 = 1/16$ (bad cascade). These probabilities add up to $1/2$. With the complementary probability of $1/2$, the second signal cancels the first, leading back to the central line. There, the game starts again with the same probabilities of $q^2 = 9/16$ or $(1-q)^2 = 1/16$ of ending in a good or bad cascade after the next two individuals, and so on. Eventually a cascade will occur. The odds that our example is a good rather than a bad cascade (with prior probabilities $p = 1/2$) are:

$$\Pr(good) = \frac{q^2}{q^2 + (1-q)^2};$$

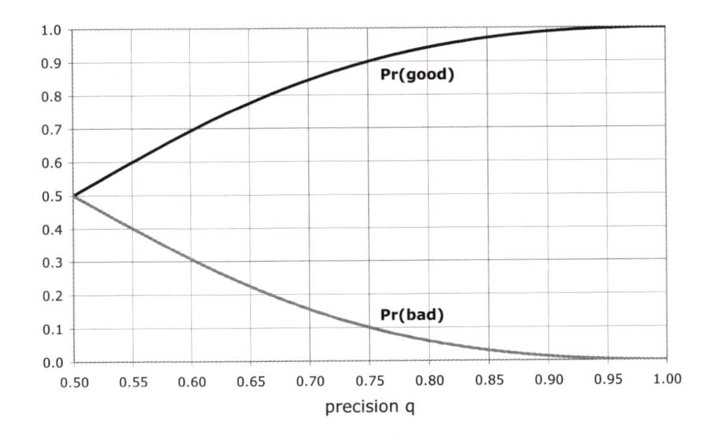

Figure 10.2 Probabilities of a "good" cascade (upper line) and of a "bad" cascade (lower line) as a function of signal precision q.

with $q = 3/4$, this is $9/10$, the probability of Restaurant A being the better place after observing two signal realizations a.

The probabilities of a good and a bad cascade in our example are illustrated in Figure 10.2.[71] The probability of a good (bad) cascade increases (decreases) with the precision of the signal q. When the precision of the signal goes towards unity (a perfect signal), the probability of a good rather than a bad cascade goes to unity as well.

Observability of actions versus observability of information

A cascade implies an inefficient use of private information. The welfare consequences of a cascade become evident from a comparison of the three possible cases:

- Signals are publicly observable (benchmark case);
- Actions, but no signals are publicly observable (cascading case);
- Neither the signals nor the actions of others are observable (isolation case).

In the benchmark case where *signals* are observable, it is public knowledge that three in four agents have signal a. The rational decision for everybody is to go to Restaurant A. The fraction of correct decisions is 1.0. In the cascading case, the fraction of correct decisions (the probability of a good cascade) in our example is 0.9. If individuals decide according to their private signals only (isolation case), the probability of hitting the right restaurant is 0.75. When individuals learn from observed *actions*, they do somewhat worse than if *signals* were observable. Nevertheless, observational learning is better than ignoring others' decisions completely.

An informational externality

Under observational learning, individuals who follow their private signal communicate private information to successors. This is a positive informational externality. However, an individual who follows predecessors rather than the private signal does not provide others with this positive externality. Even worse, a follower prevents all successors from communicating their private information, thus creating a positive externality. In our example, the single "mistake" of Individual 9 has severe consequences. It is sufficient to coordinate all successors on the wrong action to stop social learning once and for all.

The assumptions leading to a cascade

While a cascade is a logical consequence of observational learning and rational individual decisions, cascades (particularly "wrong" cascades) occur only under specific assumptions. It is important to remember the assumptions made:

- Actions are finite.
- Private information is not observable, only actions are.

The assumption of *discrete and bounded actions* is crucial for a cascade. In our example, the choice between two restaurants (with nothing in between) prevents individuals from communicating how good they think a restaurant is and how sure they are about their belief. Actions thus are a coarse filter through which individual beliefs are observed. Individual decisions are black-and-white. The history of decisions comes in five shades. The subjective probability of Restaurant A being the better place can only take the five values $9/10, 3/4, 1/2, 1/4, 1/10$. Trade in information could remove the inefficiency resulting from observational learning. Prices efficiently communicate private information (Chapter 7), as their continuous (and potentially unbounded) nature permits individuals to express their beliefs more accurately than in a discrete choice.

Given the coarse picture cast by past actions, the *non-observability of individual signals* has serious consequences. It permits cascades to occur, and it prevents them from ending as long as no exogenous event, like the arrival of new public information, breaks the cascade.

There are two further assumptions that have shaped our results:

- Only actions are observable, but not the *result* of actions (the satisfaction of guests in each of the two restaurants);
- Individuals observe an *aggregate* statistic on predecessors' decisions (the number of guests in each restaurant), but not the precise *sequence* of decisions.

These assumptions are not necessary for cascades to occur. Observability of *the results of an action* (the satisfaction of guests in a restaurant) would actually increase the probability of a cascade, which would also have a greater likelihood of being a good cascade. It is intuitively plausible that the observability of results would reduce the inefficiency of the learning process.[72]

In our example we assumed that the *sequence* in which actions occur is not observable. This is the weakest assumption that is necessary. An alternative would be the assumption that individuals can in fact observe the order in which predecessors took their decisions, rather than just the number of guests sitting in each restaurant. We would arrive at exactly the same result under this stronger assumption. In the simple model used above, the exact sequence of previous decisions is irrelevant for an individual's decision. It is only the number of decisions in favor of each the restaurants (in fact: their difference) that counts. However, even though the sequence of decisions is irrelevant for a given individual, it is *not* irrelevant for *society* as a whole. Whether a good or bad cascade starts does not only depend on the relative number of signals *a* and *b* in the aggregate. It also depends on the order in which these signals arrive. Individuals' decisions thus are *path-dependent.*

In the simple setting of our model, an observable sequence was not necessary. Yet, the order in which decisions are taken becomes important in more sophisticated models. When, for instance, some individuals, who are known as "experts", have more precise signals than others, it becomes significant *when* they decide, not only *how* they decide. This is a case of *heterogeneous signal precision*, one of the extensions of the basic model which we will discuss in the following sections.

10 C c Experts and fashion leaders

In the simple model underlying Figure 10.1, all individuals receive a signal of the same precision. In reality, signal precision is likely to be *heterogeneous*, some individuals ("experts") having better information than others. The presence of individuals known as experts has an ambiguous impact on the occurrence or the stability of cascades (see Bikhchandani, Hirshleifer and Welch, 1998; Chamley, 2004):

- When an expert moves *first*, it is likely that consecutive individuals follow the expert's action and a cascade develops immediately. Assume that in our example Individual 1 is an expert who drew two identical signals. This would immediately start a cascade.
- When an expert arrives *after* a cascade has already started, the expert may break the cascade. In our example, individuals moving after Individual 9 all go to Restaurant B. Remember that they do so on a rather weak informational basis. They only know that prior to Individual 9 there were two more signals *b* than *a*.

  Once an individual deviates from the crowd and goes to Restaurant A, the individual must have a strong reason to do so. Obviously, the individual must be an expert with (at least) two identical signals *a* (an event with probability 9/16). Here it is not necessary that the expert is publicly known as such. Yet, it is important to assume that the actual *sequence* of arrivals is observable. If only the number of guests in each restaurant were observable, the fact that the one customer in Restaurant A might be an expert would weigh little against the crowd in Restaurant B.

The observable arrival of the expert in Restaurant A breaks the cascade. The two signals *a* cancel the two former excess signals *b* and reset beliefs regarding the two restaurants to the initial (prior) value of $1/2$. The individual deciding immediately after the expert will follow the private signal. So will the next individual, and so on, until a new cascade starts.

The latter case shows that while cascades are quite inevitable under observational learning, they are also *fragile*. Experts are socially valuable groups because they help break cascades or trigger cascades in a positive direction.

Other socially valuable groups besides experts are individuals who are very certain of themselves (unless they have less precise signals than the rest of society) and non-conformists. Self-assured individuals weigh their private signals as high relative to social information implied by the behavior of others. Non-conformists derive some pleasure from acting differently from others. Like self-assured individuals they may slow down the occurrence of a cascade or may help to break it.

Fashion leaders are an interesting species. They are both experts and non-conformists. As experts, they have a "feel" for coming trends. They perceive these trends as exogenous and spend great efforts in spotting them. As non-conformists they quickly get tired of conventions. A new fashion often builds on the most despised ingredients; punk fashion, for instance, introduced garbage bags as a fashionable clothing material in the 1970s.

10 C d Wait and rush

Endogenous sequence

The presence of experts (individuals with more precise information) has a particularly strong impact when the sequence in which individuals decide is endogenous rather than given. Individuals who are uncertain wait to see others' decisions; experts tend to move first. As mentioned above, the example of experts' decisions may quickly trigger a cascade during which all non-experts ignore their own private information.

Heterogenous precision (some individuals who are known to be experts) and an endogenous sequence of events leads experts to move first and non-experts to follow. This effect is related to the problem of who should speak first in a group or organization (see Section 8 E b). One problem is that the views of some "alpha" members of a group may dominate others. In a cascade, other individuals may let the "alphas" speak first and, being impressed, absorb their point of view. As mentioned above, individuals may also disagree with the alpha view, but find it unattractive to voice their doubts. Such conformist motives or payoffs (dissent being punished) are not required for a cascade, but of course they have a tendency to contribute to the likelihood and severity of cascades.

Endogenous precision

In the basic example, individuals have signals with a given precision. In real life, individuals often can improve the precision of their information at some cost. One would think that the possibility of receiving better signals would make cascades less likely. Yet, the opposite is the case. Via actions, individuals' information becomes public knowledge. This makes individual information a public good. We showed in Chapter 6 that, as a public good, information is under-supplied. The possibility of observing predecessors' decisions makes the acquisition of costly information less attractive. The temptation to follow others' actions is even stronger than in the basic cascading model. Endogenous (and costly) signal precision leads to rampant free-riding. Cascades may form instantly as shown in Bikhchandani, Hirshleifer and Welch (1998) and Feltovich (2002).

More complicated cases involve heterogeneous precision *and* heterogeneous preferences. An individual may choose Restaurant B because it offers a vegetarian meal or because it has the better cook. Or, an agent's decision may not only have an informational externality, but also a payoff (e.g., network) externality in the sense of a coordination game. The related literature is discussed in Bikhchandani, Hirshleifer and Welch (1998).

Wait and see

Often individuals do not just have a choice among two actions, like going to one restaurant or to the other. Waiting is an alternative in many situations. An investor may invest today or just wait to see how the world looks tomorrow. In Chapter 5 we discussed a model of investing or waiting for information (Cukierman, 1980). In that model, however, an individual decides in isolation. There is no option to look at how *others* would decide.

Here the question is what happens if an investor can wait and *see* whether other investors invest or not (i.e. wait). We must assume that (like in the Cukierman model) waiting is costly, otherwise investing would be postponed forever. The outcome of such a model is driven by investor heterogeneity and by endogenous sequencing. Investors may differ with respect to their beliefs (optimists and pessimists) or with respect to their waiting cost (patient or impatient). Chamley (2004, Chs. 6–7) shows that when timing is endogenous, the most optimistic invest first. Once they invest, the mildest pessimists are likely to follow. This may "convince" the stronger pessimist, such that by the end everybody invests. Or, some individual may be pessimistic enough not to follow the less pessimistic predecessors. Then, investment stops, and once it stops, it never resumes (as only even more pessimistic investors are left). With a large number of investors the model ends with a whimper or a bang (Chamley, 2004).

Waiting games involving cascades can lead to dramatic outcomes when opportunities are limited. A few people running towards the emergency exits in a movie theater can trigger a stampede. A sufficient number of depositors queuing up at bank counters can trigger a bank run. The stumbling of a penguin can lead to a collective plunge (see Box 10.3). Such events are characterized by the coincidence of (i) observable actions with endogenous timing (leading to a cascade) and (ii) strategic complementarity (leading to a desire to coordinate). As the example shows, the anatomy of financial panic can be complicated. Although individual elements are quite well understood, their interplay in a concrete case may need careful analysis.

Box 10.3 Plunging penguins

Penguins have developed several strategies to avoid predators like the leopard seal. For example they live in groups either too small to be interesting to seals or large enough to provide safety in numbers (Ainley et al., 2005). The biggest danger for penguins is in the water. Chamley (2004, p. 1) gives the following account:

> Penguins are social animals. They live in groups above the water from which they get fish or food. Unfortunately, there is more than fish in the water. From time to time, killer whales (orcas) roam under the surface waiting for some prey. Penguins are aware of the danger and would like to have some information before taking a plunge. Indeed, any sensible penguin thinks that it would be nice if some other penguin would dive first into the water. So what is a penguin to do? Wait. Possibly some other member of the colony who is more hungry, or has other information, will go first. Is it possible that no penguin will ever go? No, because waiting becomes more costly as hunger increases. Eventually, one or more penguins will take the plunge, and, depending on the outcome, the others will either stay put or follow *en masse*. This *waiting game* is socially inefficient. It would be better if the first individual would decide to go at least a bit earlier. Actually, the penguins are well aware of this social inefficiency, which they try to remedy by pushing some poor fellow off the cliff.

This is an economist's account. Penguin researchers add some distinct flavors to the story. First, the true enemies of penguins are leopard seals, not orcas. Second, penguins may also play a coordination game. To synchronize (coordinate) individual actions they seem to motivate each other for a plunge vocally:

> Despite all this [their excellent swimming skills], penguins are genuinely afraid of the water. When chased, they will never go into the water if given a choice. They are hesitant to even wade across a small stream. When they do enter the water, it is only if there are several birds to enter at the same time. Even then, some will chicken out at the last minute.

A single penguin in the water has much less chance of escape from a leopard seal than a large group. This is the same strategy employed by all schooling fish and flocking birds. In a large group the seal has difficulty singling out any one individual for attack. Thus the Adelies wait on the beach until there is a group of ten to forty birds ready to go. The group synchronizes their entry into the water by exchanging a series of grunt-like noises. These grunts increase in intensity like the cheers at a sporting event, until the penguin in the front of the line finally takes the plunge or in some cases chickens out and lets the next guy have a go at it. Source: http://tea.armadaproject.org/wille/1.15.2000.html

Cascading as well as coordination motives may play a role in getting penguins into the water. It seems difficult to disentangle the individual mechanisms. In this respect, penguins' collective rushes are little different from similar events in the financial markets.

10 D Application: Learning in repeated games

Cascade or coordination?

A cascade looks similar to successful coordination. In both cases, at one point all individuals choose the same action. Yet, the similarity of results is superficial. In a coordination game (Chapter 9), one player's action has a *payoff* externality for another player's actions. In a cascade, one player's action has an *informational* externality for other players. The difference is illustrated by individuals choosing between an empty and a crowded place. Individuals who go to the crowded place because they like the atmosphere of a buzzing place (to be where others are) play a coordination game. Those who go to the crowded place because they take the number of other guests as a sign of quality, play a cascading game.

Technically speaking, the two kinds of game are characterized by different information structures. In coordination games, payoffs are common knowledge, but actions are not. In games leading to cascades, actions (once they have occurred) are common knowledge, but payoffs are not (individuals try to infer payoffs from others' actions). Coordination problems can arise in simple "one-shot" games, while a cascade needs a multi-period setting. One-shot coordination games involve pure *guessing*; games leading to cascades involve *learning*.

The hardest test

Of course, learning is also an issue in multi-period or repeated coordination games (Gibbons, 1992; Hirshleifer and Riley, 1992; Fudenberg and Levine, 1998). In the literature, the touchstone for agents' ability to learn in a repeated game is the prisoner's dilemma. This is for two reasons. In a one-shot prisoner's dilemma

Table 10.2 The prisoner's dilemma game

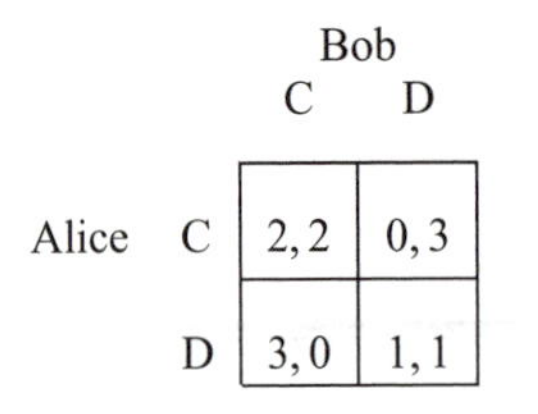

		Bob C	Bob D
Alice	C	2, 2	0, 3
	D	3, 0	1, 1

game: (i) rational agents never cooperate; (ii) communication is not credible (see the comment to Table 9.7 and the following game in Table 10.2). If agents learn to cooperate even in a prisoner's dilemma situation, learning can be trusted.

The one-shot prisoner's dilemma

Let us first have a look at the one-period prisoner's dilemma (Gibbons, 1992, Ch. 2.3). The respective game is characterized by the following payoff structure:

This game, briefly mentioned in Chapter 9, has a *unique* equilibrium $\{D, D\}$. Playing D (defect) is a *dominant strategy*. Even if one player knows that the opponent is going to play C (cooperate), playing D is more attractive. Among rational players the game ends with $\{D, D\}$, rather than with the superior outcome $\{C, C\}$.

The repeated prisoner's dilemma

Could it become worthwhile to cooperate in a prisoner's dilemma when agents know they will *repeat* the game? Cooperation by one player in round one might convince the other player to cooperate in round two, and so on. After all, regular customers in a restaurant are treated well (and leave tips). Only the one-time visitor in a "tourist trap" is served with an overpriced half-frozen pizza by an unfriendly waiter and leaves without tipping (and sometimes even without paying).

Simply playing the game twice, ten or a hundred times would not help, as long as the game has a commonly known finite number of rounds. In the last round, the game is identical to a one-shot game, with D being the dominant strategy for both players. Therefore, no player has an incentive to cooperate in the second-last round, knowing that the opponent will defect in the last round anyway. By backwards induction this argument can be applied to the third-last round, and so on, up to the first round.

Repetition might help, though, if the number of rounds is unknown to players or infinite. In this case, full cooperation (always playing C) seems to be an equilibrium. Yet, so is constant defection (always playing D). The game thus has multiple equilibria. In fact, repeating the prisoner's dilemma in an uncertain number of rounds changes it into a coordination game.

Learning from the others

Learning in a repeated prisoner's dilemma combines elements of coordination games and of cascades. Which aspect dominates depends on *who* the agents in a repeated action are:

- In a *"fixed-player world"* there are two (or a few agents) who have an identity visible to each other. In this situation it is important to learn how the opponent plays, which is typical for a coordination game. Analyzing such games can be very tricky, as each player must not only consider an opponent's past history of playing, but also his potential future reactions to the player's current decision.
- In a *"large population world"* agents are randomly matched. It is highly unlikely to meet the same agent twice (if agents are identifiable at all). Knowing the opponent or building a reputation is not possible in this kind of setting. The task of agents is to find out the proportions of different types of agents in the total population. Each agent can concentrate on what she can derive from her past experience of playing the game or from aggregate statistics on play in previous rounds. Learning the frequency of types from observing the actions of others resembles observational learning in the cascading model.

Modeling large populations

For the analysis of learning in large-population settings, three main models of learning have been developed according to Fudenberg and Levine (1998). Common to all is that in subsequent rounds one individual player is randomly matched with different players. The differences between the models arise from what players observe after a match:

- So-called *fictitious play* models assume that players only observe the outcomes of their own matches. From opponents' play, they infer the relative frequencies of different strategies. These frequencies they use in fictitious (hence the name) pre-play calculations used to coordinate on some Nash equilibria.
- In *partial best-response* models players do not only observe the outcomes of their own matches but also the outcomes of all other matches. Consequently, they can adjust their behavior in subsequent rounds to the aggregate statistic from the previous period.
- Finally, in *replicator* models, "learning" occurs on an aggregate level through survival of successful strategies, rather than through individuals becoming more knowledgable. The population consists of different types characterized by their respective strategies. The share of any type in the population grows in proportion to its success, measured by its current payoff. This leads to patterns known from biological evolution and to so-called "evolutionary equilibria". In these models, information is not processed in the form of conscious knowledge but rather passed on in the form of "genes" (innate strategies).

We shall not go into the details of the different approaches discussed in Fudenberg and Levine (1998). Our only ambition was to show that observational learning in a multi-period setting can occur in different ways. Learning from other agents' simultaneous actions about their strategies is relevant in coordination games. Learning other agents' beliefs from their actions sequentially would lead to a cascading game.

10 E Conclusions and further reading

Learning can occur within very different economic models. A realistic and interesting example of learning is "observational learning", looking at what others do. From observed actions, the beliefs of the agents can often be inferred. It is perfectly rational to use the information contained in other agents' actions.

Collectively, the attempt to learn others' beliefs from their actions is doomed to fail. At one point, the information content of observed history exceeds that of individuals' private information. A cascade starts and may continue forever. It needs a special event, like the arrival of public information or an action by a respected expert, to break a cascade.

Although cascades require special assumptions (like a discrete action space), they occur in many real situations. Even where continuous actions are available in principle (as in the stock market), individuals often seem to look at reality through a discrete filter, which is typical of a cascade. When investors buy shares because other investors bought (or claim they bought), the foundation for a cascade are laid. Cascade-like elements are present in speculative attacks (Chamley, 2004, Ch. 16) or bank runs.

For further reading we recommend Bikhchandani, Hirshleifer and Welch (1998) as an overview of cascading models and Chamley (2004) as a broader reference on cascades and herding. A more compact treatment is given in Brunnermeier (2001, Ch. 5).

Checklist: Concepts introduced in this chapter

- Observable action versus observable information
- Informational externality versus payoff externality
- Bayesian updating in a sequential decision
- Sequential learning
- Cascades
- Learning in repeated games
- The Prisoner's dilemma (also mentioned in Chapter 9)
- Fixed-player and large-population models of learning

10 F Problem sets: A bath in the crowd

For solutions see: www.alicebob.info.

Problem 10.1 As wise as before

Two restaurants (A and B) have the same prior probability of being the better place. Individuals receive a private signal with precision $q = 3/4$. The first two individuals received signal values a and b, respectively. The second individual does not know the first individual's private signal, but can observe that the first individual went to Restaurant A.

(a) What is the second individual's belief regarding the probability of A being the better place after getting the signal and seeing the first individual's choice?
(b) What were the second individual's posterior beliefs if both signals were the same (a, a) or (b, b)?

Problem 10.2 Learning from others

An individual faces two restaurants, both of which have an equal prior probability of being better. It is common knowledge that individuals have independent private signals with precision $3/4$. The individual finds 4 guests in Restaurant A and 7 guests in Restaurant B. What is the probability that B is the better restaurant?

Problem 10.3 A likely case?

As in the text, there are two restaurants with prior probability $p = 1/2$ of being the better place; individuals receive independent signals (a,b) with precision $q = 3/4$. What is the probability that the statistic of the first eight draws is $(\alpha, \beta) = (3, 5)$ if, in fact, A is the better restaurant?

Problem 10.4 The probability of a good cascade

Show that in the basic cascading model used in the text the probability of going to the better restaurant (for a large number of individuals) increases with the signal precision q. For which value of signal precision q is there a maximum difference between (i) the probability of a good cascade and (ii) the probability with which an individual who ignores others' decisions ends up at the better restaurant?

Problem 10.5 Investing with rivals

Within the same industry, large firms tend to invest when their competitors do not, while small firms invest when their competitors do (Bikhchandani, Hirshleifer and Welch, 1998, p. 165). Explain.

Problem 10.6 Convergent behavior

Bikhchandani, Hirshleifer and Welch (1998, p. 168) find that "Observational learning theory suggests that in many situations, even if payoffs are independent and people are rational, decisions tend to converge quickly but tend to be idiosyncratic and fragile."

(a) Apart from the restaurant example, what other phenomena can be explained by observational learning?
(b) What other reasons apart from observational learning might lead to convergent behavior?

Problem 10.7 The internet

Does the internet (and intranets within organizations) make cascades more or less likely? Check your answer with Bikhchandani, Hirshleifer and Welch (1998).

Problem 10.8 Crime and enforcement

Within certain city areas crime per person is significantly higher than in other areas.

(a) Does that mean that people living in these areas haven a disposition for crime? Explain.
(b) In face of observational learning what policies would you introduce?
(c) In what other context can you find similar social interactions or peer-group effects?

11 The macroeconomics of information

11 A Introduction

> In a world where information is relatively scarce, and where problems for decision are few and simple, information is almost always a positive good. In a world where attention is a major scarce resource, information may be an expensive luxury, for it may turn our attention from what is important to what is unimportant.

This quote by Herbert A. Simon (1978) is a nice illustration of the fact that individuals usually face a large number of decisions to make and an even greater amount of information to process. Many of these decisions will also affect the future choice set of the individual, and—when aggregating the decisions of all people in an economy—the market conditions under which these future decisions have to be made. If an individual chooses to consume more and save less, she will have fewer resources to live on in the future. If all the citizens in a country do the same, the economy will have a lower capital stock in the future, which affects both its production possibility and future interest rates. Ultimately this has an impact on the decisions we make tomorrow.

Information and information processing have always been of great importance in macroeconomics. How much individuals consume and save, and what firms decide about investment or hiring strategies is crucial for the economy. Such decisions have a strong forward-looking component that makes any information about the future course of the economy very valuable. It also allows economic agents to incorporate the impact of economic policy measures when attempting to make an optimal decision.

Traditional Keynesian models, for example, are based on adaptive expectations in which today's expectations are a simple function of past expectations and currently available information. Since the failure of these models to explain real world phenomena and as a consequence of the so-called Lucas critique, rational expectations have become *the* information processing paradigm in modern macroeconomics. They are the only logically consistent way to formulate expectations.

But problems begin once rational expectations are implemented. To start with, many realistic processes on which the individuals have to form expectations, such as asset prices, would require an infinite amount of information processing capacity (Shannon, 1948). More significantly, the models using full-information rational expectations cannot explain some important macroeconomic phenomena, such as the sluggish response of consumption to monetary shocks and long-term deviations from purchasing power parity.[73] Moreover, if current prices reflect future prices, as is the case in asset and exchange rate markets, it will be important to anticipate not only the values of the underlying economic variables, but also the expectations of other market participants—a classic example of Keynes' beauty contest ((Keynes 1936, p. 156), discussed in Chapter 8).

The problem is often not only expectation formation *per se*, but the challenge of information processing. Some observed macroeconomic phenomena are consistent with *rational expectations* if information is costly to obtain or to process. Given the complexity of information acquisition, people may rationally decide to stay uninformed.

11 B Main ideas: Who acquires information and why?

Almost all economic decisions involve taking actions today for which the payoff tomorrow is uncertain. This is particularly true in macroeconomics with its strong intertemporal dimension. If a person decides how much to save, for example, she must take into account that the return on her savings is uncertain and also that built-up inflation may erode the value of her assets. A firm that hires new workers during a boom faces the probability that an upcoming recession may make these additional workers redundant. If it is costly to fire workers, the risk of such a negative outcome in the future will also affect today's decisions.

As expectations about the future value of economic variables are so important, the working of macroeconomic models depends crucially on how these incorporate expectations or, more generally, information processing. Two key points stand out:[74]

1 The way individuals *acquire* and *handle* information to make the necessary decisions. This also includes the *type* of information individuals use.
2 The way *expectations* are formed about the future value of key variables. In many circumstances it is not possible to pin down a single value of a variable in the future. The prediction then comes in the form of a probability distribution.

The present chapter mainly focuses, like the rest of the book, on the first point. Nonetheless, it is worthwhile to pause for a moment here to think about the second point. Throughout the book, we tacitly assume that people form their expectations sensibly, based on the information they have and/or are willing to acquire. Putting it differently, expectations are formed in line with the utility maximization behavior of individuals. In fact, the bulk of our analysis would not make much sense without this requirement, which corresponds to what is usually called the *weak form of rational expectations* (see Box 11.1).

Returning to information acquisition and processing, the cost of information can play an important role in macroeconomic models. If individuals or firms incur the direct costs of information acquisition, they face a trade-off between bearing the necessary costs to obtain the relevant information, or saving it and deciding under incomplete information (Chapter 5). As a consequence, an individual or a firm might rationally decide not to update information at a given moment, if the costs of buying the information exceed the present value of planning errors. This behavior is called *rational inattentiveness* and will be explained in more detail below. One point worth noting here is the heterogeneity of market expectations. Although individuals have access to basically the same information set, costly information leads them to acquire and process information at different times. This creates different individual information sets which form the basis for any decisions they may make.

Costly information violates one of the assumptions behind the *Efficient Market Hypothesis* (see Chapter 7). Markets are (informationally) efficient, if prices always fully reflect all available and relevant information. The EMH holds under three conditions:

- There are no transaction costs;
- Information is available without cost;
- Investors have rational expectations and use the same model.

The violation of the second of these assumptions leads to rational inattention. Unfortunately, costly information about the economic fundamentals also leads to a second problem. In a world of costly information acquisition there is often substantial *disagreement about expectations* of macroeconomic variables.

A striking example of this disagreement is the huge span of inflation expectations,[75] which are unlikely to be due to information differences *per se*. As an example, Figure 11.1 depicts the evolution of inflation expectations of Polish households since 1992. While there has been considerable uncertainty during the high-inflation transition period in the 1990s, persistent disagreement on the course of inflation after the turn of the century is surprising. It is very likely that these different perceptions also feed into the decisions made by households.

The heterogeneity of market expectations is particularly critical, if current prices reflect future prices. Important examples are the general price level, asset prices and exchange rates. If beliefs are non-homogenous, information processing is much more complex. One needs to forecast not only future values of a variable, but also the expectations of other market participants.[76] The latter will do the same, of course, and one wants to form the expectation of the expectation of future prices. By iterating this strategy, one ends up having to forecast the expectation of the expectation of the expectation of ... and so on. It is not surprising that Keynes regarded financial markets as a sort of beauty contest. We will look at the implications of *higher-order expectations* (introduced in Chapter 8) below.

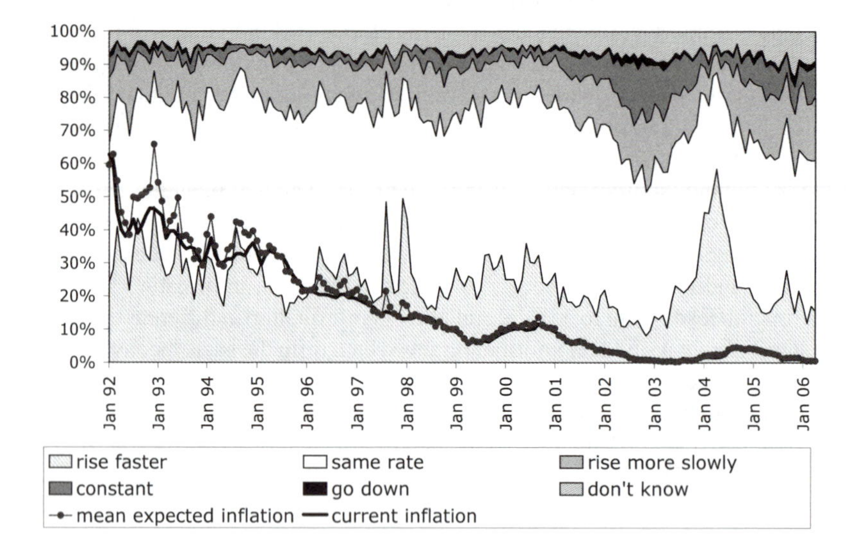

Figure 11.1 Inflation expectations of private households in Poland since 1992. The shaded areas represent the fractions of the different forecasts given by the surveyed households. The observed and expected mean inflation rates are depicted as solid and dotted lines, respectively.

Source: Ipsos survey data; calculations by the National Bank of Poland.

In reality, information processing in macroeconomic contexts is likely to be a combination of both *costly information acquisition* and *higher-order expectations*: Those who trade (because they have incurred the cost of updating their information about fundamentals) do not only have to take into account the value of those fundamentals, but also the behavior of individuals with different expectations and/or those who are temporarily inattentive. Needless to say, this is extremely hard to model and to date has been little explored.

Box 11.1 The rational expectations hypothesis (REH)

The Rational Expectations Hypothesis (REH) had its origin in a seminal paper by John Muth in 1961. Similar notions about rational expectation formation had been put forward by other economists, even before Muth. Nonetheless the work by the latter had by far the greatest influence on the economics profession. Although initially considered as an expectation formation in the microeconomic context, it had the most substantial impact in macroeconomics.

The REH came close to a revolution in the field and was vividly and controversially debated for a long time. Barro (1984, p. 179) even started his critical appraisal of rational expectations with the following phrases:

> One of the cleverest features of the rational expectations revolution was the application of the term 'rational'. Thereby, the opponents of this approach were forced into the defensive position of either being irrational or of modeling others as irrational, neither of which are comfortable positions for an economist.

The REH basically states that people form their expectations sensibly based on the information they have or are willing to acquire. It can be considered to be in the spirit of Muth in its strictest sense, or as a more generally applicable paradigm of information processing. One usually differentiates two versions:

Weak form of REH: In forming their forecasts about the future value of variables, rational individuals make the most efficient use of all publicly available information about the factors that, in their view, determine those variables.

In a formal way, the weak REH can be written as:

$$E_t(X_{t+1}) = E(X_{t+1}|\Omega_t),$$

where Ω_t is the information available at the time of the forecast t.

Strong form of REH: In addition to the efficient use of available information as already contained in the weak from of the REH, individuals also use the correct form of the underlying economic model. The forecast error is thus essentially random with mean zero, serially uncorrelated over time, and has the lowest possible variance.

More formally, we can write:

$$E_t(X_{t+1}) = X_{t+1} + \epsilon_{t+1},$$

where ϵ is a mean zero error term that is uncorrelated to the information set available at t when the expectations are formed.

When the rational expectations revolution started, orthodox monetarists used adaptive expectations to derive the expectations-augmented Phillips curve. Adaptive expectations of future values of economic variables imply that individuals only use past information, in particular past values of the variables concerned. A consequence of this backward-looking approach is that forecasts may be repeatedly wrong over a long period.

An example shall demonstrate the *difference between adaptive and rational expectations*, and also illustrate why the issue was most hotly debated in macroeconomics.

Let us consider throwing a dice several times. After having rolled the dice 10 times without a single occurrence of a 6 (an event with a probability of 16 per cent, after all), adaptive expectations would give a very low weight to throwing a 6 in the next round, although the probability of a 6 is, of course, still 1/6 in the next round. In such well-defined microeconomic situations adaptive expectations obviously make little sense.

However, what does not seem to be a sensible expectation formation in the context of the dice example is more difficult to ignore in a macroeconomic context. Not only are people forced to form expectations about a stochastic and complicated future, but they generally do not know the underlying economic theory, unlike in the very simple dice example. Under these conditions, forming expectations solely based on past records of economic variables does not seem too far-fetched.

There are situations in which the reliance on past values is clearly "wrong". Think of an announcement of the central bank to target a lower inflation rate, for example. Here rational expectations come into play. Individuals might not know the true underlying model, as in the weak form of the REH, but should incorporate all available information into their optimal planning.

Today a wide consensus has emerged with regard to the use of rational expectations in macroeconomic models. Like the market clearing real business cycle economists, modern Keynesian macroeconomists generally depart from similar notions of optimizing behavior, and thus rational expectations. Instead, they place greater weight on market imperfections, frictions, transaction and information costs.

Nonetheless, everybody would acknowledge that in real life people do not have all the information they would want when making decisions. As we have argued throughout the book, information has benefits and costs. It is therefore not efficient or rational to use all the information that is potentially available. Expectations can be simultaneously rational and quite inaccurate.

11 C Theory: Information is imperfect and costly

Information may not only be noisy, but also costly to acquire and process. We present two different approaches to incorporating imperfect and costly information into macroeconomic models. These have received the most attention in recent years and represent state-of-the-art research in macroeconomics.

Rational Inattention

The *Rational Inattention* strand of the literature explicitly models the fact that information processing is costly. Rational individuals take this into account and may decide not to purchase all available information or to use limited brain capacity to process the news. Following the latter idea, Sims (2003) and Luo (2005) emphasize capacity constraints of human brains to process information. Reis (forthcoming), on the other hand, argues that the costs of acquiring information reduce the number of planning dates at which information is fully updated. All individuals basically have the same information at their disposal, but not all use it. In other words, there is no interaction between the expectations of market participants. When Alice chooses her optimal strategy she ignores Bob, which also means that she does not infer any additional information from Bob's behavior.

Forecasting the forecasts of others (higher-order expectations)

The *Higher-Order Expectations* (HOE) setup is driven by the interaction of market participants' expectations. Not everybody has the same information, so individuals try to guess what other market participants do, at least on average. For Bob's optimal response it is important what he thinks Alice's information might be. Such interaction is especially important in situations in which prices depend on the expectation of future prices, such as foreign exchange rates and asset pricing.

11 C a Inattentive agents

Consumers-savers and producers alike are affected by costly information processing. At each point in time individuals will have to decide whether they want to incur the direct costs of information acquisition or bear the indirect costs of a plan based on incomplete information. Following the analysis of Reis (forthcoming and 2006), we present the decision problems of consumers and (in fewer details) producers separately.

Inattentive consumers-savers

Alice and Bob still have a few thousand dollars they made (after amicably pooling all profits and losses) from dealing in Hedgeville Hopper shares. "Let's spend the money on something special!" Bob says. "Shouldn't we build up some reserves?" Alice objects, adding: "You see, about once a year I go through all the hassle of working out how much I can spend. Having fixed my monthly spending budget, I save the rest." "It's not that I don't save! I even have a savings target, but I gladly spend the rest," Bob counters, and somewhat timidly admits: "I thought we might use the money for—a joint vacation." "Oh, I see . . .," Alice says and, smiling, replies: “That’s fine by me: I’ll remember a nice vacation for the rest of my life. Therefore, it’s not consumption, it’s an investment! That means I’ll not in fact spend the money, I’ll just convert it into memories. This is perfectly consistent with my budget.”

Let us start with a rather intuitive—albeit somewhat complicated—formulation of Alice’s and Bob’s problem. Both maximize their expected intertemporal utility based on costly information. The objective function can be written as:

$$max_{z_t,D}E_0\left[\sum_{i=0}^{\infty}\sum_{t=D(i)}^{D(i+1)}\beta^t u(c_t)dt\right],$$

where c_t denotes consumption, $u(\cdot)$ instantaneous utility, $\beta < 1$ the discount factor, z_t the decision variable (either consumption or saving), and $D(\cdot)$ planning dates.

What looks rather complicated is in fact a matter of simple intuition. In contrast to standard intertemporal models, Alice and Bob not only choose their optimal consumption or savings path, but also their *planning dates* $D(i)$ in an optimal way, in the knowledge that information is costly. The beginning of the planning horizon is $t = D(0) = 0$. Between $D(0)$ and $D(1)$—and more generally between $D(i)$ and $D(i+1)$—they do not update their information and follow their initial consumption or savings plan regardless of any potential changes in the economy. At the first chosen planning date, $t = D(1)$, Alice and Bob acquire new information, and update their optimal plans. In a completely analogous fashion they then optimally determine the future points in time at which they will update their information as well as the consumption or savings schedule to be followed between two adjacent planning dates.

How exactly does information cost enter the analysis? If planning were costless, an individual would continuously revise her plans. With costly information, the cost of continuous planning would eat up the very funds she has at her disposal to allocate. Planning expenditure reduces the funds available for either consumption or saving. The simplest way of modeling this is to assume that the planning costs K reduce asset holdings at planning dates D when the information is updated. Formally, the budget constraints obtained at planning date $D(i)$ then are:

$$\begin{aligned} s_{t+1} &= (1+r)s_t + y_t - c_t - K \quad \text{if } t \in \{D(i)|i = 0, 1, \ldots, \infty\}, \\ s_{t+1} &= (1+r)s_t + y_t - c_t \quad \text{else.} \end{aligned}$$

Alice’s and Bob’s decision variable z_t can be either consumption c_t or saving s_t. This means that they make a plan for either c_t or s_t and adjust the other variable to their income y_t (which is subject to shocks). Alice chooses her *consumption* (and leaves savings as a residual), while Bob chooses his *savings* (and consumes whatever is left).

A crucial assumption in this setting is that, between planning dates, individuals receive no new information. In particular, individuals do not even know the income they have accrued since the last planning date. Within our simplified setting it is obviously hard to defend why information about a person's past and/or current income should be so difficult to obtain. Yet, in reality, the delayed availability of information on variables, such as after-tax income or the stock and yield of total accumulated assets, may make it difficult to assess an individual's net income and/or the underlying state of the economy.

Alice's decision variable is consumption

Alice's income follows a random process (see Box 11.2), which can be viewed as representing an underlying business cycle. Income can either be 100 or 120 units, and in each period the probability of receiving the same income as in the previous period is 75 per cent (though this is not important). Planning reduces her assets by 20 units. For simplicity's sake, we assume that there is no interest on savings and consumer debt.

Box 11.2 Markov Process

A *Markov Process* is a convenient way to model a stochastic process with some persistence, such as the business cycle, but without the need to know the full history of the process. Persistence means that the current state depends on the previous state of the economy. If the economy is in recession, for example, the probability of a recession is still high in the next period.

Another important feature of processes is that the probabilities of future states only depend on the current state, but not on how the current state had been reached or how long it had been in place. In that respect, the process has no memory. For our business cycle example, the chance to come out of a recession does not depend on how long the recession has already lasted. This structure facilitates information acquisition considerably.

The most simple Process is characterized by two possible outcomes Y_H and Y_L, the so-called states. The probabilities of staying in the same state are q and p, respectively. With a probability $1-q$, the state changes from Y_H to Y_L, with a probability $1-p$ from Y_L to Y_H. A Markov Process, which is also depicted in the figure below, can conveniently be represented by the so-called transition matrix:

	$Y(t+1) = Y_H$	$Y(t+1) = Y_L$
$Y(t) = Y_H$	q	1-q
$Y(t) = Y_L$	1-p	p

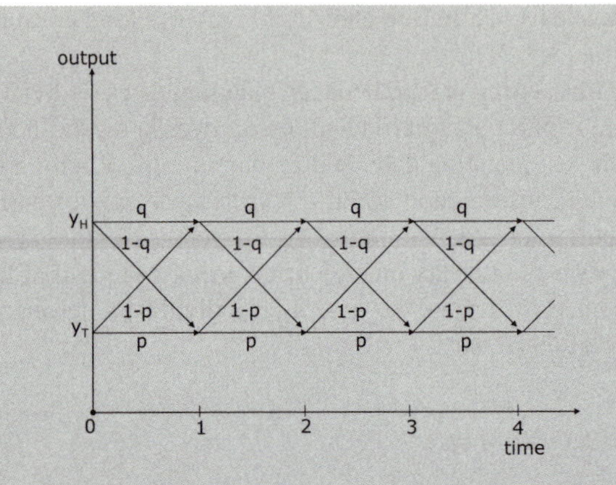

If p and q equal $1/2$, the process has no persistence, i.e. the probability of a boom Y^H in the next period is independent of the state of the economy today. The more these probabilities deviate from $1/2$, the longer the economy remains in the same state.

The second figure illustrates a realization of such a Process with two states and $p = q = 1/2$. Interpreting Y^T as recessions and Y^H as booms, the whole path maps a business cycle. Although this specification has no built-in persistence, it already resembles, to some degree, a pattern of booms and recessions.

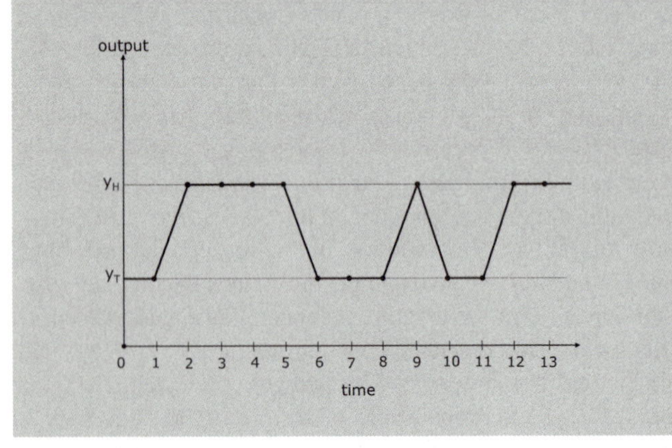

We have assumed that Alice is an inattentive consumer who spends a planned amount and leaves the rest in the bank. Figure 11.2 illustrates the effects of her strategy over the first 48 quarters. Note that the figure is not based on either optimal planning dates or optimal consumption/savings plans. It is purely drawn to illustrate the economic effects of Alice's rational inattention, assuming for expositional reasons that Alice mechanically plans every second year (every 8 quarters) and varies her consumption by arbitrary amounts.

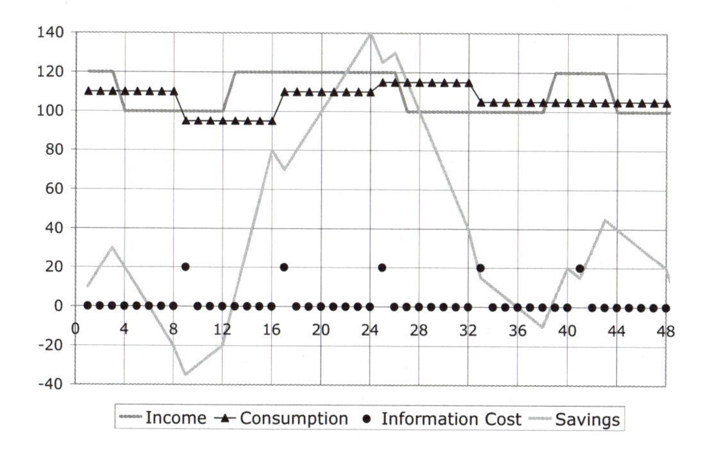

Figure 11.2 Inattentive consumer: An example of Alice's possible savings and consumption plan with a fixed planning horizon of eight quarters.

Starting in a boom, Alice fixes her quarterly consumption at 110 units (which corresponds to the long-run average of her income) and decides to re-plan in quarter 9. At that time, Alice expends the 20 units of planning costs and realizes that the economy is in a recession and that her current savings are actually negative. She thus decides to reduce her consumption to 95 for a while. Eight periods later she finds herself in a boom and in a comfortable asset position. This makes her increase her consumption. The situation is even better in period 25, so the spending spell is maintained for a while. At the next planning date, however, Alice's savings are almost depleted and so she reduces her consumption to a level slightly above the recession income. Although she finds herself again in a boom in quarter 41, she maintains her low consumption in order to boost her savings further.

Costly planning and inattentiveness affect the dynamics of Alice's consumption *and* its level. The longer she does not react to shocks, the greater her risk of ending up with a high level of debt, which in turn may seriously limit her consumption opportunities in the future. If Alice's utility function implies a decreasing marginal utility of consumption (as the commonly used utility functions in macroeconomics do), such low levels of consumption represent a suboptimal intertemporal allocation of resources. As a consequence, the desired *precautionary* savings as a safeguard against negative shocks are higher if she plans at longer intervals. Inattentiveness can lead to sizeable swings in an individual's savings. The optimal length of a non-planning interval is a trade-off between the cost of being inattentive (not reacting to potential shocks) and the planning costs.

Reis (forthcoming) also shows that the interval between planning dates decreases with the underlying risk (of Alice's income process, for example) and the degree of risk aversion. The higher the interest rate, the shorter the length of inattentiveness.

A high interest rate magnifies the impact of a wrong savings decision, rendering savings even more volatile. *Between* planning dates, i.e. within a planning period, Alice behaves as if living under perfect certainty.

Bob's decision variable is savings

Now consider Bob, whose income and planning costs are the same as Alice's. In contrast to Alice, however, Bob starts out as an inattentive saver who puts aside a planned amount of savings, and spends the rest. Figure 11.3 illustrates Bob's possible strategy over the first 48 quarters. Again, the figure does not necessarily represent the optimal plan, but is drawn to illustrate the impact of Bob's behavior.

Initially Bob, during an economic boom, decides to save 5 units per time and to re-plan in period 9. Having accumulated a small buffer stock, he reduces his regular savings to 2.5 units per period. Unfortunately he soon finds out that small amounts of regular savings are eaten up by planning costs. Bob might extend the time between planning dates, or even more simply, decide not to plan any more and consume whatever he earns, i.e. follow a savings plan of zero units forever.

As soon as Bob decides to stop planning and to save an amount of zero each period, consumption absorbs all income shocks, as can be seen in Figure 11.3. The marginal propensity to consume is equal to one and consumption is as volatile as income. Bob and others who behave in a similar way are often called hand-to-mouth consumers.

Reis (forthcoming) shows that as long as the planning costs are not too small, the inattentive saver never changes his plans. To illustrate this, consider the even simpler special case with serially uncorrelated income shocks: If a person rationally chooses not to update his information this period, he would not do it next period

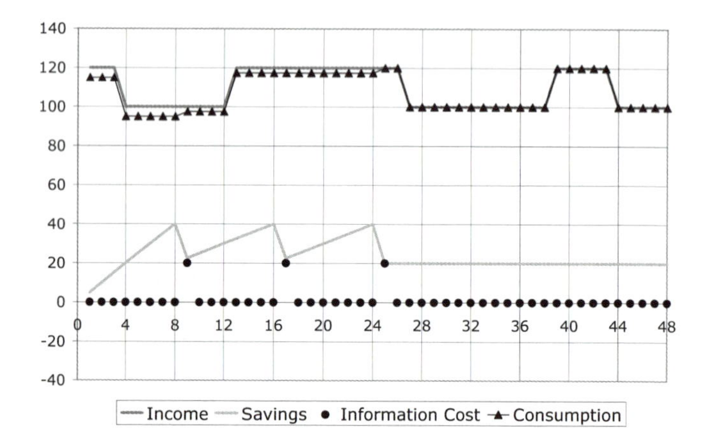

Figure 11.3 Inattentive saver: Bob's savings and consumption over time with a fixed planning horizon of eight quarters.

either. As the next period's assets are predetermined and therefore non-stochastic, the planning problem is unchanged. Inattentive savers are rational non-planners (Reis, forthcoming).

The difference between inattentive consumption and saving

A comparison between the consumption and savings profiles of Alice and Bob suggests that an apparently minor difference in the planning strategy implies quite striking differences in outcomes: Alice's asset holdings fluctuate much more than Bob's, but her consumption is somewhat smoother than his. It seems as if planning is more important in Alice's case, as it prevents a bad outcome in the form of debt, for example. So, the last question is what factors determine whether a person is an inattentive saver or an inattentive consumer?

Unfortunately, this question is too difficult to answer analytically. But it comes as no surprise that the size of the planning costs is crucial. The larger the costs, the lower the impact on inattentive savers, but inattentive consumers find themselves worse off. If the costs are *above* a certain threshold (seemingly the case for Bob), individuals only choose savings plans. If they are *below* (seemingly the case for Alice), individuals only follow consumption plans.

Why is inattention important for the macroeconomy?

Rational inattention models have been developed to explain macroeconomic phenomena that are difficult to reconcile with full information rational expectation models. One such phenomenon, large variations in inflation forecasts, will be discussed below. Two other puzzles concern aggregate consumption: The first is the so-called *excess sensitivity puzzle* in which consumption reacts too strongly to predicted changes in income. The second is the *excess smoothness puzzle* in which consumption reacts too little to permanent changes in income. At first sight the two puzzles seem to contradict each other, but they in fact address different aspects of the income-consumption relationship. A person can simultaneously react too strongly to winning the lottery and not enough to a permanent change in income, such as an increase in retirement benefits.

Reis' rational inattention framework helps to explain both puzzles. Individuals only gradually update their plans. As a consequence, costly news trickles slowly into the economy. Recall that high-cost individuals are inattentive savers (hand-to-mouth), while lower-cost individuals are inattentive consumers. Let us look at two cases:

- A *predicted*, but *temporary* increase in income (after a period of good weather, for example): Aggregate consumption is excessively sensitive as hand-to-mouth consumers do not save the windfall gain, but consume whatever they receive. But even rational consumers contribute to the excess sensitivity.

As not everybody updates his or her information at any point in time, income shocks in the past affect current consumption growth.

- A *permanent* increase in income (after an increase in output growth, for example): This should immediately increase consumption in a full-information environment. But since under information processing constraints only a fraction of the agents react contemporaneously to changes in permanent income, consumption will be smoother than income.

In reality, planning dates are often adjusted under special circumstances, such as a natural disaster. In these situations, the costs of information processing relative to the implied mistakes of non-planning would be small and justify a re-optimization of resources for almost everybody. A useful extension of the rational inattention setup model would, therefore, be to allow all individuals to observe some extraordinary events as they occur.

Inattentive producers (Reis, 2006)

Households are not the only important players in macroeconomic models. Firms' decisions to produce goods, invest capital and hire workers play an equally important role for the economy. In particular, firms set prices. These largely influence the aggregate output of a country and the prevailing overall price level. Looking at individual goods prices, one feature stands out: Prices seem to be much less volatile than what would be compatible with fully flexible prices. Based on this observation, sticky prices have long been successfully incorporated into macroeconomic models. Nonetheless, the microeconomic foundation for sticky prices has always posed some difficulties.

That information might be key to understanding this sluggish adjustment of prices has also been posited by Radner (1993), based on the observation that even in manufacturing firms, a considerable fraction of the staff is employed to carry out managerial tasks. These people essentially acquire, absorb and process information as a basis for making the right decisions.

In a similar spirit to the inattentive consumer above, Reis (2006) has developed a model of rationally inattentive producers. Again, information about market conditions is costly to process. Not only is it costly to acquire the pieces of information relevant to estimate the demand for the products, but it also takes time and energy to compile these pieces into the prices to be charged to the customers. A manufacturer facing such costs may decide to update information only sporadically and be inattentive to any news arriving between planning dates. Once the information is updated, the manufacturer has to decide whether to set prices or quantities (similar to the consumption/savings decision). To illustrate these simultaneous decisions, Reis considers the example of a fictional baker:

> Her first decision is on which variable to write a plan on: price or quantity. If she sets a price for her bread, she will keep the oven burning and bread coming out as long as customers are walking through the door. If instead she

chooses to produce a certain amount of bread, she then gives it to a seller. This seller takes the bread to the market and distributes it among homes and shops, charging whatever positive price is necessary to sell all the bread today, since by the end of the day the bread becomes stale and worthless. Finally the seller returns the sales proceeds to the baker. Facing this choice of prices versus quantities, the baker forms an expectation of her profits under the two alternative and chooses the most profitable one.

As in the example of the inattentive consumer, who follows either a consumption or a savings plan, the baker first has to determine the optimal planning strategy. Reis shows for a variety of demand and cost functions that setting the price is likely to be the preferred strategy, and is consistent with the empirical evidence. The second decision the baker has to take is when to change plans given the structure of uncertainty and her information acquisition and processing costs. There is again a trade-off between the costs of not following an optimal plan, and the costs of acquiring new information. Reis (2006) also demonstrates that the aggregate inattentive producer economy matches the data on the real effects of inflation very well. Unlike in the inattentive consumer setup, there is also a problem that needs to be addressed: In principle, some information about current demand conditions is virtually costless to acquire. The baker can easily observe when she runs out of bread or how much bread is left over at the end of the day. It would be easy to adjust the price or the quantity to align the supply of bread with the observed demand. On the other hand, the example of the baker is probably not the most important one for modern economies. Reis argues that even low information-processing costs are sufficient to generate substantial inattentiveness.

11 C b Forecasting the forecasts of others

Our second example where information has an impact on the economy are higher-order expectations. Higher-order expectations become relevant, when individuals' optimal decisions depend on expectations about other individuals' optimal decisions, and vice versa. Unlike in the case of rational inattention, where heterogeneous information was an *ex post* outcome (of costly information), in the case of higher-order expectations, individuals hold heterogenous beliefs *ex ante*, from the very beginning.

The law of iterated expectations . . .

When all market participants have the same information set, the *law of iterated expectations* holds, i.e. the expectation of the expectation of a variable equals the expectation of the variable. Equilibrium prices that depend on future prices can then be solved using the law of iterated expectations: Today's expectation of tomorrow's expectation of a future variable is the same as today's expectation of this future variable.

$$E_t E_{t+i} X_{t+j} = E_t X_{t+j}, \quad \forall\, i \geq 0,\, j \geq i.$$

Many (macroeconomic) prices are of a forward-looking nature. Here are three examples, which we state without going into details:

- The general price level: In the simple form of a money demand equation $m_t - p_t = -\eta E_t(p_{t+1} - p_t)$, current prices p_t not only reflect current conditions such as the money supply m_t, but also future prices. The expected inflation rate $E_t(p_{t+1} - p_t)$ is an important determinant of the real money demand $m_t - p_t$, and hence of the price level p_t.
- Exchange rates: The (uncovered) interest rate parity $E_t s_{t+1} - s_t = i_t - i_t^*$ predicts that the expected depreciation of a currency $E_t s_{t+1} - s_t$, where s_t is the exchange rate, is equal to the interest rate differential between the home country and the foreign country, $i_t - i_t^*$;
- Stock prices: $P_t = \frac{1}{1+r} E_t(P_{t+1} + D_{t+1})$.
  Future stock prices P_{t+1} are an important factor in the determination of today's prices. Future prices reflect expectations about the future stream of dividends D. In fact, if one substitutes out future prices in the equation above,[77] the current price of a stock is basically the present value of expected future dividends.

... and its limits

When prices depend on future prices, the expectations of market participants become a crucial determinant of these prices. If there is some heterogeneity of information among the individuals, the "correct" way to compute the forecasts of future variables is less straightforward than it would be under homogeneous expectations. "Forecasting the Forecasts of Others"[78] is then not only a nice game like Keynes' beauty contest, but an essential strategy for successfully dealing in the market.

An illustration

To illustrate the concept of higher-order expectations, let us consider the following example, taken from Allen, Morris and Shin (2006):

- θ is a random variable. As we will use inflation forecasting as an application below, it is useful to think of θ as the future inflation rate, given the current monetary policy. The uncertainty about its true value comes from potential changes in the underlying macroeconomic fundamentals, such as labor productivity or simply the oil price.

We will assume that θ is drawn from a Normal distribution $N(0, 1/\alpha)$.[79] The parameter α is the inverse of the variance and is called the *precision* of the distribution. We will interpret α as the precision of the *public signal*.[80]

- $E_{i,t}(\theta)$ denotes the expectation of individual i of θ (the inflation rate) at time t.
- $\overline{E}_t(\theta)$ is the average expectation of θ at time t.
- $E_t^*(\theta)$ denotes the public expectation, i.e. the expectation of the inflation rate θ given publicly available information only.

The *law of iterated expectations* holds for individual and public expectations,

$$E_{i,t}E_{i,t+1}(\theta) = E_{i,t}(\theta)$$
$$E_t^* E_{t+1}^*(\theta) = E_t^*(\theta),$$

but usually not, as we will see below, for *average expectations*, if information is asymmetric:

$$\overline{E}_t\overline{E}_{t+1}(\theta) \neq \overline{E}_t(\theta)$$

We now assume that each individual observes a signal x_i about the future inflation rate, which is distributed as $N(\theta, 1/\beta)$. The distribution of the signal is symmetric around the true value of the future inflation rate, i.e. the realization of the random variable. The inverse of the signal's variance β is again called the precision of the signal. Figure 11.4 illustrates this environment with the unconditional distribution of the future inflation rate as the dashed line, together with the conditional distribution of the signal, where the random variable takes on a value of 2, as the solid black line.

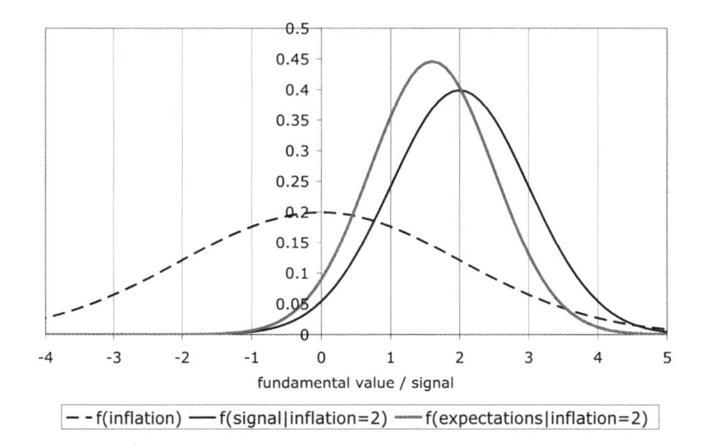

Figure 11.4 The unconditional distribution of the random variable "f(inflation)" together with the private signal's distribution, conditional on a possible realization of an inflation rate of 2 per cent, can be used to derive the distribution of market expectations "f(expectations|inflation=2)".

To simplify matters, we consider a setting without learning. Time subscripts can then be dropped. In the assumed case of a Normal distribution, it can be shown using the continuous distribution version of Bayes' Rule that:

$$E_i(\theta) = \frac{\alpha y + \beta x_i}{\alpha + \beta}.$$

This implies that even a person who receives a correct signal about the true value of the inflation rate, $x_i = \theta$, would expect a value between the unconditional mean of the future inflation rate, y, and the observed private signal. The reason is the following. Compare two inflation rates that are of equal distance from the true value of the random variable: $\theta \pm d$. Given the true value, these two inflation rates are equally likely, as expressed by the solid black curve in Figure 11.4. But if the true value of the signal is unknown and $\theta > 0$, an inflation rate of $\theta + d$ seems less likely than one of $\theta - d$, as it is further away from the expected value of the future inflation rate. This relationship is represented by the solid grey line in Figure 11.4.

In terms of Bayes' Rule we can explain the bias as follows: We know the distribution of the signal based on the realized value of the random variable $f(x_i \mid \theta)$ (the solid black curve in Figure 11.4), but unfortunately not the random variable, in this case inflation. To infer market expectations, we need to know the expectation of this random inflation rate, given the signal $E(\theta \mid x_i)$. Given the distribution of the signal, the distribution of market expectations (= expectations of individuals receiving different realizations of the signal) is depicted by the solid grey line in Figure 11.4.

The average market expectation is the expected value of this distribution, therefore

$$\overline{E}(\theta) = \frac{\alpha y + \beta \theta}{\alpha + \beta}:$$

Individuals have to take into account that other individuals might have different expectations of the random variable. An individual's expectation of market expectations can then be computed as:

$$E_i\left(\overline{E}(\theta)\right) = \frac{\alpha y + \beta E_i(\theta)}{\alpha + \beta} = \frac{\alpha y + \beta\left(\frac{\alpha y + \beta x_i}{\alpha + \beta}\right)}{\alpha + \beta}$$
$$= \left(1 - \left(\frac{\beta}{\alpha + \beta}\right)^2\right) y + \left(\frac{\beta}{\alpha + \beta}\right)^2 x_i$$

which yields the corresponding average expectation of the average expectation:

$$\overline{E}\left(\overline{E}(\theta)\right) = \left(1 - \left(\frac{\beta}{\alpha + \beta}\right)^2\right) y + \left(\frac{\beta}{\alpha + \beta}\right)^2 \theta$$

Iterating this procedure k times yields:

$$\overline{E}^k(\theta) = \left(1 - \left(\frac{\beta}{\alpha+\beta}\right)^k\right) y + \left(\frac{\beta}{\alpha+\beta}\right)^k \theta$$

The expectation of the expectation . . . of inflation is biased towards public information, that is, the expected value of the inflation rate in the absence of private signals. The higher the precision (β) of the private signal about the realized inflation rate, the closer the expectation is to the true value of the random variable. For $k \to \infty$, the expectation of the expectation of . . . of the expectation converges to the expectation of the random variable conditional on public information only, $\overline{E}^k(\theta) \to y$. This implies that after several rounds of guessing, the private signal becomes worthless, although it is, on average, correct. Iterated expectations over several periods compress information to potentially rather unhelpful public information.

Is this merely a theoretical gimmick or can we find an empirically relevant counterpart to this exercise? Given that inflation expectations of private households differ quite a bit, as we have illustrated in Figure 11.1, Morris and Shin (2005) think that the issue is also relevant for the conduct of monetary policy, as our next application shows.

11 D Application: Central bank transparency

Figure 11.1 above has clearly demonstrated that there is considerable heterogeneity of private forecasts of even the most important macroeconomic variables. Nonetheless, not many researchers have addressed this issue so far. Mankiw, Reis and Wolfers (2003) start their famous paper on the dispersion of inflation expectations in US survey data with the following paragraph (emphasis added):

> At least since Milton Friedman's renowned presidential address to the American Economic Association in 1968, expected inflation has played a central role in the analysis of monetary policy and the business cycle. How much expectations matter, whether they are adaptive or rational, how quickly they respond to changes in the policy regime, and many related issues have generated heated debate and numerous research studies. Yet throughout this time, one obvious fact is routinely ignored: *Not everybody has the same expectation.*

Unavoidable disagreement

Mankiw et al. (2003) also demonstrate that not only private households, but also professional economists and forecasters disagree about expected future inflation. Moreover, the disagreement shows substantial variation over time. Forecasting inflation is not an easy task as future prices depend not only on macroeconomic fundamentals, but also on the policy of the central bank and on the expectation of other market participants. The observed disagreement might, therefore, not be so surprising, after all. Nonetheless, it has important consequences for monetary policy.

Inflation targeting

In recent years, many central banks have adopted a new policy called *inflation targeting.*[81] Ben Bernanke, the Chairman of the Board of Governors of the Federal Reserve System, explained this policy as follows (Bernanke (2003), emphasis added):

> Inflation targeting, at least in its best-practice form, consists of two parts: A policy framework of constrained discretion and *a communication strategy that attempts to focus expectations and explain the policy framework to the public.* Together, these two elements promote both price stability and well-anchored inflation expectations; the latter in turn facilitates more effective stabilization of output and employment.

The role of communication ...

An important building block of the new policy is that the monetary authority announces its own expectation about the future evolution of prices. Even central banks that have not explicitly adopted inflation targeting put much more emphasis on communicating their expectations to the public than in the past. As a consequence of well-anchored inflation expectations, as Bernanke (2003) puts it, monetary policy is believed to be more effective. Many leading economists consider the management of expectations as the most important task of monetary policy. Not everybody agrees with this assessment, however, as the debate initiated by Morris and Shin (2002, 2005),and Allen et al. (2006) shows.

... and public signals

As we have seen in Section 11 C b, the existence of a public signal may bias expectations even if the private sector's signals were correct and more precise on average than the central bank's information. Households and producers can use two sources of information to forecast prices: The publicly available announcements of the monetary authority and their private assessment of market conditions. The correct first-order estimate of market expectations is a precision-weighted combination of the two sources of information. The more rounds of the guessing game that are played, the more weight is placed on the central bank's forecast. Ultimately, the private signals of market participants become useless and market prices cease to reflect the aggregated information of the private sector.

Learning from prices when prices reflect managed expectations?

The key problem is that even the central bank has to learn about the state of the economy. It has no independent clock to read off the time, so to speak. Its information comes, at least partially, from observing market prices and transactions. These market conditions and prices in turn depend on what the central bank communicates to the economic agents. As a consequence, the information value of prices is weakened and more transparency may actually be bad for the economy. Morris and Shin (2005) describe this situation as a tension between managing information and learning from market expectations.[82] In their view, the monetary authority cannot manipulate prices by an appropriate policy and, at the same time hope that prices reflect market information. Morris and Shin (2005, p. 42) even warn that too much transparency can be harmful for the economy, implying the possibility of price bubbles:

> If the central bank does not recognize that signal values are impaired in a world of managed expectations, it may be lulled into a false sense of security when in fact imbalances are building in the economy.

Is communication bad?

Does this mean that communication is bad for the conduct of monetary policy? Morris and Shin (2005) argue that too much transparency can indeed be detrimental, but many other economists, notably in the field of monetary policy, disagree, as the heated debate subsequent to the presentation of Morris and Shin (2005) demonstrates.[83]

One of the criticisms is that a clear distinction should be made between the information disseminated about inevitable *policy measures* and the release of information about the *state of the economy*. In fact, an important feature of the Morris and Shin (2005) analysis is that it only offers a binary choice between full transparency (= communicating a public signal) and full opaqueness, although transparency in reality comes in different shades. Gosselin, Lotz and Wyplosz (2006), among others, argue that a central bank has little choice but to disclose at least *some* information as it has to set the interest rate. However, it can still decide to release or withhold the fundamental information in its possession, which forms the basis for interest rate decisions.

The optimal policy is still debatable. According to Gosselin et al. (2006), the optimal policy of the central bank depends on the precision of public and private information. If the precision of the two sources of information is unknown, revealing all the fundamental information is optimal: Putting all the cards on the table prevents the private sector from misinterpreting the signal the central bank sent out in the form of interest rates.

Morris and Shin (2005) also state that central bank transparency reduces the signal value of prices. In his discussion of the paper, Christopher Sims argues that any policy that is successful in reducing the fluctuations in the price level is bound to make the signal value contained in the average price level less informative. But this is not an undesired outcome, as only relative prices matter for the optimal allocation of resources. Stable average prices eliminate the uncertainty about the general price level and facilitate information gathering about local market conditions. New evidence, provided by Gürkaynak, Levin and Swanson (2006) based on the Swedish experience, suggest that inflation targeting may indeed help to anchor private sector expectations of future inflation.

Central bank transparency as accountability

Last but not least, economic efficiency is just one dimension of the debate on the optimal transparency of a central bank. As Alan Blinder mentions in the discussion (cited from Morris and Shin 2005, p. 59), openness *per se* has an important social value in a democracy:

> As a public body, the central bank should be accountable to the public, and transparency is an important element of accountability.

The interesting contributions of Morris and Shin (2002, 2005), therefore, highlight one facet of central bank transparency, but not all relevant aspects. As in many other situations, the economics of information is important, but other non-economic aspects not discussed in this book may offer equally valuable insights.

11 E Conclusions and further reading

Heterogeneous beliefs can be a consequence of truly private information, information processing constraints or genuinely different forecasting models. In this chapter we have offered a glimpse of the exciting new literature on the importance of information acquisition and processing in macroeconomics. We would like to underline, however, that information and expectation formation have always played a crucial role in macroeconomics. The new and important aspect is, as Mankiw et al. (2003) put it: "*Not everybody has the same expectation.*"

The initial direction of research primarily focuses on the cost of information for individual consumers and producers. Faced with such costs, consumers or producers may decide not to acquire new information. In that respect, individuals who decide not to purchase the information are considered as rational, but inattentive. Taken together, the behavior of these agents may lead to a sluggish response in terms of consumption or to other phenomena which are difficult to explain using standard macroeconomic models.

While the rational inattention literature does not assume any heterogeneity in the underlying information set, but (potentially different) costs to acquire and process new information, private information is at the root of literature on higher-order expectations. As macroeconomic variables often depend on expectations about future prices, individuals are not only forced to forecast the value of fundamental variables, but also to guess what other market participants anticipate. The resulting guessing game may have many rounds, making a public signal such as a Central Bank's announcements and actions a valuable, but also potentially ambivalent anchor.

In reality, information acquisition and processing is a combination of both aspects: If it does not pay to obtain new information at any point in time, those who *do* acquire it are confronted with the problem of guessing what the uninformed market participants think. A first attempt to do this in the context of exchange rate dynamics has been made by Bacchetta and van Wincoop (2006). Needless to say, this is technically very demanding and well beyond the reach of this book.

Checklist: Concepts introduced in this chapter

- Inattentive agents
- Heterogeneous expectations
- Decision variables under rational inattention
- Rational expectations
- Efficient markets hypothesis
- Higher-order expectations
- The law of iterated expectations
- Convergence of iterated expectation to public signal
- Central bank communication
- Markov Process

11 F Problem sets: As time goes by

For solutions see: www.alicebob.info.

Problem 11.1 Inflation expectations in Poland

Since the early 1990s the inflation rate in Poland has dramatically fallen, as shown in Figure 11.1. Nonetheless, the divergence of individual perceptions about the course of future prices seems very persistent.

(a) What might be the reasons for such a substantial heterogeneity of expectations? Can you conclude something about the precision of the private signals individuals use?

(b) Discuss the importance of inflation forecasts and higher-order expectations in the Polish context, based on Morris and Shin (2005) and the debate discussed in Section 11 D.

(c) For students with a background and interest in macroeconomics: The price level today crucially depends on the expectation of future prices as we have seen in Section 11 C b, and represented in the money demand equation $m_t - p_t = -\eta E_t(p_{t+1} - p_t)$. Can you see any evidence of this relationship in Figure 11.1?

Problem 11.2 Telling your type

Two individuals show you an account of their monthly consumption expenditures and savings for the last ten years. Could you read from the figures who is an inattentive consumer and who is an inattentive saver?

Problem 11.3 An example of inattentive consumers

Let us consider an individual with the following intertemporal utility function:

$$U_0 = E_0 \sum_{t=0}^{\infty} \beta^t \left[u(c_t) + 0.9u(d_t)\right],$$

where $\beta < 1$ is the discount rate. The two different consumption goods c and d are perishable and are thus not storable for future consumption. The price of one unit of c and d is one. Instantaneous utility $u(\cdot)$ of the two goods can be tabulated as

$x\ (x \in \{c, d\})$	0	1	2	3	4	5	≥ 6
u(x)	0	4	7	9	10	10	10

So if two units of c and four units of d are consumed, the period t utility would be $u(c_t) + 0.9u(d_t) = 7 + 0.9 \times 10 = 16$.

Both goods can only be consumed as integer values and it is not possible to consume negative amounts. It is also assumed that within a period, either good must be purchased in one visit to the shop, excluding the possibility that individuals go back to the shop in case they have left-over income.

In each period individuals receive a non-storable serially uncorrelated income Z_t with

$$Z_t = \begin{cases} 2 & \text{with} \quad P = \frac{1}{2} \\ 6 & \text{with} \quad P = \frac{1}{2} \end{cases}$$

For various reasons, it is assumed that *prior* to the purchase of the first good the income can only be observed at a cost K. As a consequence an individual has two options:

- She pays K and gets to know the exact amount of her income before she goes shopping
- She only learns her income after having purchased the first good. This is tantamount to following an optimal plan that fixes the amount of consumption in one of the two goods, spending the rest on the other good.

(a) Describe the optimization problem, including the budget constraints.

(b) For which of the two goods does the individual want to fix the amount to consume under uncertainty? (Hint: Draw a table for all possible combinations of quantities of c and d for the two income levels and determine which strategy delivers the highest expected utility. Recall that it is not possible to consume negative amounts of either good.)

(c) Determine the information costs that would make the individual indifferent between acquiring the information about the stochastic income or not. (Hint: Compute the expected utility in case the income is revealed before the first good is purchased.)

(d) Determine the optimal planning horizon of the individual. (Hint: The plan is adjusted every period or never at all. Why?)

(e) How would the optimization problem of the individual look like if the stochastic income was serially correlated (Answer in words only!)?

Problem 11.4 Central bank communication

In recent years many central banks put emphasis on a communication strategy. In line with this observation several economists argue that the central bank should be as transparent as possible.

(a) What is the rationale for their argument?

(b) Assume that you get a normally distributed private signal of $x_i = 2$ for inflation with precision $\beta = 3/4$. A normally distributed public signal with precision $\alpha = 1/4$ takes the value $y = 0$. Calculate the average expectations for one, two and three iterations. Are the values biased towards the public signal?

(c) In view of your calculation from above: Can transparency be bad for conducting monetary policy taking into account a publicly available signal from the central bank?

(d) Can you think of further reasons for and against transparency?

Part III

Asymmetric information

12 The winner's curse

12 A Introduction

In the year 193 AD one of the most remarkable items ever went under the hammer: the Roman Empire (see Box 12.1). The auction was won by the boldest bidder, a certain Didius Julianus. Yet, his experience would be repeated by thousands of auction winners after him: Regret at making the winning bid. Many winners of auctions or auction-like contests for items with an objective yet unknown value discover that, with hindsight, they should have bid less. Too often they find themselves with an item worth less than they thought. Of course, overpaying may be the result of auction room fever. However, the phenomenon of regretful winners is particularly frequent in environments where business-like sobriety should prevail: In sales of oil fields, spectrum licences, or corporate takeover bids. The phenomenon of regretful winners—losing money by winning an auction—is known as the *winner's curse*.

Box 12.1 Auctioning off the Empire

In 193 AD the praetorian Guard killed Roman emperor Pertinax. This made them the *de facto* owners of the empire. Unwilling to run the Empire by themselves, the Guard decided to do away with it. A historian reports:

> Then ensued a most disgraceful business and one unworthy of Rome. For, just as if it had been in some market or auction room, both the City and its entire empire were auctioned off. The sellers were the ones who had slain their emperor, and the would-be buyers were Sulpicianus and Julianus, who vied to outbid each other, one from the inside, the other from the outside [of the praetorian camp]. They gradually raised their bids up to twenty thousand sesterces per soldier. Some of the soldiers would carry word to Julianus, 'Sulpicianus offers so much; how much more do you make it?' And to Sulpicianus in turn, 'Julianus promises so much; how much do you raise him?' Sulpicianus would have won

> the day, being inside and being prefect of the city and also the first to name the twenty thousand, had not Julianus raised his bid no longer by a small amount but by five thousand at one time, both shouting it in a loud voice and also indicating the amount with his fingers. So the soldiers, captivated by this excessive bid and at the same time fearing that Sulpicianus might avenge Pertinax (an idea Julianus put into their heads), received Julianus inside and declared him emperor.
>
> (*Dio's Roman History, book LXXIV*; quoted from Temin and Klemperer (2001))

Didius Julianus won by a strategy known today as "jump bidding" (Temin and Klemperer, 2001). With his bold move he succeeded in discouraging his competitor. Yet, having spent an incredible fortune, Julianus could not enjoy the fruits for long. He was beheaded a mere 66 days later. Consequently, Didius Julianus would later become the patron saint of auctioneers. Temin and Klemperer (2001) suggest he was the first reported victim of the "winner's curse".

Fundamentally, the "winner's curse" is easy to understand. In competitive bidding, some buyers normally underestimate the value of an item, while others overestimate it. The highest bidder usually is among those who overestimate. Therefore, there is a good chance that the "winner" pays too much for the item and ends up being the loser.

The winner's curse arises when at least one contender naïvely assumes he or she is better informed than the totality of the other contenders—which far too often is simply not the case (and cannot be true for all). Of course, rational individuals anticipate the winner's curse and take precautions: They bid less than they think an item is worth. The winner's curse then becomes the seller's curse. This may explain why issuers leave considerable amounts of "money on the table" in initial public offerings (IPOs). The winner's curse also has implications for the price system. Paradoxically, due to the winner's curse, rationing demand may under certain circumstances increase, rather than decrease the price of a good. The winner's curse even extends to the classroom: Typically, larger audiences ask fewer questions. Each individual might argue: "If my question were good, someone else would have already asked it".

In this chapter we introduce the winner's curse and its statistical roots. In particular, we discuss two applications, one specialized and one broad: The money left on the table in IPOs and the impact of the winner's curse on the price system (the "rationing paradox"). The chapter is a prelude to Chapter 13 on asymmetric information, where even more serious information problems will be discussed.

12 B Main ideas: How to lose by winning

Alice has invited Bob and Charles for tea. In the course of conversation Bob asks: "Alice, how old are you actually?" "That's a rude question to ask, Bob," Alice replies, "but I'll tell you, at a price! Each of you must make a secret bid about my age, bidding one dollar for each of my years. The one who makes the highest bid will pay that dollar amount; in return, I promise to pay him a dollar amount equal to my age."

Bob and Charles are both uncertain about Alice's age. Bob thinks she is 30 plus/minus 5 years. Charles thinks she is 26 plus/minus 5 years. Of course, they cannot discuss the issue in front of Alice. How much should they bid?

Let us first examine the most natural strategy: Bidding one's own best guess. This would lead to Bob winning the auction (bidding \$30 against \$26). What would his profit be? As long as we do not know Alice's age, we cannot tell. However, we can make a plausible guess. Assuming that Bob and Charles are about right in the average (a group knowing more than its individual members) Alice would be around 28. With a bid of \$30, Bob would win the auction but lose \$2!

This is, in simple terms, the winner's curse. In an auction for a good with the same value which is unknown to everyone, naïvely bidding one's best guess leads to an expected loss. The "winner" of the auction, i.e. the bidder with the highest guess, is exactly the one who overestimates the value most. Yet, neither Bob nor Charles are naïve.

Bob reflects: "I have to take two things into consideration. Charles maybe thinks that Alice is older (or younger) than I do. In the former case, he is probably going to outbid me. This means that I should focus on the possibility that he thinks Alice is younger than I do, as this is the only one which gives me a chance of winning the auction. In this case, Charles, like me, thinks Alice is younger than 30. If I bid 30, I may be right; but he too may be right. In that case I would win the auction but lose money. So, I should have a safety margin. I'll bid \$27."

After reasoning like Bob, Charles, who thinks Alice is 26, bids \$22.

Alice reveals that she is in fact 28. All are happy. Bob wins the auction and makes a profit of \$1; Charles does not mind, given his low guess he could not have won anyway. And Alice, not at all disappointed about losing \$1, smiles: "I'm flattered that you think I'm so young! Although, Bob, I notice that Charles is more of a gentleman than you are!"

This example summarizes the two main ideas of this chapter:

- Bidding one's own best guess for an item with (uncertain) common value leads to an expected loss.
- As a consequence, rational bidders underbid their best private guess for the value of the item.

Box 12.2 Private and common value

Private and common value are important notions in economics, particularly in auction theory. The two notions capture two different dimensions (Wolfstetter, 1999):

- whether a good has the same or different value for different individuals;
- whether the value of a good (for one or all individuals) is private or public information.

The benchmark for both comparisons is a good that has the same (common), observable value for everyone, like a crisp, new US$50 bill. By contrast:

- A *private value* good has a different value for different individuals, but each knows his/her own private valuations.
- A *common value* good has the same value for all individuals, but they do not know this value. They only have imperfect private signals about the true value.

Examples of private value goods are collectors' items which are sought not with a view to their potential market value but for the private satisfaction they give their owners, like works of art or collectors' pieces. Examples of common value goods are oilfields, spectrum licences and takeover targets, i.e. assets that are sought for their (yet uncertain) commercial value rather than for their owners' personal satisfaction. While the distinction between private and common value is clear in theory, in practice many goods combine characteristics of both.

The winner's curse has been discovered in practice. The first explicit account we know of is Capen, Clapp and Campbell (1971). These authors were puzzled by the experience of oil companies who bid for petroleum lease tracts in the Gulf of Mexico in the 1960s. Most companies found that successful bids were much too high. Capen, Clapp and Campbell (1971) compute the bids that would have maximized companies' profits; for twelve out of eighteen companies they find results that are up to 15 per cent lower than actual bids. The authors conclude: "If one wins a tract against two or three others, he may feel fine about his good fortune. But how should he feel if he won against 50 others? Ill." This is a very nice formulation of what we will try to formalize somewhat in the theory section.[84]

The winner's curse has also been quoted as a potential explanation of excessive prices paid in corporate takeovers. The presumption in a corporate takeover would be that a firm (the bidder) spots an undervalued firm (the target) and tries to buy it by making a bid to existing shareholders. If the bid is successful and the target is integrated into the acquiring firm, the market value of the new company should exceed the sum of the market values of the bidder and the target prior to the takeover (more precisely: prior to the announcement). Just after the announcement, both the market value of the target and of the bidder should increase. However, empirical evidence does not confirm these hypotheses. While the value of targets on average rises with the announcement of a bid, the value of the bidders often does not. The aggregate value of the new companies, on average, is roughly equal to the sum of prior values of the individual firms (target and bidder), at least in "normal" times. During a financial frenzy, acquiring firms overpay. Moeller, Schlingenmann and Stulz (2005) show for the US that acquisitions (i) almost broke even in the 1980–1990 period (−US$4 bn), (ii) were slightly profitable from 1991–1997 (+US$24 bn), but (iii) led to wealth destruction on a massive scale (−US$240 bn) during the merger wave that accompanied the internet boom 1998–2001. In the first two periods, the winner's curse dealt only a soft blow: Acquiring companies did not lose (in the average), but they hardly benefited from the "synergies" perceived prior to the mergers. In the last period, the winner's curse hit hard: The most optimistic bids were far above the average market perception of target companies' values.

Box 12.3 The winner's curse in practice

The winner's curse lurks in many contexts including the following:

- Oil field auctions
- Spectrum auctions
- Sale of broadcasting rights
- Initial public offerings (IPOs)
- Corporate takeovers
- Real estate transactions
- Standing in a (short) queue
- Asking questions in the classroom

12 C Theory: The importance of conditional expectations

12 C a Informational assumptions

The winner's curse arises in situations exhibiting three characteristics:

- *Common value:* An item has an uncertain common value for all potential contenders (see Box 12.2).
- *Heterogeneous information:* Individuals have imperfect private signals about the true value of the item.
- *Unbiased signals:* Individual signals are about right in the average.

These are necessary conditions, although one further ingredient is needed: Individuals only fall victim to the winner's curse if they are not fully rational.

12 C b The problem

The notion that is important to the understanding of the winner's curse is the *conditional expectation* of a variable. Assume that a seller offers an (indivisible) object to the highest bidder from a number of $n > 1$ potential buyers. The true value of the object, V, is identical to each of the n bidders. However, V is not directly observable for potential buyers. Each bidder has a private signal s_i for the true value V. Signals are independently drawn from the same distribution. We denote the highest signal by s_1. Bidder i has signal:

$$s_i = V + \varepsilon_i,$$

where the error terms ε_i have an expected mean of zero.

The expectation conditional on a signal

The expectation of V for a bidder who got signal s_i thus is the signal value:

$$E(V|s_i) = s_i.$$

This is the expectation conditional on having received signal s_i, but *not conditional* on signal s_i being the *highest* of all signals (i.e. on bidder i winning the auction).

The expectation conditional on winning

Of course, bidder i does not know whether his signal is the highest or not. But it might be, and what if it were? Similarly, if signal s_i is the highest of all signals ($s_i = s_1$), it exceeds the average signal (unless all signals trivially happen to be equal). This is nothing other than the fact that the tallest girl in the class is taller than the average girl.

The expectation of V conditional on having received signal s_i, *and conditional on having the highest signal* is:

$$\hat{V} = E(V|s_i = s_1) < s_i.$$

In words: The expected value is lower than indicated by the signal.

Losing winners

What happens if bidders bid their expectations unconditional of winning, s_i? First, the bidder with the highest signal will win the auction.[85] Second, the winner will probably overpay. It is unlikely that the most optimistic bidder has a signal lower than the true value of the object. The higher the number of bidders, the greater the probability that the highest signal exceeds the true value V, and the higher the loss to the "winner" of the auction.

To put it simply, bidding one's private signal value (for a common value object) leads to an expected loss. A bidder's signal value is the best estimate of the unknown true value as long as the bidder may or may not win the auction. Once the bidder allows for the possibility of winning the auction, the signal provides an overly high estimate of the object's true value. This is true, even though no bidder knows in advance who has the highest signal. Either a bidder does not win the auction, or he does. The former case is irrelevant; but in the latter case, bidding too much (signal value) is dangerous.

It is important to distinguish once again between:

- the true value V (which is not known to individual bidders),
- a bidder's expectation of the value of the object, V, conditional only on the signal s_i,
- a bidder's expectation of the value of the object, $\hat{V}$, conditional on the signal s_i *and* on winning the auction which is lower than signal s_i ($\hat{V} < s_i$)).

Quantifying the winner's curse

The difference between a bidder's signal and the expected value of the object conditional on winning the auction can be seen as a measure of a potential winner's curse. It measures the discount from the signal value that is necessary to arrive at the best estimate of the true value of the object in case of winning the auction. To compute this difference we need to know the actual distribution of signals. For illustration purposes, let us assume that signals are distributed uniformly between $\underline{s}$ and $\overline{s} = \underline{s} + \Delta$. While Δ is common knowledge, $\underline{s}$ and $\overline{s}$ are not known to individuals.

What we are looking for is the difference between the expected highest signal and the true value of V. With uniformly distributed signals, the expected signal of the highest of n bidders is $\underline{s} + \Delta n/(n+1)$ (see Box 5.2). The expected true value of V is $\underline{s} + \Delta/2$. The difference thus is:

$$E(L) = E(s_i|s_i > s_{j \neq i}) - E(V) = \frac{n-1}{2(n+1)}\Delta. \tag{12.1}$$

This is the loss bidder i expects from bidding the signal conditional on winning the auction. In other words, this is the amount by which bidder i should underbid his signal to avoid the winner's curse.

12 C c A numerical example

Assumptions

We will illustrate the expected loss from the winner's curse, as given in (12.1), by means of a numerical example. We will stick to the assumption that bidders are completely naïve and bid the value indicated by their private signals. We also keep the assumption of uniformly distributed signals; for simplicity's sake we normalize the width of the distribution to $\Delta = 1$.

Expected

As (12.1) shows, the expected loss of this naïve strategy increases as the number of bidders grows. This is illustrated by the increasing line $E(L)$ in Figure 12.1. With $n = 1$, $E(L) = 0$: A lonely bidder could bid the signal value without an expected loss, though this would be very naïve (see below). As the number of bidders rises, so does $E(L)$, and quite steeply; it takes only a handful of bidders to produce a severe winner's curse. At the limit where $n \to \infty$ $E(L) =$ approaches 1/2 with

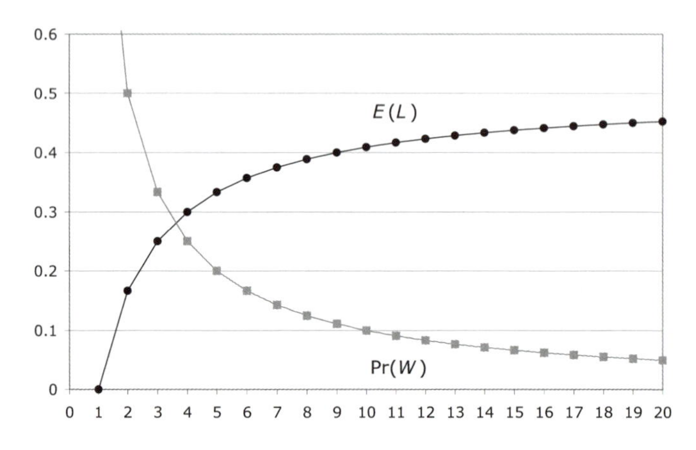

Figure 12.1 The expected loss of a naïve bidder, $E(L)$, illustrated for uniformly distributed signals, increases as the number of bidders rises. The probability of having the highest signal (and winning the auction), $\Pr(W)$, falls as the number of bidders decreases.

a large number of uniformly distributed bidders, the highest signal overestimates the true value of an item by an expected amount equal to half the width of the signal distribution. The figure thus nicely illustrates that one may feel good after winning against two or three others, but ill after winning against 50 others (Capen, Clapp and Campbell, 1971); in fact, 10 others may be more than enough.

Safety in numbers?

With an increasing number of competitors, a bidder faces an ever more severe winner's curse in the event of his winning the auction. At the same time, the probability of actually winning the auction, $1/n$ (having the highest signal of all bidders), decreases. This is illustrated by line $P(1)$, indicating the probability $1/n$ of having the highest signal: This probability, starting at $P(1) = 1$ for a lonely bidder quickly falls and approaches zero, as n rises.

Bidding in a large crowd can therefore be both very dangerous and fairly safe. It is dangerous to win, but an individual bidder is almost sure not to win. What is the combined effect in the sense of the unconditional expected loss of bidding naïvely? The unconditional expected loss is equal to the probability of having the highest signal (winning the auction if bidders bid signal values) *times* the expected loss conditional on winning. It is found by dividing (12.1) by n. The result is a function *decreasing* in n (for $n \geq 3$). In other words, the expected loss of bidding naïvely falls as the number of bidders drops. Yet, this should not lull bidders into a false sense of security. If a naïve bidder happens to win in a crowd, the winner's curse is likely to strike hard.

12 C d Underbidding in auctions

The winner's curse feeds on bidders' limited rationality. A rational bidder anticipates the winner's curse and underbids a private signal. Individuals often underbid what they think is the auctioned object's value. Consequently, underbidding can be the result of two fundamentally different strategies:

- *"Shaving"*: Underbidding the private signal value in a common (uncertain) value auction;
- *"Shading"*: Underbidding the perceived value in a first price auction (private or common value).

Shaving

The first strategy is an attempt to avoid the winner's curse. The individual with the highest signal normally "wins" the auction, but has an over-optimistic signal. Therefore, underbidding the signal (shaving) is necessary to avoid expected loss. The optimal amount of underbidding for the winner's curse *increases* as the number of bidders rise.

Shading

The second strategy is altogether different. Assume an individual knows the value of a good. The question then is: Should the individual bid that value? In a first price auction, where the winning bidder pays the effective bid, the answer is "no". Bidding the perceived value means that he only breaks even: The successful bidder would pay the full true or perceived value of the good. It is better to underbid ("shade") the perceived value. The optimal bid strikes a balance between the chance of winning the auction (bidding close to the value) and the chance of profit in the event of winning (underbidding) (Milgrom, 1989). The optimal level of underbidding for profit maximization *decreases* as the number of bidders falls.

On their way home from their evening with Alice, Bob complains to Charles. "Charles, that was unfair! You bid $22, although you knew perfectly well that Alice was older. You were just playing Mr Nice Guy!" "No, Bob, you're being unfair," Charles counters, "I just tried to make the best of the auction. I guessed that Alice was 26. But I was afraid that, if I'd win, this would be too high. I thought that $24 was a bid that would not lose if it won. But I didn't just want to break even. That's why I went down to $22. That would have left me a margin of some $2, if I'd won. This wasn't to keep Alice happy. In fact, I didn't expect you to bid so aggressively."

If all bidders are rational, they can count on others to underbid rationally as well. Yet, in the presence of naïve bidders the winner's curse gives rise to a certain dilemma for rational bidders. A rational bid (one that avoids the winner's curse) has little chance of succeeding. It is difficult to be cautious *and* win the auction. In the presence of naïve bidders, the probability of winning with a rational bid would be lower than indicated by the decreasing line $\Pr(W)$ in Figure 12.1.

One possible response for rational bidders is to spread information. Any public information on the true value of an object reduces value uncertainty, and thus the potential severity of the winner's curse. A similar strategy is to signal one's own rationality. The ever more frequent use of auction consultants may have a special merit in this context. Bringing in a consultant sends a clear signal to potential competitors: "We do understand the winner's curse" (= we will not bid as aggressively as you may fear; hence you may also bid less aggressively).

Sellers are likely to have an ambiguous view of the winner's curse. Facing naïve bidders who tend to overpay is a great situation to find yourself in. However, if sophisticated bidders with high signals are scared off, the seller may even suffer from the winner's curse. If the latter effect is likely to dominate, a seller may try to disseminate information in order to reduce the winner's curse.

Box 12.4 *Google* and the winner's curse

In August 2004, *Google*, one of the leading internet search engines, went public. The company sold 19.6 million shares in an auction-like procedure. In the prospectus issued prior to the auction, Google devoted several pages to reasons why investors might overestimate the company's value. The list also included a warning about the winner's curse:

> **Risks Related to the Auction Process for Our Offering**
> ***The auction process for our public offering may result in a phenomenon known as the "winner's curse," and, as a result, investors may experience significant losses.***
> The auction process for our initial public offering may result in a phenomenon known as the "winner's curse." At the conclusion of the auction, bidders that receive allocations of shares in this offering (successful bidders) may infer that there is little incremental demand for our shares above or equal to the initial public offering price. As a result, successful bidders may conclude that they paid too much for our shares and could seek to immediately sell their shares to limit their losses should our stock price decline. In this situation, other investors that did not submit successful bids may wait for this selling to be completed, resulting in reduced demand for our Class A common stock in the public market and a significant decline in our stock price. Therefore, we caution investors that submitting successful bids and receiving allocations may be followed by a significant decline in the value of their investment in our Class A common stock shortly after our offering.
>
> To the extent our auction process results in a lower level of participation by professional long-term investors and a higher level of participation by retail investors than is normal for initial public offerings, our stock price may decrease from the initial public offering price and be more volatile.

With hindsight, concerns about the winner's curse were slightly exaggerated. The auction process resulted in an issue price of US$85.00 per share. In the average bidders received 74.2 per cent of the shares they had subscribed. The trade in shares opened at a price of US$100.01.

One strategy for the seller is to "insure" bidders against the winner's curse. This can be achieved by tying the winning bidder's payment to the value realized from the object (its true value). For example, the right to broadcast the summer Olympics held in Seoul was auctioned against a bid price *plus* a royalty payment (from buyer to seller) that increased with the size of the television audience. Another way of insuring bidders against the winner's curse occurs in repeated transactions. The underpricing of initial public offerings (IPOs) may provide investors with such insurance, allowing them to bid more aggressively in the average.

12 D Application: The underpricing of IPOs

12 D a Empirical underpricing

Start-up firms are often financed by venture capitalists. At some point in time a firm, if it has not failed in infancy, is well established and looks for a broader financial basis. This is the time to go public. Going public means selling a company's shares to a large number of investors through an initial public offering (IPO). The issuing technique still dominant in many countries is underwriting the whole issue by a bank or a group of banks at a fixed price. The banks then offer the issue at that price to the public. The issuer pays banks a fixed fee.

A big puzzle with IPOs is that issue prices are normally "too low": First-day closing market prices exceed the issue price, often by a large margin. The issuing firms sell their shares for less than they seem to be worth. "Why Don't Issuers Get Upset About Leaving Money on the Table in IPOs?" Loughran and Ritter (2002) ask. The authors compute that from 1980 to 2003 the average IPO was underpriced by 18.7 per cent, or US$9.1 million in monetary terms. This is approximately twice the fees paid to investment bankers and represents a substantial indirect cost to the issuing firm.

IPO underpricing is very robust across countries. Over time, the degree of underpricing is subject to some fluctuations. Hot periods (strong underpricing) alternate with cooler ones (weak underpricing), but underpricing is always there; periods of sustained overpricing are non-existent. Loughran and Ritter (2002) report for different sub-periods an average underpricing of 7.3% (1980–89), 14.8% (1990–98), 65.0% (1999–2000), 11.7% (2001–03).

The puzzling aspect is that issuers do not seem to mind. One possible explanation is the winner's curse: Issuers have to underprice securities, because banks would not underwrite otherwise, and investors would not subscribe the issue.

12 D b A potential explanation

The starting point is the asymmetry in the commitment of an underwriting bank or a subscribing investor. An underwriting bank has a say in fixing the issue price with the issuer; but once the price is fixed, the bank has to fulfil its promise: It has to buy the full issue at the agreed price, whether the issue is a success with investors or a failure. If the issue is a success, the bank sells the shares. If the issue fails in the market, however, the bank finds itself with overpaid shares on its books. Similarly, an investor who subscribes to one hundred shares has to buy all one hundred if demand from other investors is low. If the issue turns out to be "hot", the investor may count himself lucky if he receives fifty.

In the following example (from Jay Ritter's homepage[86]) an investor is faced with two IPOs which are fairly priced in the average:

> If you are an investor who is asking for an allocation of 1,000 shares, and there are two IPOs: One is overpriced, the other is underpriced by 20%. Do you get out even? Probably not. When the offering is overpriced, few other people will have asked for shares and you typically get all 1,000 shares. When the offering is underpriced, half the world will have asked for shares, and typically you rarely receive any shares. (The situation becomes even worse when underwriters distribute the "good shares" only among their friends, and you are not one of them.) For the sake of argument, assume you receive shares only half the time. Then, your expected return on these two offerings would be
>
> $$50\% * (+20\%) + 100\% * (-20\%) = -10\%$$
>
> Consequently, an investor who participates in an offering that is fairly priced is likely to make a negative return and thus would not wish to participate. To encourage investors to participate, issuers must set a lower price for the IPO. It will look as if the average IPO left money on the table, but the typical investor cannot profit from this money.

12 D c Available cures

Alternative formats

One would assume that competition induces issuers and banks to search for ways to reduce underpricing. One option would be to change the underwriting mechanism: Issuers could auction off the shares themselves. Indeed, there are some banks who try IPO auctions. However, the winner's curse is also present in an auction. Another option is that banks work on a "best effort" commitment, rather than on a fixed underwriting basis. This is sometimes used for smaller issues. Yet, for bigger issues the underwriting procedures observed in practice still dominate. They seem to be reasonable approximations of an optimal procedure, although they do not solve all of the informational problems arising from going public.

Discretionary allocation

There may be ways to minimize underpricing within the traditional fixed underwriting procedures. Sherman (2000) shows that a discretionary allocation of shares by underwriting banks can partially insure investors against the winner's curse. Within a long-run relationship the underwriting bank can allocate shares in hot issues to investors who had asked for (and received) allocations in earlier overpriced issues. The expectations that overpriced allocations will be compensated by future underpriced allocations allows investors to issue more aggressive share allocation demands, thus lowering the funding cost for issuers. We could call this the "uninformed investor insurance view". This view may explain some average degree of underpricing.

A somewhat complementary theory suggests that banks allocate shares not to uninformed investors, but to sophisticated investors. Participation of sophisticated, professional or institutional investors provides underwriting banks with important information; often such investors signal their interest before the issue price is fixed. Allocation of hot issues thus compensates them for information revealed during the underwriting process. We may call this the "information revelation view".

Rival explanations

Loughran and Ritter (2002) subsume the investor insurance and the information revelation views under the "academic" view, as these are typically found in explanations examined in academic papers. They confront the academic view with rival explanations. One is the "pitchbook view", the view most popular among investment bankers who do the "road shows" for IPOs: Underpriced issues are allocated to long-term buy and hold investors, who provide a stable ownership. A similar explanation is that, via underpricing, issuers pay banks for analyst coverage. Loughran and Ritter point out that coverage by highly-rated analysts has become a major competitive factor in the underwriting business.

In recent years, some less savory explanations have gained ground not only in the press but also in academic papers. Loughran and Ritter call these the "profit-sharing view" and the "corruption view". The former refers to the observation that investors who obtained allocations of underpriced shares often generate considerable commission income for the underwriting banks (or vice versa). Banks and investors thus collude and share profits from underpriced issues. The latter view builds upon several cases in which hot issues were partly allocated to managers of the issuing firms.

In both the profit-sharing and the corruption view underpricing is not only a precondition for profit-sharing or corruption, but also a consequence. If investors anticipate that banks allocate the best shares to some special clients or to management, they face an aggravated winner's curse which, in turn, requires additional underpricing. Such explanations of the winner's curse do not fundamentally contradict the academic view.

Beyond revenue maximization

After all, the explanation of underpricing may also be found in "behavioral" motives. We have implicitly assumed that in an IPO issuers maximize expected revenue. Yet, they may not. This is suggested by standard terminology: Most issuers (and, less surprisingly, their banks) deem an issue with high first-day returns (i.e. with a high degree of underpricing) as a "successful issue". Management is positively surprised by the market's valuation of the firm. While we all like positive surprises, it is not quite rational to forego money by underpricing simply in order to be positively "surprised".

All potential explanations of why underpricing exists and why issuers do not complain more, can cite some evidence in their favor. An IPO involves several information problems between issuer and underwriter, between underwriter and the different categories of investors, between investors and issuer, and finally between managers and owners of the issuing firm. It remains to be seen which issuing procedure is optimal in the sense of minimizing the aggregate deadweight cost of these information problems. One attempt to formulate an optimal underwriting mechanism is offered in Biais et al. (2002).

12 E Application: Prices and the winner's curse

Contesting the obvious

At least since the times of Adam Smith, economists took it almost for granted that increasing the supply of a good or restricting the demand for it would reduce its equilibrium price. Alas, our profession manages to disagree even on this seemingly simple issue.

The claim that higher demand may in fact *decrease* the equilibrium price was made by Bulow and Klemperer (2002). Their point is quite simple for a reader familiar with the winner's curse. We have shown that the winner's curse becomes more severe as the number of participants increases. As illustrated in Figure 12.1 above, the expected loss from bidding one's signal increases with the number of bidders. A higher expected number of bidders may thus lead individual bidders to be more cautious. Bajari and Hortaçsu (2003) found that the average bid in eBay auctions decreases by almost 4 per cent with any additional bidder.

From the point of view of the seller, the number of bidders has an ambiguous influence. On the one hand, it exacerbates the winner's curse and leads to lower bids. On the other hand, as the number of bidders increases, so too does the expected value of the highest signal. The effect which dominates depends on signal distributions. Bulow and Klemperer (2002) show, however, that based on fairly plausible assumptions, there is a number of bidders $\hat{n}$, beyond which the expected seller revenue falls with each additional bidder. The results also depends on whether bidders are symmetrical or asymmetrical.

Where rationing may pay

Bulow and Klemperer (2002) suggest that their findings are most relevant for financial markets, as financial assets are typical common value goods. In air wave spectrum auctions, for instance, revenue (per license) was often higher, when two licenses, rather than one, were offered simultaneously. Conversely, in (fixed price) initial public offerings (IPOs), restricting supply to guarantee excess demand may be more profitable than finding the market clearing price.

A dangerous case

Finally, Bulow and Klemperer (2002) also warn of a case in which the winner's curse is particularly dangerous: In "almost-common-value" auctions. Adding a tiny amount of private value to a common value auction is not a negligible deviation from the common value assumption. On the contrary, if an object, above its common value, yields a (commonly known) small private benefit to one or to a number of the bidders, this exacerbates the winner's curse dramatically. In such an auction, all other bidders know that they would only win the auction if they overbid a competitor by at least his private benefit. This requires a fairly high signal and a correspondingly high risk of winner's curse. Bidders thus have to reduce their bids more than in a standard common value auction. Even worse, the bidders with private benefit anticipate that their competitors will be cautious. This permits them to reduce their own bids, too. In the end, the seller of the object may wind up with a disappointing return.

12 F Conclusions and further reading

Whoever competes for an object—a petroleum lease, a spectrum licence, a indexspectrum license, a takeover target, or just a family home—may end up asking: "How come, that it was exactly me who made the highest bid? What did the others know that I did not?" This reflects the winner's curse, the danger that the highest bidder for an item of common but unknown value is overly optimistic.

Of course, like all economic wisdom, the idea of a winner's curse can be taken too far. A groom, panicking before the altar, suddenly thinks to himself: "Why exactly did I end up with her?" However, he may find comfort in the fact that his bride may have also asked herself a symmetrical question. This paradox was formulated in the immortal quote attributed to Groucho Marx (1890–1977): "I don't care to belong to a club that accepts people like me as members."

Who should beware of the winner's curse? We have shown that the winner's curse occurs in many contexts, particularly in financial markets where goods typically have a common value. The potential victims are primarily those competing for common value goods; whether the selling procedure is explicitly called an auction or not seems less relevant. Pure private value goods seem free from the winner's curse. Most disturbingly, though, adding a pinch of private value to a common value good exacerbates the winner's curse, rather than mitigating it.

We have only given a basic introduction, abstracting from much interesting detail. Several texts provide students with more detail and a deeper analysis. An easy-to-read introduction to the winner's curse is Thaler (1992). A more formal treatment is offered in Wolfstetter (1999, Sec. 8.6). The application to auctions and the price system is developed in Bulow and Klemperer (2002). A primer in auctions is Milgrom (1989).

Checklist: Concepts introduced in this chapter

- Expectation of value conditional on winning an auction
- Private versus common value
- The winner's curse

- Underbidding in auctions (shaving versus shading)
- Initial public offering (IPO) underpricing
- Almost-common-value auctions

12 G Problem sets: Cursing winners

For solutions see: www.alicebob.info.

Problem 12.1 Misleading signals

An individual i has an imperfect signal s_i for the value V of an object, with:

$$s_i = V + \varepsilon_i$$

(a) What is the expected signal if the distribution of the error term ε has zero mean?
(b) What is the individual's expectation of V after receiving signal s_i?
(c) Assume that n individuals have signals on V drawn independently from the same distribution. How does the difference between the highest signal and V vary with n?

Problem 12.2 Winner's curse in a Uniform distribution

Assume that in a common value auction there are n bidders. Bidders have individual signals about the true value of the item to be auctioned. Signals are independently drawn from a uniform distribution with width Δ.

(a) Draw a graph plotting the expected loss against n.
(b) What is the expected loss of a bidder who bids signal value s_i if (i) $n = 1$, (ii) $n = 2$, (iii) $n \to \infty$?

Problem 12.3 Bidding and the number of bidders

When the number of bidders in a common value auction increases, bidders have three options: (i) bidding more, (ii) bidding an unchanged amount, (ii) bidding less. Make a point for each of the three options.

Problem 12.4 Takeover bid

This problem is adapted from Thaler (1992). Company A ("acquirer") considers acquiring Company T ("target"). The value of a share of Company T depends directly on the return from an oil drilling project. It is uniformly distributed in the range from 0 to 100 (all values are equally likely). Company A does not know the return from the project (and hence the share value) when it makes its tender offer. The board of Company T learns about project outcome after Company A has made its offer. The board accepts the offer if it is higher than the value of Company T. Company T will be worth 50 per cent more in the hands of Company A than under its current management. How much should Company A bid per share of Company T?

Problem 12.5 IPO

Greycells.com for the first time offers its shares for sale to the public. A bank has underwritten the issue at a fixed price. Investors are invited to subscribe. After studying the prospectus you decide that you will subscribe for 100 shares. Yet, before sending your subscription, you read in the financial press: "Greycells.com seems to be a real hot issue. We estimate that the issue will be oversubscribed by 100 per cent."

(a) How would you revise the number of shares you subscribe: (i) less than 100, (ii) 100 (as planned before), (iii) between 100 and 200, (iv) 200, (v) more than 200?
(b) Plot the amount you subscribe against the overall subscription ratio (subscribed divided by available shares), r, you expect (in a range of $0 < r < 5$).

Problem 12.6 Transparency

Would you advise eBay to disclose the number of bidders in an ongoing auction or not?

Problem 12.7 Ali and Baba

Ali and Baba have done a service to the caliph. The grateful caliph calls a servant who brings two identical looking leather bags, each containing a certain number of gold coins. The caliph hands over one bag to Ali, the other to Baba. They are asked to look into their personal bag without disclosing anything about its content.

The caliph tells them that they may of course exchange the two bags, provided that both (secretly) agree to such a swap. He announces that the individual who ends up with the higher number of dinars will become his new vizier. Finally, he reveals that one bag contains exactly twice as many dinars as the other, the two bags being drawn from a series with 1, 2, 4, 8, 16, and 32 dinars, respectively.

(a) Assume Ali finds four dinars in his bag: What is the expected number of dinars in Baba's bag?
(b) Should he agree to swap?
(c) Should he expect Baba to agree?
(d) Will they agree on a swap?
(e) Who is more likely to become the new vizier?

13 Information and selection

13 A Introduction

In the 1990s a Swiss insurance company[87] launched a niche product: marriage insurance. Children younger than 10 could be insured (by their parents) against the risk of early marriage. The company offered a maximum coverage of CHF 100,000 for a marriage taking place before the age of 25, with some benefit reduction for marriage before the age of 19. Pricing was based on expectations derived from Swiss marriage statistics. In Switzerland the probability of being married by the age 25 is 17 per cent for men and 29 per cent for women, respectively. Of course, the price included a safety and a profit margin.

Was the policy profitable? Not quite. The vast majority of parents seemingly did not have sleepless nights about the possibility that their children might marry early. Among majority groups, the policy barely sold. Yet, the early marriage insurance policy proved extremely popular among ethnic and religious communities where marriage at a tender age is the rule. The policy thus was not bought as coverage against a damage, but rather as an opportunity to embellish a lucky event. As a consequence, the early marriage ratio among the insured, i.e. the fraction of insurance policies where the company had to pay out, was a stunning 99 per cent! As a representative of the company put it: "There was a technical loss on each and every policy sold. At least for once it was easy to calculate the necessary provisions. We just had to multiply coverage by the number of policies."

There is an old saying in the insurance business: "The scent of the premium should never mask the smell of risk". When a deal looks too attractive, something may be wrong with it. In Chapter 12 we argued that the opportunity to buy a good means that *competing buyers* may find it less attractive. Here, we look at a similar if even more dangerous case: The opportunity to buy a good presumes that the *counterparty* finds the deal attractive. But the counterparty, the seller of a car, say, may know something about the vehicle a potential buyer does not. After all, the seller has owned it for some time and is likely to know it better than anyone else. Overbidding the seller, so to say, may be even more dangerous than overbidding competing buyers.

Asymmetric information, as economists call a difference in information between the two sides in a contract, was first analyzed in the 1970s by researchers like George Akerlof, A. Michael Spence, and Joseph Stiglitz. For their work these researchers were honored with the 2001 Nobel prize for economics. The importance of their contribution may seem obvious with hindsight—to contemporaries it was not: The path-breaking article on the used car market, "The market for lemons" by George Akerlof, was rejected by one journal after another until the *Quarterly Journal of Economics* accepted it in 1970—seven years after it had been written. Today "adverse selection" is a well known concept to first-year students and to many others who never took a formal course in economics.

Indeed, life is full of situations of asymmetric information. The economics of information asymmetries explains why people who eat well at home deliberately feed on junk as tourists, why borrowers on local Third World markets have to pay skyrocketing rates, or why for students a diploma is valuable even if the courses are boring.

Asymmetric information is an area where the simple rule "more information is always better" does not hold. More private information often leads to market failure due to so-called "adverse selection". But even more public information may destroy welfare, as we will demonstrate in an Application on the so-called insurance destruction effect (13 D).

An important example of an information-sensitive market is the market for annuities, the subject of our second Application (13 E). This market raises important policy questions, like: Should the poor subsidize the rich? What information should a pension fund use (or be allowed to use)? How do we distinguish between adverse selection and beneficial sorting? As far as we know, the latter question has been largely neglected in the literature to date.

In the present chapter we focus on the type of information asymmetry that exists already *before* deals are proposed and made, the so-called "adverse selection" case. A related type of asymmetry arises when one party acquires relevant information *after* a deal has been made, though this case will be discussed in Chapter 15. A fairly different case of information asymmetry relates to actions, rather than to facts. After a contract is signed, one agent can choose an *action* unobservable to the contract party. This leads to so-called "moral hazard", a phenomenon to be discussed in Chapter 16.

13 B Main ideas: When information prevents trading

It has become somewhat easy to cheat. On internet auction sites large numbers of used or rare goods are traded. Typically, these are of unknown quality to potential buyers. Nothing seems to prevent a seller from overstating the quality or condition of his article. Given that buyers and sellers are anonymous, a disappointed buyer cannot initiate legal proceedings against the seller. While successful bidders are normally supposed to pre-pay, a seller may even choose not to send the article at all.[88]

In traditional markets, trading parties are rarely anonymous. Yet, information asymmetries remain numerous. Often the seller knows more than the buyer. For example, restaurants, car sellers, computer manufacturers, doctors, lawyers or investment advisors all have an information advantage over their clients. In other areas buyers are better informed than sellers, insurance markets being the classic example.

The Romans were well aware of such problems and coined the phrase *caveat emptor* (buyer beware). Yet, if buyers are cautious they may not buy at all. As Riley (2001) states, the invisible hand of the market only works when "agents have the same information about the nature of the goods traded." George Akerlof, with his "market for lemons" model, addressed the question of whether markets still work in the presence of a small amount of hidden information. His answer is the main topic of this chapter.

Alice and Bob want to go on vacation and therefore need a car. Fred, known as a clever guy, is ready to sell his car for $10,000. But Alice and Bob conclude from Fred's offer that the car's value is *at most* $10,000—otherwise Fred would not sell—and probably less. They consider a counter-offer of $8,000 assuming that a discount of $2,000 would sufficiently protect them against the quality risk. Yet, Alice cautions: "Fred will only accept if the value of the car is even less than $8,000! If we don't want to lose money, we must go even lower!" "Right," says Bob, "but where does this leave us? If we don't want to overpay, we have to offer more than the value of a lemon. But then we'll only get the car if it is indeed a lemon!"

Under hidden information, all but the worst qualities may disappear from the market. This effect is called "adverse selection". Selection effects become important whenever a uniform price is offered for goods of different quality. The adverse selection problem is more severe than the winner's curse, which was treated in the preceding chapter. The winner's curse arises because individuals have *different* information, none of which is necessarily superior or inferior *ex ante*. Here, one agent, in our example Fred, has in fact *superior* information. Potential buyers can avoid the winner's curse if they all underbid their private signal by some safety margin. By contrast, under asymmetric information a buyer cannot acquire a car at a fair price unless the car is a "lemon".

Both adverse selection and the winner's curse illustrate the importance of higher-order information, which we discussed in Chapter 8. It is not only important which facts the agents know (here the quality of Fred's car), but equally what they know about each other's knowledge. Alice and Bob are rather lucky: At least they *know* that Fred knows his car better than anyone else. Unlike naïve buyers, they can try to protect themselves against overpaying. Yet, the price for not overpaying may be high.

"I refuse to go on vacation in a heap of junk," Alice says. "But," Bob replies, "I do want to go on vacation with you!". After some further reflection they agree on the following: They would not like to travel in a vehicle worth less than $5,000. Nor would they want to stay at home. They cannot fully eliminate either risk. But they feel that offering $8,000 is a good compromise. Such an offer reduces the scope for overpaying; at the same time it leaves them with a good chance of acquiring a more or less decent car.

The gains from trade, in particular the fact that Alice and Bob may want the car more urgently than Fred himself, make a deal possible. Some inefficiency remains, though: Fred may not sell his car (if it is worth more than $8,000), or the vehicle may be a lemon (worth less than $5,000). The solution is only a "second-best", the best Alice and Bob could achieve under asymmetric information. Under perfect information—when the car's value is observable—there is a "first-best" outcome: If the car satisfies their quality standard, Alice and Bob simply offer a price equal to the car's value. At this price Fred is ready to sell.

There are two reasons why markets work despite obvious opportunities for well-informed parties to cheat their less-informed counterparties. One reason is the *potential gains from trade*. When Bob wants to go on vacation with Alice he takes the risk of overpaying for a car or for a hotel room. When he wakes up with a terrible toothache during the trip, his main worry is not that he does not know the quality of the local dentists, but getting to a dentist in the first place. His "gain from trade", getting rid of the pain, is just too important.

The other reason why markets work in the face of information asymmetries is, roughly speaking, that *sellers can pledge their reputation*. Participants in internet auction sites can collect ratings, often called feedback scores, from their trading partners. A seller with a better rating is more likely to earn higher revenues from future auctions and more likely to stay in business in the first place (Bajari and Hortaçsu, 2004). In traditional markets, firms offer goods of standardized quality and develop brand names. A brand name is a pledge to the customer. It is valuable because buyers are ready to overpay for standard qualities relative to goods of unknown quality. A brand name loses value, though, once the seller is unable to maintain quality standards. When tourists eat fast-food, they know that they pay a bit too much for what it is; but at least they know what they are getting. This is moving closer to the more general topic of contracting which is treated in a separate chapter (Chapter 14).

The analysis of asymmetric information also casts some light on the functioning of markets, which we have already discussed in Chapter 7. There, we have argued that trading is driven by differences in opinion; bulls buying shares from bears, bears selling to bulls—until the price is such that both camps are equally strong. Here we show that differential information may actually prevent trade.

This is the idea behind some of the no-trade theorems mentioned in Chapter 7. If information is distributed asymmetrically and traders *know*, then markets are likely to fail. Once more, we can use the Nathan Rothschild legend as an example: Given that the other traders knew that Rothschild knew more, why did they stupidly trade naïvely? How come that in reality financial markets, far from breaking down, deal with enormous sums every day? The standard explanation is the same as in Alice's and Bob's vacation trip: There are some traders who are ready to lose a small amount in expected terms because they have a personal motive for trading. One such motive is a need for liquidity or its counterpart, excess cash, both creating a potential gain from trading. It is these gains that keep markets going.

Hidden information, the subject of the present chapter, is only one form of information asymmetry. The other is hidden action (see Chapter 16). To clarify the difference, we shall use the example of health insurance. When individuals seek insurance, they have hidden information about their health. This may lead to adverse selection: Individuals in poor health tend to buy more insurance, thus crowding out individuals in good health. Once individuals are insured, they can take hidden actions—going to the doctor too often or asking for unnecessary expensive treatment. Such behavior is called moral hazard. In reality, the potential for hidden information and hidden action is considerable: Gardiol, Geoffard and Grandchamp (2005), for example, find that in the Swiss health insurance system coverage and health care expenditures are correlated. A decrease in the co-payment rate (the part of expenditures paid by the insured individual) from 100 per cent to 10 per cent increases the marginal demand for health care by about 90 per cent; a decrease in the co-payment rate from 100 per cent to 0 increases it by about 150 per cent. Both hidden information (selection effects) and hidden action (incentive effects) contribute to this result. The correlation between insurance coverage and health care expenditures can be broken down into two effects: 75 per cent of the correlation may be attributed to selection, and 25 per cent to incentives (moral hazard).

13 C Theory: The market for lemons

13 C a Modeling hidden information

Assume that a buyer goes to the used car market, offers a price and buys all cars the sellers are ready to sell at that price (or picks one of them at random). Between seller and buyer there is an asymmetry of information. The seller of a car knows the car better than a potential buyer. We assume that the seller has hidden information about the quality of the car when a buyer appears on the market. The buyer learns of the quality of a car only after it has been bought. This sequence of events is represented in Figure 13.1.

The distribution of information between buyer and seller is illustrated in Figure 13.2. The seller knows in which of six quality categories a car falls, while the buyer cannot tell the quality of a car. The figure is an extreme case of Figure 3.3

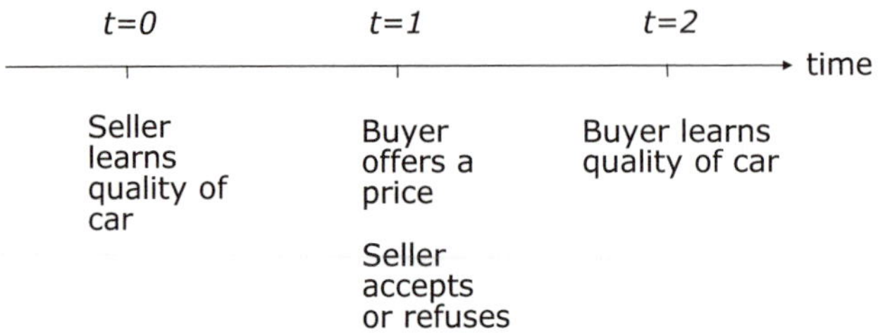

t=0 t=1 t=2 time

Seller learns quality of car

Buyer offers a price

Seller accepts or refuses

Buyer learns quality of car

Figure 13.1 The time line in the market for lemons model: The buyer who offers a price does not know the quality of the car, but knows that the seller is more likely to sell if the car is of poor quality.

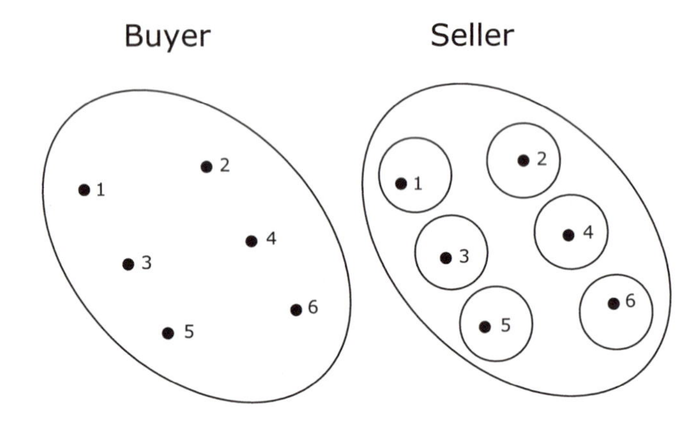

Figure 13.2 Hidden quality: The buyer only knows the potential qualities, while the seller knows the true quality.

from Chapter 3. Here, the seller can recognize each quality category, while the buyer merely knows the potential qualities and their respective frequency, but has no idea of the actual quality of any particular car.

This looks like a home match for the seller. Yet, this is not the case. The information structure, as given in Figure 13.2, is common knowledge between buyer and seller (see Chapter 8). In other words, the buyer *knows* that the seller knows the quality of the car (and the seller knows the buyer knows, etc.). This higher-order information about who knows what is just as important as the fundamental information about the quality of the car. As it turns out, the seller's information advantage may in fact be a handicap: buyers are not naïve; and a seller who is known to know cannot sell a poor quality car for the value of a high quality car. On the contrary, even for a high quality vehicle the seller may not be able to obtain more than the value of a poor quality car.

13 C b Adverse selection

Partial market failure

To illustrate the impact of the seller's information advantage, assume that in the used car example, there are just two different qualities of car with an equal share in the car population: One half is worth \$5,000, the other \$10,000. A seller privately knows the actual quality of a car, while a potential buyer does not. Figure 13.3 shows the respective information structures. These structures, including the prior probabilities of the two qualities (\$5,000 and \$10,000), is common knowledge. So is the seller's information "advantage".

Somewhat unrealistically, we assume like Akerlof (1970) that buyers make take-it-or-leave-it offers to sellers: an offer is either accepted or rejected forever. To guess the outcome of this car market, we first need the seller's best response function. The best response of a seller to a price p offered by a buyer is: Accept if the price p at least equals the car's value, V, i.e. if $p \geq V$, otherwise reject it. This yields the buyer's expected value function. The expected value $E[V(p)]$ of a car coming forth at price p is the average value of all cars offered at that price. As cars sold at p have value $V \leq p$, the expected value is $E[V(p)] = E[V \mid V \leq p]$.

The expected value is the step function in Figure 13.4. It reflects the seller's best response plus an assumption on the shares of different qualities in the aggregate car pool. Assume a buyer is bold enough to offer a price of \$10,000 or even above. At such a price all cars will be offered for sale. Their average value will be equal to the average from the whole car pool, i.e. \$7,500. Once the buyer lowers his bid below \$10,000, the better quality cars will cease to be offered, the expected quality offered falling abruptly to \$5,000. For a bid of less than \$5,000 no car will be offered.

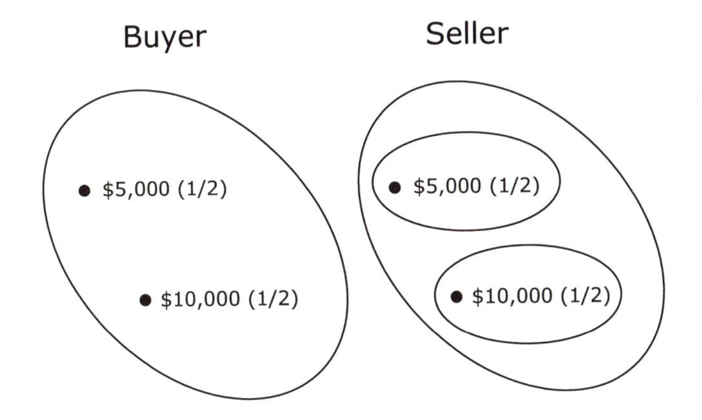

Figure 13.3 Hidden information with two qualities (\$5,000 and \$10,000). The true quality is observable to the seller, while for the buyer both qualities have equal probability.

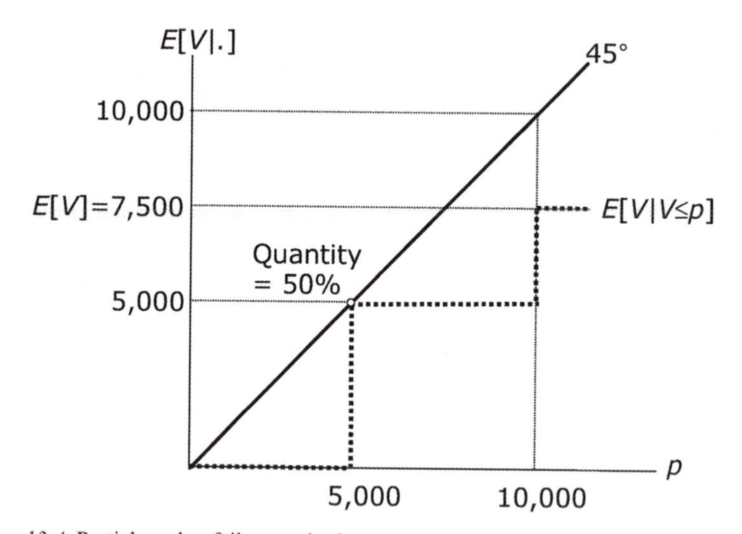

Figure 13.4 Partial market failure: only the worst of two qualities is traded.

The expected value line can be seen as the "supply" curve. It describes the expected value, that is the average quality, offered as a function of the price (unlike a standard supply curve which describes the quantity offered as a function of price). The corresponding "demand" curve describes the minimum expected value required to satisfy a buyer. The buyer agrees to a deal if the expected value is at least equal to the price. The minimum value thus is given by the line $E[V] = p$, the 45°-line in Figure 13.4.

The market is in equilibrium at $p = E[V] = 5,000$. Only the poor quality cars are sold. The high quality cars are not sold at this price; a buyer announcing a price that would bring the high quality cars to the market would make a loss. Exactly half of the car pool will be sold. The other, better half will not be sold. We will now show that in the case of continuous qualities the outcome is even more dramatic.

Market breakdown

Assume that car values are uniformly distributed between \$5,000 and \$10,000, i.e. $V \sim U[5,000, 10,000]$. The expected value line is given in Figure 13.5. It has a segment with $E[V] = 0$ below \$5,000, increases in a straight line between \$5,000 and \$10,000, and is again flat above \$10,000. The logic is still the same: If a buyer announces a price of \$10,000, for example, all cars will be offered, with an average quality of \$7,500. If the buyer lowers the price somewhat, the best quality cars will be withdrawn from the market. With each reduction in price, cars in successive quality categories will be withdrawn. The market equilibrium is reached at a price of \$5,000, where only the very worst vehicle worth \$5,000 will come to market. The fraction of cars sold is almost nil.

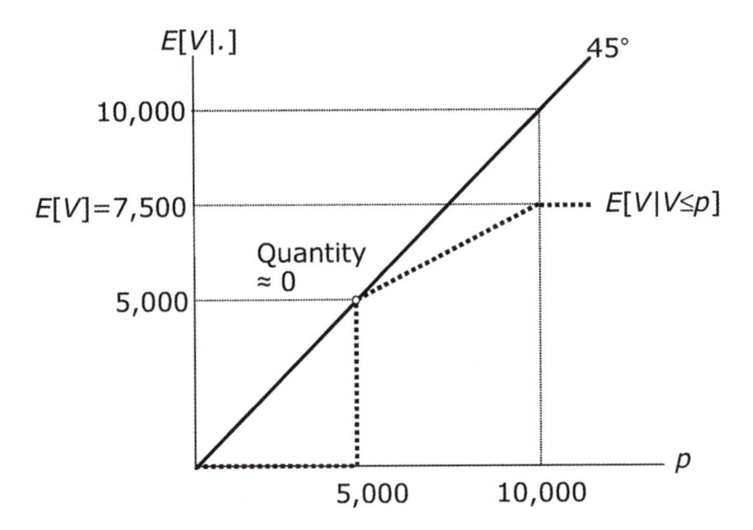

Figure 13.5 Market breakdown with a quality continuum.

The complete market failure in this context is not really a problem: Cars have the same common value for sellers (their actual owners) and for potential buyers. From a welfare point of view it is irrelevant who owns a particular car. Market failure *per se* is not inefficient. A market outcome is only inefficient when a car is not sold, although it is worth more to a potential buyer than to the present owner. This is the subject of our next example.

Adverse selection

We maintain the assumption of uniformly distributed qualities. However, we assume that cars are worth more in the hands of potential buyers than in the hands of their owners. Such an assumption is perhaps more plausible if one thinks of labor, rather than cars. Most workers are more productive in a firm, i.e. when they sell their labor, than at home. Yet, as a tribute to Akerlof's classic contribution we shall stay with the car analogy. For a labor market framing, see Mas-Colell, Whinston and Green (1995) or Wolfstetter (1999).

The value of a car to a buyer, V, now differs from its value to a seller, V_S. We denote the buyer's differential valuation by a, that is $V = aV_S$ $(a > 1)$. The seller's best response still is to sell if the price offered exceeds the seller's valuation. Cars come to the market if, and only if, $p \geq V_S$. A potential buyer offering price p faces an expected value function of $E[V \mid p] = E[aV_S \mid V_S \leq p]$.

For illustrative purposes, we set $a = 1.2$. Buyers value cars 20 per cent higher than sellers. This shifts the expected value function, the value as seen through the eyes of a buyer, upwards, as illustrated in Figure 13.6. If a buyer declares a price

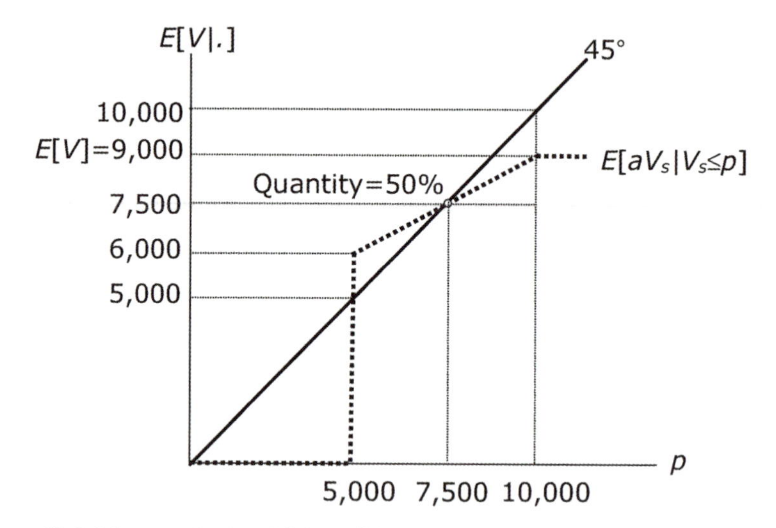

Figure 13.6 Adverse selection: High qualities remain unsold despite potential gains from trade.

of $10,000, all cars come to market. Their average value for sellers is $7,500, but for a buyer it is *a* times that value, i.e. $9,000. Lowering the offered price continuously excludes a broader range of qualities from the market, until only the worst car is offered at $5,000. To a buyer it is worth $6,000.

The market equilibrium is at a price of $7,500. To sellers, the average value of cars sold is $E[V_S] = \$6,250$. All cars in the lower half of the quality range are sold. Their average value to buyers is $E[V] = \$7,500$. Although the market does not break down completely (as in the above example), the outcome is inefficient. The better half of cars remains unsold, although buyers would value them higher than their present owners. This phenomenon is called *adverse selection*.

13 C c The gains from trade

Winners and losers

Who wins and who loses from information asymmetry? To answer this question we shall look at the adverse selection equilibrium from the perspective of the potential *gains from trade*. Gains from trade are created when an object is sold to a buyer who values it higher than the seller.[89] In the presence of informational asymmetry, gains from trade play an ambiguous role, with potential gains becoming partly lost. This is why we talk of market failure. Note that the no-trading result is not a problem if there are no gains from trade in the first place. The other part of potential gains are realized; these are the reason why, even under asymmetric information, markets exist after all. However, the realized gains are not distributed evenly. Sellers of low quality goods acquire relatively high shares in the realized gains of trade; their presence creates a negative externality for all sellers of high quality goods (who cannot sell) and to sellers of median quality goods (who get less than they would under observable quality).

A bit of arithmetics may clarify the issue. For simplicity's sake, we normalize the number of cars to unity or 100 per cent; their aggregate seller value in this case is equal to the average seller value of $7,500. In the example where $a = 1.2$, the total potential gains from trade are 20 per cent of the aggregate seller value, or $1,500.

Under full information, the gains from trade in the present setting—where buyers make take-it-or-leave-it offers—would go to buyers. Buyers who observe qualities can infer their owners' valuations and offer the minimum price at which an owner sells. For a car worth $6,000 to a buyer, the buyer would bid the seller's worth of $5,000 and will seal the deal.

In the asymmetric information equilibrium, gains from trade are distributed unevenly across the quality spectrum:

- Cars in the *upper* quality category (cars with a seller's worth of $7,500–10,000) are *not sold*. The potential gains from trading these cars are lost. The aggregate loss amounts to $875 (20% of 50% of $8,750); this is more than half of the potential gains from trade.
- Cars in the *lower* quality category (cars with seller's worth of $5,000–7,500) are *sold*. The gain from trade on these cars is $625 (20% of 50% of $6,250). As buyers pay $7,500 for an expected quality of $7,500, the gains from trade accrue with sellers. Within the group of successful sellers, sellers of relatively poor quality cars do better than sellers of higher quality cars:
    - The seller of the best quality sold (seller's worth of $7,500) earns no benefit from trading; buyers gain $1,500;
    - The seller of the median (= average) quality sold (seller's worth of $6,250) gains exactly 20 per cent (a times the price); buyers break even;
    - The seller of the worst quality (seller's worth of $5,000) earns a 50 per cent profit, more than the gains from trading this quality under full information; buyers lose $1,500.

As these figures show, there is a redistribution of the gains from trade among the sellers of different qualities. The sellers of "lemons" do best, as they can "hide" behind successful sellers. In general, sellers of below-average quality goods appropriate some gains from trade on the above-average quality goods.

Can information become a burden?

In the above example, the cost of information asymmetry fell fully on potential buyers. This extreme effect is a consequence of our assumption that buyers make take-it-or-leave-it offers to sellers and therefore would skim off the whole gains from trade under full information. Once we assume that market power is less one-sided, car owners may indeed suffer from their information “advantage”.

To the owners of high quality cars, their private knowledge can indeed become a burden. Their information advantage is in fact a handicap. This looks like an exception to the rule that more information is always better. Can information become a burden even when it is a basis for action (as distinct from the psychological motives discussed in Chapters 4 and 17)? The answer is no. An individual is strictly better off with information than without—*ceteris paribus*. The burden is not knowledge *per se*, but the fact that others know about the information advantage. To use the language of Chapter 8, it is always an advantage to have some fundamental information, but the problem lies with *higher-order information*, i.e. with what others believe an individual knows. Being thought to know may hurt an owner if he knows the car’s quality—but it is likely to hurt him even more if he does not know the quality. While the secret insider does best, the known insider does less well, and the (unjustly) suspected insider is to be pitied.

In the morning paper, Bob reads an advertisement: “Fertility test for men—results within 5 minutes”. He pauses for a moment. Would he want to know? Just for himself, yes, he might. Better know the truth and adapt one’s life accordingly, he thinks. But what about Alice? Would he manage to keep a negative result secret from her? Probably not. And would he still have a chance with her? “Better not to test!” he concludes.

Gains from trade and qualities: How adverse is selection?

Our preceding analysis was based on two very simplistic assumptions:

- Uniform distributions of qualities;
- Constant gains from trade, a, across qualities.

If we modify these assumptions, the adverse selection problem becomes more or less severe. The impact of both factors, quality distribution and gains from trade, are ambiguous. A higher fraction of high quality cars or a higher potential for gains from trade (a) reduces the range of qualities remaining unsold, but increases the efficiency losses on these unsold qualities. A practical example may illustrate the impact of these two factors.

The example of life insurance

In life insurance, the gains from trade tend to outweigh the potential losses from adverse selection. In industrial countries a small minority of adults are "short-lived" in the sense that they are likely to die before the age of 65. Even if these short-lived individuals have a hunch and tend to buy more life insurance than the rest of the population, a life insurer is faced with mostly good risks. The gains of trade—the increase in utility from taking out insurance—are sufficiently large to attract many people who think they will survive. The bad risks can hide behind the good risks; premia in their presence will be somewhat higher than in their absence, but not so high as to repel the good risks.

In reality, individuals do not only differ with respect to mortality, but also with respect to the potential gains from trade. Individuals with a high income tend to have a higher demand for life insurance. In addition, mortality risk and gains from trade are correlated: Richer individuals also tend to live longer. Such real world complications will play a role in our Application 13 E.

Sorting versus pooling

We have shown that when agents are heterogeneous and privately know their individual "type", adverse selection can prevent markets from functioning. This is a particularly inefficient outcome, when adverse selection prevents insurance of a contingency such as bad health or living too long. We will discuss (Section 13 E) why health insurance and annuities markets seem to form informational asymmetries, namely because individuals have a good idea about their health status and their remaining life expectancy. This may hinder risk-averse individuals from reaping the potential gains from trade, or from pooling their risks in the insurance example.

There are two responses to potential adverse selection in insurance markets:

1 Making insurance compulsory;
2 Prohibiting the use of some information.

Making insurance compulsory for individuals avoids the best risks not taking out insurance. The insurer can price the insurance policy on the basis of the average damage probability of the whole population without making a loss.

Prohibiting the use of information prevents insurers from asking applicants questions about family history or from demanding medical tests, with a view to sorting applicants and applying "risk-based" rates to different types.

It is important to distinguish two different motives for mandatory insurance and for forbidding the use of information:

- **The "insurance destruction effect"**
  The availability of information on individuals' risk characteristics may prevent insurance due to market failure. This is true in the case where individuals have (and are known to have) *private* information (the lemons' market case). But insurance may also fail when information is *public*. Individuals may then be able to insure at "fair" rates (given their risk type), but not against the

risk of falling into a high-risk category. We will analyze this effect below (Section 13 D).

- **Redistribution among good and bad risks**
  In most circumstances, notably in health insurance and life insurance, the good risks are well-off, whereas the bad risks are rather poor. Forbidding the use of information or mandatory pooling are similar to a subsidy from the good to the bad risks, i.e. a transfer from the rich to the poor. Interestingly, there are exceptions where the good risks are poor and the bad risks rich. One case we will discuss below (Section 13 E) is annuities/pension funds.

13 D Application: The insurance destruction effect

Bob has invited three friends for a round of cards. While Bob is dealing, Alice arrives, plate in hand. "I'll give you four pizza slices. If you all manage to keep quiet for one minute, you'll each get a piece of pizza," she says. "Otherwise, I'll give all four slices to the same guy . . . " All four remain silent, suggesting they have agreed to the one slice each arrangement. Then Alice adds ". . . to the one who has, say, the Queen of Spades." "Me! That's me!" Eugene cries, brandishing the Queen of Spades in the air and quickly cashing in the four pizza slices. His friends complain about the unjust outcome until Alice offers four more pizza slices. "The rules are the same," she says, "except that this time, if someone breaks the silence, I'll give all four slices to . . . " "Stop! Don't tell us!" Eugene interrupts, "We're better off not knowing!"

Destructive information?

Is it possible that a group of people would prefer not to have a piece of information? As the example shows, the answer is clearly yes. In an uninformed group, risk-sharing arrangements (one pizza slice per person) are possible which are not attainable, once some information on outcomes is released. This is the insurance destruction effect or Hirshleifer Effect, as described in Hirshleifer and Riley (1992) and Brunnermeier (2001).

Does the insurance destruction effect mean that information can have a negative value? The simple answer is no. The individual choice in the above example is not between having the information or not. The choice is between *everybody* having the information or *nobody*. An individual who privately acquires a piece of information is still better off. An individual who acquires a piece of information together with all (or sufficiently many) others, may be worse off. As our example shows, it may not even make much of a difference whether the information remains private to the individual recipients (only Eugene knows he has the Queen of Spades) or is made public.

We will try to show the insurance destruction effect within a simple model. In this model, we take individuals' information as exogenous; they receive a test result without having asked for it. We do not look at the wider issue of individuals' incentives to *gather* (free or costly) information. Such incentives are analyzed in Doherty and Thistle (1996).

A simple model

In a population some individuals have a high innate probability (a "gene") to develop a certain disease. The probability of others contracting the disease is low. Fraction g of the population belongs to the high-risk group that carries the pathogenic gene and develops the disease with probability $\pi > \frac{1}{2}$.[90] The $1-g$ low-risk individuals contract the disease with probability $1-\pi$. Population statistics are summarized in the following table:

Size	Risk group	Pr(healthy)	Pr(sick)
$1-g$	Low risk	$\pi > \frac{1}{2}$	$1-\pi$
g	High risk	$1-\pi$	π

In the terms of Figure 13.1 used above, the time line of events is the following: Individuals have an initial wealth of Y (which they may use later to buy insurance). In $t = 0$, individuals learn their types. In $t = 1$, individuals can trade (buy insurance). In $t = 2$, individuals develop the disease or remain healthy. Health costs are $d = \{0, D\}$ for the healthy and the sick, respectively. All these assumptions, including the different probabilities, are common knowledge.

We assume that there is an insurer who offers insurance at competitive (zero profit) rates. The insurer offers to pay the damage amount D to those affected by the disease. The premium charged under competitive conditions is equal to the expected health costs of the applicant.

Insured individuals pay the premium, but can recover their cost D from the insurer should they go on to develop the disease. Those who remain uninsured save the premium, but receive nothing should they contract the disease. In addition, they suffer a psychological cost of not being insured. We model this cost as a simple percentage v of their expected cost of sickness. This is a crude device to capture risk aversion while keeping utility functions linear. Note that the expected cost of sickness depends on individuals' beliefs regarding their types.

No information

First we look at the benchmark case in which individuals do not know whether they belong to the low-risk or to the high-risk category. This is the case when genetic testing is not possible. The insurer who cannot distinguish individuals faces a population with an average risk:

$$\bar{\pi} = [g\pi + (1-g)(1-\pi)]$$

The insurer charges a premium $z = \bar{\pi} D$. An individual buys insurance if:

$$z < (1 + v)\bar{\pi} D,$$

that is, if the premium is smaller than the expected damage plus the disutility of being uninsured. As $z = \bar{\pi} D$, all individuals buy insurance. The saved deadweight cost vD represents the gains from trade. Individuals' expected utility is:

$$U = Y - \bar{\pi} D.$$

This utility level is reached (i) when both categories are *unobservable* to the insurer as well as to individuals themselves, or (ii) when neither individuals nor insurers are allowed to use their information about categories, i.e. under *mandatory* insurance (see below).

Public information

The other extreme is public information about risk categories. This is the case where genetic test results are made public (or where the presence or absence of a gene is obvious, as in the case having blue or brown eyes). The insurer charges different premia to high- and low-risk individuals respectively:

$$z_L = (1 - \pi)D,$$
$$z_H = \pi D.$$

At these terms both types buy insurance, as the expected gains exceed the premia:

$$(1 + v)(1 - \pi)D - z_L = v(1 - \pi)D > 0,$$
$$(1 + v)\pi D - z_H = v\pi D > 0.$$

Expected utility $\hat{U}$ to individuals with observable types are:

$$\hat{U}_L = Y - (1 - \pi)D,$$
$$\hat{U}_H = Y - \pi D.$$

While both types buy insurance, they obtain fairly different terms. The degree of "discrimination" depends on π, the probability that a carrier of the bad gene develops the disease. If $\pi = 0.5$, both groups do equally well. But the higher π is, the greater is the difference in expected utility between the two groups. The insurance fund only insures against developing the disease, *given* someone's type. Unlike insurance under unobservable types, the insurance fund does not insure against carrying the bad gene. This is the insurance destruction effect of public information.

Asymmetric information: Pooling equilibrium

Asymmetric information lies between the two extremes of no information and of public information. It means that (i) individuals know their types, and (ii) the insurer cannot distinguish types but knows that individuals can. As we have shown above, adverse selection may occur under asymmetric information. An insurer charging premia for average risk must be careful not to end up with only bad risks.

The insurer has two strategies available: Insuring both categories (pooling) or insuring only the high risks (separation).[91] We first look at pooling. A pooling premium z^{LH} must lie in the interval between the minimum that yields non-negative profit to the insurer and the maximum that is still acceptable to low-risk types:

$$[g\pi + (1-g)(1-\pi)]D \leq z^{LH} \leq (1+v)\,(1-\pi)D.$$

A pooling equilibrium exists if this condition can be satisfied. The condition can be simplified to:

$$\frac{v}{g} \geq \frac{2\pi - 1}{1-\pi}. \tag{13.1}$$

High risk aversion, roughly measured by v, facilitates pooling, while a high frequency of the gene, g, makes pooling unlikely. Most detrimental for pooling equilibria is a high π, i.e. a high incidence of the disease among carriers of the gene.

If a pooling equilibrium exists the competitive insurance premium:

$$z^{LH} = [g\pi + (1-g)(1-\pi)]D = \bar{\pi}D,$$

reflects the average risk of the population. Under condition (13.1) both types insure.

The pooling equilibrium is in all respects identical to the no-information equilibrium discussed above. Although individuals know their types, they have no opportunity to make use of their information. As in the no-information equilibrium, both risk types have a utility level

$$U_L^{LH} = U_H^{LH} = Y - \bar{\pi}D.$$

Carriers of the gene do better than under public information, while non-carriers do worse: $U_H^{LH} > \hat{U}_H$, while $U_L^{LH} < \hat{U}_L$: High-risk types can hide in the total population, passing some of their insurance costs to the low risks. This is the same externality we found among successful sellers in the "lemons' market" example.

Asymmetric information: Separating equilibrium

The second option to the insurer under asymmetric information is to insure only high-risk individuals. It is the sole available option, when condition (13.1) is not met and no equilibrium exists in which both types buy insurance. The insurer is then unable to insure the low-risk category without making a loss. The competitive (zero-profit) premium for high-risk individuals only is $z_H^H = \pi D$, the same as under public information. The two types obtain utility levels:

$$\begin{aligned} \breve{U}_L &= Y - (1 - v)(1 - \pi)D, \\ \breve{U}_H &= Y - \pi D. \end{aligned}$$

As in the public information case, carriers of the gene appear to be insured, but they pay for their risk type with high premia. With a high incidence of the disease among carriers of the gene, π, there is not much of an insurance effect. With $\pi \to 1$, the carriers of the gene merely pre-finance the cost of the disease, rather than sharing it with healthy individuals. While the separating equilibrium is expensive to the g high-risk types, it is not wholly satisfactory to the $1 - g$ low-risk individuals either. These remain uninsured and incur a deadweight loss of $(1 - \pi)D$, representing the adverse selection effect.

Mandatory insurance

There are several reasons to look at mandatory insurance:

- A pooling equilibrium might not exist, and the high premium for high-risk individuals in a separating equilibrium may be considered socially unacceptable.
- A pooling equilibrium might exist, but the insurer may prefer a separating equilibrium (due to imperfect competition, for example).
- High-risk individuals may in fact not be able to finance the insurance premium in a separating equilibrium; this is the case if $Y < \pi D$.
- Individuals might prefer to be insured *ex ante*, i.e. before they know their types.

Mandatory insurance means that everybody has to buy insurance. We also assume that the insurer cannot (or is not allowed to) distinguish risk categories, and that insurance is available at the competitive rate $\bar{z} = \bar{\pi} D$ for all individuals. Under mandatory insurance, as in the no-information setting, the outcome is identical to the pooling equilibrium. Mandatory insurance thus implements the pooling equilibrium when it would not exist (when (13.1) is not met) or when an insurer would not offer a pooling contract.

Utility comparison . . .

Which of the three possible scenarios would individuals prefer if they could choose: (i) public information, (ii) a pooling equilibrium (resulting from either asymmetric information, from no-information, or from mandatory insurance), or (iii) a separating equilibrium?

Table 13.1 Utility levels in different equilibria as a function of individual types

	Public information	*Pooling equilibrium*	*Separating equilibrium*
Ex ante	$Y - \bar{\pi}D$	$Y - \bar{\pi}D$	$Y - [\bar{\pi} + v(1-g)(1-\pi)]D$
Low risk	$Y - (1-\pi)D$	$Y - \bar{\pi}D$	$Y - (1+v)(1-\pi)D$
High risk	$Y - \pi D$	$Y - \bar{\pi}D$	$Y - \pi D$

The answer is likely to depend on what an individual knows about her risk type. We therefore distinguish between *ex ante*, low-risk and high-risk. The *ex ante* view would correspond to a decision to be taken *before* individuals are tested and discover their risk. The results of a utility comparison are summarized in Table 13.1 (recall that $(1-\pi) < \bar{\pi} < \pi$).

... before testing

As Table 13.1 shows, *ex ante*, all individuals would want to *pool*, either voluntarily or under mandatory insurance. Pooling permits them to insure unconditionally against the gene and the disease. Under pooling, individuals' utility becomes independent of their types.

From the *ex ante* view, the *public information* equilibrium appears to be equivalent to pooling. Yet, it is only equivalent in terms of *expected* utility, but not in terms of *realized* utility. In the public information equilibrium an individual's payoff, *ex ante*, is risky (as the risk of carrying the gene is not insured), while in a pooling equilibrium it is safe. If individuals dislike risk, the pooling equilibrium dominates the public information equilibrium. It would be Pareto-optimal to suppress public information or make insurance mandatory.

From an *ex ante* perspective, the outcome to avoid is the *separating equilibrium*. In the separating equilibrium there is a deadweight loss, as low-risk individuals go uninsured.[92] The possibility of testing thus may eliminate the ability to insure. When no pooling equilibrium exists, mandatory insurance Pareto dominates voluntary insurance.

... after testing

Once individuals are tested, their interests differ:

- *Low-risk* individuals prefer the public information setting. The ranking between the pooling and the separating equilibrium (in which low-risk individuals remain uninsured) is ambiguous. Low-risk individuals prefer the pooling equilibrium if one exists, i.e. if (13.1) is met. Low-risk individuals would agree to mandatory insurance if a pooling equilibrium exists but is not implemented by the insurer.

- *High-risk* individuals prefer a pooling equilibrium to either a separating equilibrium or public information. They benefit from mandatory insurance whenever a pooling equilibrium does not exist or is not implemented.

Conclusion: Genetic testing and insurance destruction

As the utility comparison showed, mandatory insurance or, equivalently, the prohibition of genetic testing can be Pareto improvements from the *ex ante* view of individuals who do not yet know their risk type (who have not tested). The critical condition is that the market fails to pool individuals. Such failure is caused by the insurance destruction effect and by adverse selection. As long as individuals do not know whether they carry a pathogenic gene, they are better off if no test is available, if the test is forbidden or if insurance is mandatory.

These results apply, it should be stressed, in the case where there is no available treatment against the disease. If a treatment is available, mandatory insurance is obviously preferable to a ban on genetic testing. Preventing insurance companies from using test results, as a "softer" form of pooling, is unlikely to help. If the market equilibrium is not characterized by pooling anyway, individuals will self-select into low-risk and high-risk groups, the former remaining uninsured, the latter paying high premia—if they can afford them.

The insurance destruction effect may become increasingly important, as genetic testing becomes available for a growing number of yet incurable conditions like Huntington's disease or Alzheimer's. The higher the fraction of carriers of identified genes in the population (g) and the more fatal the condition (π), the lower the chances that the pooling of risks will survive in a free market (see (13.1)). *Ex ante*, there is a case for mandatory insurance of genetic disorders. Mandatory insurance would also permit testing without fear of adverse insurance effects. Unfortunately, *ex post* it may be difficult to find a majority for mandatory insurance. Once individuals know they are in the low-risk category (as the majority of the population is), they are likely to resist mandatory insurance, and opt out of voluntary insurance, too.

Box 13.1 Prophesy without prevention

In 1968, Alice and Nancy Wexler were in their twenties when their mother Leonore was diagnosed with Huntington's disease. Huntington's disease, affecting one in about 20,000, is an incurable, debilitating and fatal neurological disorder caused by a dominant gene. The Wexler sisters faced a 50-50 chance of carrying the critical gene and some day contracting the disease themselves. They found the uncertainty quite unbearable. Their father Milton, a clinical psychologist and psychoanalyst, and the two sisters decided to devote the rest of their lives to studying the disease. Eventually their efforts led, in 1983, to the discovery of its genetic markers and the possibility, finally, of testing.

This discovery placed them in a dilemma: Should they find out if they were to develop the disease? Later, in an interview, Nancy Wexler gave their answer: "When we were trying to develop a test, we assumed we'd both take it. Then once the test existed, we were thinking about it differently. Our family talked an enormous amount about the consequences. Even if you live at risk all of your life, and you've thought about it and cried about it, there's a certain amount of denial that helps you get through the day. Being tested can take that away" (Avins, 1999).

Whether or not to be tested is only one of the tormenting questions for the potential carriers of such genes. Another is: Given a test is positive or a relative is diagnosed with the disease, what should he or she do? For example, should a surgeon or an airline captain stop practising immediately, when they find out they carry the gene? And: What can and what shall be kept secret? In the Wexler sisters' case, that issue was eventually resolved when Alice, in 1996, published her book *Mapping Fate*, a family memoir (Wexler, 1996).

Many others are struggling with similar questions. Since 1983 genes have been located for many other diseases or crippling conditions like cystic fibrosis, manic depression, and Alzheimer's disease. Each year the list grows longer. As Nancy Wexler put it: "We are in the problematic position of being able to identify those who carry genes for fatal or crippling conditions without being able to prevent or treat the diseases" (Wexler, 1989).

The possibility of genetic testing certainly raises the most difficult questions for those who are afraid to test positively or have, in fact, done so. Yet, for all others, the existence of tests may mean that there will be no insurance against the consequences of those diseases, at least not in a free market. We ought therefore to think just as hard about the economics of our genes as the Wexler sisters (see Box 13.1) thought about Huntington's disease.

13 E Application: Annuities

13 E a Life is uncertain

In his autobiography, Robert Graves (1929) reports the following example of risk-sharing in the trenches during World War I (1914–1918)[93]:

> Beaumont had been telling how he had won five pounds' worth of francs in the sweepstake after the Rue de Bois show [battle]: a sweepstake of the sort that leaves no bitterness behind it. Before a show, the platoon pools all its available cash and the survivors divide it afterwards. Those who are killed

> can't complain, the wounded would have given far more than that to escape as they have, and the unwounded regard the money as a consolation prize for still being here.

Ordinary life is not quite as dangerous as serving on the front line during World War I. Still, life is uncertain; typically, and fortunately, we do not know how long we will live. We may benefit from something similar to the sweepstake mechanism described above which made sure that resources left behind by the soldiers killed in action were made available to their luckier comrades.

What is an annuity?

Consider an insurance product called an *annuity*. The buyer of an annuity pays an amount of money (the annuity capital) in exchange for the promise of receiving a regular benefit payment as long as (s)he lives. The annuity automatically expires at death even when the annuity capital has not been "used up". But the annuity continues to be paid should its holder live much longer than expected, even after the paid-in capital has been depleted. Provided a large number of individuals buy annuity contracts, their mortality risk can be pooled. The above example of risk sharing in the trenches, in essence, was an annuity contract. Those who survived shared the money of those who died early. This is the formula, roughly speaking, of modern pension funds.

Annuity versus life insurance

An annuity is also the opposite of life insurance. Under an annuity contract, the insurance premium is a capital stock that will be translated into periodic payments until death. Under a life insurance contract, by contrast, periodic payments are due until death after which the survivors receive a capital payment specified in the life insurance contract.

Annuities and life insurance are not free from hidden information, as the buyers of annuity contracts presumably know their life expectancy (as distinct from their effective life span) much better than the seller. Having parents in their 90s is a good indicator of a long life. So is a high income—wealth is health—at least in expected terms. Economic theory predicts that people with a low life expectancy are more likely to buy life insurance, while those with a long life expectancy should purchase annuity contracts.

Here we focus on selection in the presence of existing schemes, rather than on the design of the optimal contract to address potential selection. We start with a simple stylized annuity contract in the absence of mortality differences, and then discuss the consequences of hidden information.

13 E b The annuity contract

Assumptions

Let us first assume that all individuals have the same life expectancy. The easiest way to capture this in a two-period model is by the *survival probability* $\Psi < 1$ to reach the second period. We also suppose that insurance markets are competitive, leading to zero profits in equilibrium. Unlike individual agents, insurance companies are risk-neutral and can operate at no costs. The risk-free market interest rate is denoted by r.

If the number of people is large enough, the fraction of individuals who survive corresponds to the survival rate of a single individual Ψ. For example, if the probability of survival is 60 per cent for each person, it follows that 60 per cent of all people survive. Mathematically, this regularity is called the *Law of Large Numbers*, but the claim looks intuitive even without proof.

The *annuity contract* is similar to the one described by Robert Graves during World War I. The invested capital is pooled and invested in the market at the prevailing market interest rate. Those who die receive nothing and the survivors divide the capital and its return according to their respective investment share.

The annuity contract

Let us now look at the annuity contract in more formal terms. In exchange for a capital payment of 1 dollar in the first period, an individual receives:

- $1 + r^A$ dollars in case of survival,
- 0 in case of death.

In equilibrium, the zero profit assumption requires that the expected return of the annuity must be equal to the market return r. The formal condition can be written as

$$\overbrace{\Psi \times (1 + r^A)}^{\text{survival}} + \overbrace{(1 - \Psi) \times 0}^{\text{death}} = (1 + r)$$

$$\Rightarrow \quad 1 + r^A = \frac{1 + r}{\Psi} > 1 + r$$

In the event of survival, returns on the annuity are higher than an investment at the market interest rate r. If a person derives no utility from leaving behind assets in the event of death, he will always invest his entire savings in an annuity contract.

If $\Psi = 1/2$ and $r = 0$, for example, the buyer of an annuity can consume twice as much in the event of survival as without such insurance. The higher consumption of survivors is “financed” by their deceased peers. And of course, there are no wasted resources if you take the annuitants’ viewpoint (or no unintended bequests if you take that of the annuitants’ heirs).

13 E c Who should buy an annuity contract?

Short- and long-lived

The assumption of homogenous mortality rates in the population is obviously unrealistic. Some differences in life expectancy can be observed (such as differences in survival rates between men and women). Others are not observable or cannot not be contracted upon. To illustrate the consequences of heterogeneous, but unobservable mortality, let us consider two types of individuals:

- *Long-lived* individuals have a survival probability of $\overline{\Psi} = \frac{2}{3}$, and represent 50 per cent of the population.
- *Short-lived* individuals have a survival rate of $\underline{\Psi} = \frac{1}{3}$ and make up for the remaining 50 per cent of the population.

The average survival probability in the population is $\frac{\overline{\Psi}+\underline{\Psi}}{2} = \frac{1}{2}$.

The money value of an annuity

To compute the "value" of an annuity, let us introduce the concept of a "money's worth ratio" (MWR):

The money's worth ratio of an annuity is the expected present value of the annuity payments relative to the money invested in the annuity contract. An MWR of 1 says that for each dollar invested you get an expected present value of 1 in the form of annuities.

In our simple two-period model the MWR can be computed as follows:

$$\begin{aligned} \text{MWR} &= \Psi \frac{1}{1+r}(1+r^A) \\ &= \Psi \frac{1}{1+r} \frac{(1+r)}{\tilde{\Psi}} = \frac{\Psi}{\tilde{\Psi}} \end{aligned}$$

where Ψ corresponds to the individual survival probability, and $\tilde{\Psi}$ is the relevant rate on which the insurance company bases its computations. In a perfectly competitive market with uniform survival rates, $\tilde{\Psi} = \Psi$, and the MWR is 1:

$$\text{MWR} = \Psi \frac{1}{1+r}(1+r^A) = \Psi \frac{1}{1+r} \frac{1+r}{\Psi} = 1$$

Let us assume that the insurance company only observes the average survival rate $\tilde{\Psi} = \frac{1}{2}$. The implied MWRs for the two groups of the population are then:

- *Long-lived:* MWR $= \frac{2/3}{1/2} = \frac{4}{3} > 1$;
- *Short-lived:* MWR $= \frac{1/3}{1/2} = \frac{2}{3} < 1$.

What is the optimal response of the individuals? The long-lived will purchase an annuity in any case. For the short-lived the situation is less obvious. For each invested dollar, the individual only gets back $2/3$ dollar and the annuity looks like a bad deal. But what is her alternative? If she invests in the market asset, the present value of her consumption is merely $1/3$, as in $2/3$ of the cases the savings are wasted because of death. So the $2/3$ is still better than the expected present value of the return from investing in the market asset.

Should one purchase an unfair annuity?

At first sight, we have the puzzling result that even an "unfair" contract increases an individual's welfare. Moreover, we have not even taken into account that an annuity provides an insurance against outliving one's assets in old age. If individuals are risk averse, the case for an annuity becomes even stronger. Yet, in reality we do observe adverse selection in annuity markets, as Friedman and Warshawsky (1990) and Finkelstein and Poterba (2004), among many others, have clearly demonstrated. People choosing to buy an annuity have a considerably higher life expectancy than the average population. So, here are some reasons why this might be the case:

- We have assumed that the insurance company has no operating costs. Even a small fee on annuities can render the previously beneficial contract unattractive for short-lived individuals. See Problem 13.5 for an illustration.
- We have also assumed that the individual derives no utility from unused savings in the event of death. In reality, many people also have a bequest motive and would like to leave behind some assets. The optimal degree of annuitization for people with a desire to leave a bequest involves a trade-off between the risk of dying too early (in which case a full annuitization means a very low bequest) and the risk of living too long and thus outliving one's assets.
- The demand for an annuity depends on the entire mortality profile, and not just on a single number Ψ. As an (unrealistic) illustration consider a person with a below average, but known life span. Her demand for an annuity will be much smaller (in fact zero) than that of an individual with the same, but uncertain life expectancy: She has no need for insurance and does not risk to leave unintended bequests behind.
- There are also psychological reasons for the low demand for annuities. In general, people do not like to hand over control of their savings, as would be the case for an annuity. They may also find it unfair that assets remain with the insurer in the event of premature death.

If there *is* adverse selection, the market outcome is inefficient: The short-lived are driven out of the market (like the high quality cars) and cannot adequately insure against longevity. In many funded pension schemes, the problem is "solved" by compulsion. In essence, the pension benefits only depend on the paid-in capital, but not the individual's life expectancy. As in other mandated insurance schemes, the good risks subsidize the bad risks. But there is a fundamental difference with, say, health insurance, as we will argue in the following section.

13 E d When construction workers subsidize professors

Mandatory annuity schemes redistribute from the short-lived (the good risks) to the long-lived (the bad risks), just as a mandatory health insurance does from the healthy (the good risks) to the sickly (the bad risks). There is a very important difference between health insurance and longevity insurance, however. In the health insurance case, the good risks are the lucky ones (because they are healthy), whereas in an annuity scheme the good risks are the unlucky ones (because they have a shorter life expectancy). That is also the reason why most compulsory health insurance systems try to avoid risk selection (even where risk is observable, the pooling of risks is beneficial to the more disadvantaged).

When the lucky ones are the bad risks

The good risks in the annuity contract are the short-lived, that is those who are doing badly. The bad risks, on the other hand, are doing well as they can expect to live longer. The pooling of risks leads to a transfer to annuitants with an *ex ante* higher life expectancy. As longevity is also positively correlated with income and education, funded pension systems redistribute resources to employees with a high level of education and income, such as professors, at the expense of blue-collar workers with low education and income levels.[94] The well-offs (= bad risks) receive much more than they had paid in, the less well-offs (= good risks) much less.

A matter of justice?

Should construction workers really subsidize professors? After all, information is at least partly observable. Why not take it into account and enforce a system that allows for the sorting of risks? This is an analytically difficult question. An answer depends, among other things, on both the *individuals' utility function* and the *social welfare function*. The individual utility function is important to assess the value of the insurance against outliving one's assets. The social welfare function, which aggregates individual utilities, specifies how much weight should be given to different subgroups of the population.

In a Rawlsian social welfare function, for example, society's welfare depends on the utility of the least well-off individual. In this case, sorting is clearly better than pooling, because it increases the welfare of the short-lived. In the utilitarian case, in which each individual's utility is given equal weight, the outcome is ambiguous: A transfer from the short-lived to the long-lived increases the inequality of a society and thus decreases social welfare. But it also increases the value of the insurance against longevity. As the long-lived have a greater weight in the social welfare function due to their higher life expectancy, the social welfare is raised, and the net effect is unclear. In any case, the higher the inequality aversion of a society, the more likely it is that sorting would lead to a better outcome than pooling, as will be illustrated in Problem 13.3. The importance of the social welfare function shows that the "ranking" of the two options is not possible without a value judgment.

Is it real?

How much do construction workers suffer in existing funded pension schemes? In many countries, funded pension plans are tied in some way or another to the employer. As the workforce of a firm is generally more homogeneous than the whole population, there is already some implicit sorting in the scheme. Professors are not likely to be covered by the same employer pension plan as construction workers. In fact, construction worker pension plans in Switzerland, for example, offer more generous early retirement options, as these can be "financed" by a shorter average life expectancy of the covered workers.

13 E e Selection with more than one dimension

Annuities versus life insurance

Annuities protect individuals from the risk of living too long (longevity) and outliving one's assets, while its opposite, life insurance, protects individuals and especially their dependants from the risk of early death. We have argued above that the annuity and the life insurance markets are potentially both plagued by adverse selection. The theory predicts that individuals with a higher than average life expectancy should be more likely to purchase an annuity while the buyers of life insurance can be expected to have a shorter life than the general population.

In fact, evidence from several different countries has shown that this is indeed the case for annuity contracts, but surprisingly not for life insurance policies. Annuity buyers are longer-lived, i.e. they represent a bad risk compared to the general population. But the buyers of life insurance, which insures the opposite of a longevity risk, are *also* longer-lived than the average comparable population and therefore a good risk. Does this mean that adverse selection effects are unimportant? Or, is there any way to reconcile these puzzling findings with economic theory?

Risk and preference as dimensions of type

In a recent paper, Finkelstein and McGarry (2006) offer an interesting explanation of such apparent selection differences. They argue that there is an additional dimension of private information, based on individual preferences. In their words:

> Two types of people purchase insurance: individuals with private information that they are high risk and individuals with private information that they have high insurance preferences. *Ex post*, the former are higher risk than insurance companies expect, while the latter are lower risk.

This is akin to the distinction made above between (i) the distribution of types and (ii) the potential gains from trade as the two determinants of how adverse selection will be under asymmetric information. Finkelstein and McGarry remind us of two things. First, individuals may have private information not only on their type, but also on their potential gain from trade. Second, there may be a correlation between type and gains from trade. In life insurance, wealthy individuals, due to their lower mortality risk would tend to opt out of a pooling arrangement, were it not for their higher demand for insurance.

Preference may offset asymmetric information

Finkelstein and McGarry use data on the long-term care insurance market to demonstrate the existence of such multiple dimensions of private information. By using subjective assessments of individuals' odds to enter a nursing home, they show that there is clear evidence of *asymmetric information* in long-term care insurance. However, the same data does not provide any evidence for *adverse selection*. There is no positive correlation between individuals' insurance coverage and their risk experience as would be the case if selection mattered. The authors conclude:

> A resolution to this apparent puzzle may be that individuals have private information about a second determinant of insurance purchase, their preferences for insurance coverage, as well as private information about their risk type. If individuals with private information that they have strong tastes for insurance are lower risk than the insurance company would predict, private information about risk type and private information about insurance preferences can operate in offsetting directions to produce an equilibrium in which those with more insurance are not more likely to experience risk.

The importance of correlation

The selection puzzle in the cases of life insurance and annuities can potentially be explained by the possibility that people with a high life expectancy also have a stronger preference for insurance. While there is no direct empirical evidence of such a correlation between longevity and insurance preference, it does not seem implausible. After all, life expectancy does not only depend on our genes, but also on how prudent we are. So if we value personal security highly, we are at the same time more likely to purchase insurance and to live longer. Insurance preference and risk go in the same direction for annuity contracts, but in opposite directions for life insurance. For the insurance company the preference-selection dimension reinforces the problem of adverse selection in the former case, but alleviates it in the latter.

13 F Conclusions and further reading

It is dangerous to trade with a better informed party. Therefore, markets function badly, whenever one party is known or suspected to have an information advantage. Some goods, often those of the best quality, are not sold; others are overpaid. At the limit, a market may break down completely.

The reason why markets do not break down are the potential gains from trade. The potential buyer may need some good more urgently than the seller, or the seller may need cash more urgently than the buyer. When such gains from trade exceed the expected loss from trading with a better informed party, deals take place despite the information asymmetry. Gains from trade have an ambiguous influence, though. They increase the scope for trading, but they also mean a higher deadweight loss on the units that are not sold due to hidden information.

The present chapter also casts some further light on Chapter 7. There we argued that the market aggregates individuals' private information. Informed individuals find profitable trading opportunities, thereby giving away their information. But why should non-informed individuals trade? In most of Chapter 7 we assumed that non-informed individuals are just naïve and do not expect others to be better informed. The present chapter suggests a more plausible explanation: For some individuals there might be gains from trade distinct from the trading profit. Individuals may, for example, sell (buy) securities because they need (have excess) liquid cash. In finance, such traders with private trading motives are often called noise traders.[95] They are essential to avoid "no trade" scenarios. If a private trading motive is stronger than the potential fear of losing in a deal with an insider, trade occurs despite asymmetric information.

Problems of self-selection are particularly important in the insurance sector. The insurance destruction effect is an example where more (private or public) information may reduce welfare from an *ex ante* point of view. In health insurance, for example, the availability of tests for conditions like the presence of a faulty gene may destroy the possibility of insurance against carrying the gene. The ever growing possibilities of genetic testing are a mixed blessing, therefore.

But even without formal tests, selection problems exist. Individuals can make a good guess of their life expectancy or of their health risks. Puzzlingly, such private information leads to adverse selection in some markets (annuities and health insurance), but not in closely related markets, such as life insurance. The explanation seems to lie in a correlation between individual risk and insurance demand. When the better risks have a high demand for insurance, self-selection may be beneficial rather than adverse. When the bad risks have a lower demand for insurance, or lower wealth, difficult problems of public policy arise. One example is the subsidization of bad risks (long-lived professors) by the good risks (short-lived construction workers) in a pension fund.

There is a wealth of literature, theoretical and empirical, on hidden information. For a more detailed treatment of the theory we suggest Laffont (1993) and Mas-Colell, Whinston and Green (1995). Wolfstetter (1999) shows the links between hidden information, general microeconomics, and auction theory. Bolton and Dewatripont (2005) treat hidden information as a background for optimal contracting, our next subject to be discussed in Chapter 14. In addition to these theoretical treatments, there are countless empirical studies in areas like health economics, labor economics, insurance, banking, and others.

Checklist: Concepts introduced in this chapter

- Hidden information
- Market failure
- Adverse selection

- Realized and unrealized gains from trade
- Pooling and separating (sorting) equilibria
- Insurance destruction effect (Hirshleifer Effect)
- Annuities
- Money's worth ratio
- Private information versus insurance preferences

13 G Problem sets: Buying the cat in a bag

For solutions see: www.alicebob.info.

Problem 13.1 Getting a lemon
Assume that (i) the qualities of cars are unobservable and uniformly distributed in the interval between \$0 and \$10,000 and (ii) the value of a car to a buyer (in \$) is: $5,000 + 0.5V_S$. Buyers make take-it-or-leave-it offers.

(a) What is the equilibrium price and quantity?
(b) Is there adverse selection?

Problem 13.2 Gains from trade and adverse selection
Assume that (i) qualities of cars are unobservable and uniformly distributed in the interval between \$5,000 and \$10,000 and (ii) the value to buyers is k times the value to sellers, where $1 < k < 2$ is a constant.

(a) Draw a graph with the price bid by buyers (x-axis) and the expected value (to buyers) of cars offered.
(b) How does the social loss due to asymmetric information change with different values for k?

Draw a graph with the price bid by buyers (x-axis) and the value (to sellers and to buyers, respectively)

Problem 13.3 Professor versus plumber
This problem is based on Wolfstetter (1999, p. 248). Assume that (i) the productivity of a university professor is negatively correlated with his productivity as a self-employed plumber, (ii) universities pay productivity wages. Show that there are too many professors.

Problem 13.4 Life insurance versus pension funds
Explain the difference between life insurance and retirement insurance.

(a) What risks do they insure against?
(b) What types of people are good risks and what types of people are bad risks for the insurers?

(c) Which of both types of insurance is more exposed to asymmetric information?

Problem 13.5 Annuity contract with operating cost

Assume that, as in Section 13 E, the economy consists of two types of individuals differing in their probability, p to die old, rather than young. Half the population is *short-lived* ($p = 1/3$) and the other half *long-lived* ($p = 2/3$). Individuals are able to insure against longevity. However each insurance contract involves a fee f for each dollar invested.

(a) Calculate the fee that makes short-lived indifferent between buying the insurance or not.
(b) How does the result change if only $1/4$ of the individuals are short-lived?

Problem 13.6 Should construction workers subsidize professors?

Reconsider the example of long-lived (professors) and short-lived (construction workers) individuals as presented in Section 13 E c. Assume again that there are as many professors as construction workers in the population.

The second period utility function of individual i is given by $u_i = \Psi_i \ln(\text{annuity}_i)$, where annuity_i is the annuity stream resulting from investing one dollar into an annuity contract.

The social welfare function is given by $S = \sum_i \frac{u_i^{1-e}-1}{1-e}$, where e is the degree of inequality aversion of a society. For $e = 0$, the social welfare function is utilitarian, for $e = 1$ it is logarithmic, and for $e = \infty$ it is Rawlsian (S is equal to the utility of the worst-off individual in society).

(a) Compute the social welfare of the population under pooling and sorting as a function of e.
(b) What is the watershed value of e between social optimality of sorting and pooling? (Hint: This is best done in a spreadsheet.)
(c) In the social welfare function S above, professors get a higher weight as the utility of the annuity is weighted by the survival probability Ψ. How would your answers to the previous two questions change if u_i were not weighted by Ψ_i, i.e., $u_i = \ln(\text{annuity}_i)$?

Problem 13.7 Endogenous quality

In a one-period economy, you are the producer of a car with unobservable quality. Your cost of producing a car with a quality worth q to a buyer is $q/2$. The possible range of qualities is from \$5,000 to \$10,000. You sell in perfect competition to other producers.

(a) What quality do you offer if you do not know what qualities your competitors offer?
(b) What qualities are offered in equilibrium?

14 Optimal contracts

14 A Introduction

On Boxing Day 2004 a tsunami caused by an earthquake west of Sumatra struck the coasts of several countries. There were 300,000 casualties, but also a few miracles. A small baby, believed to be three or four months old, was found alive among the rubble and corpses after the waves hit Sri Lanka. Covered in mud but otherwise healthy, he was the 81st admission to the local hospital. The boy was thus referred to as "Baby 81".

No fewer than nine desperate Sri Lankan families claimed Baby 81 was theirs. The police had to separate the putative mothers, who turned up every day to stake their claim. One man threatened to kill himself and his wife outside the nursery of Kalmunai Base hospital unless they were given their baby. The couple had to be arrested for attacking the police. Finally, DNA tests revealed that the distraught couple were the true parents of the baby, Abilass Jeyarajah, born 19 October 2004.

The Jeyarajahs were not the first parents who had to fight for their child. There is the much cited biblical story of King Solomon (see Box 14.1). In most cases where one person has an incentive to lie, an objective test, like DNA analysis, is not available or too costly. Yet, game theorists and economists have developed methods to transfer information credibly from one party to another. These methods, an "economic lie detector", are the subject of the present chapter.

Economists prefer to speak of optimal mechanisms or contracts, rather than of a lie detector, when they discuss methods that help parties transfer asymmetric information. The field has flourished over the last twenty years. Its importance is highlighted by at least two Nobel prize winners in economics (A. Michael Spence and Joseph Stiglitz). These have shown that private information can be credibly transmitted among parties: An uninformed party can separate informed parties by "screening", while an informed party can credibly reveal information by "signaling".

The present chapter introduces one of the key ideas in information economics: The transfer of information via contracting. Such transfers become important when markets fail. In previous chapters we argued that markets aggregate (heterogeneous) information. Yet, they may fail when information is asymmetric (as discussed in Chapter 13). Here we will try to explain that despite informational asymmetries, there is hope for the agents involved. Asymmetric information may muddy the waters between parties, but contracting provides an—often quite elegant—bridge.

The basic ideas are fairly intuitive. Our leading examples will be from the world of insurance. Typically, an insurer faces individuals with different risk characteristics which are unobservable for a third party. We will look at the insurer's scope for screening customers for hidden information both in a monopolistic and a competitive setting.

Screening customers requires that a buyer or seller can make offers differing in one more dimension than just with respect to price. In the Theory section we will use quantity as a second instrument. Offering different price–quantity bundles is commonly referred to as *non-linear pricing*. Yet, the second instrument could also be quality. Price–quality discrimination is the topic of our first Application. A second Application illustrates the role of financial constraints in contracting. It is devoted to subordinated debt, an instrument that plays an important role in contemporary discussions on banking supervision.

Box 14.1 Solomon's Judgement

King Solomon (Kings III, 3) was consulted by two women, each claiming to be a baby's true mother. Solomon listened to each of them. Then he raised his sword and announced he would cut the child in half if none of the women were to renounce her claim. One woman kept firm; the other renounced, begging Solomon to spare the baby. Solomon decided that she who preferred to leave the child to her opponent rather than have it killed was the true mother.

German dramatist Bertold Brecht used the story in *Der Kaukasische Kreidekreis* (The Caucasian Chalk Circle, 1945): The child was placed into a chalk circle and the mothers had to pull it out by its arms. The winner would keep the child. One woman let go early. The judge (not a wise king, but a bit of a drunken rascal) declared her to be the true mother. Was she? Biologically, she was not. The rich and hard-hearted biological mother had given the child away to be brought up by another women. That "true" loving mother, who had raised the child as if it were her own, was the one who let it go and thus finally kept it.

Game theorists have found out that Solomon's scheme (in both versions) was not entirely watertight (Farrell, 1987): The false mother could have done better by imitating the true mother's strategy. If both mothers had renounced, Solomon's scheme would not have worked.

Yet, several authors have pointed out that Solomon's technique could be improved to become a true lie detector. The trick is to find a "side payment" that is cheaper for the true than for the false mother. Selling the child in an auction is a possibility; but it might not work if the true mother is poor and the false mother rich. Solomon could have asked the mothers to make non-monetary bids: Level of risk to their own lives or numbers of years in jail they would incur in order to save the child's life. The true mother would probably have made the higher bid.

14 B Main ideas: The economic lie detector

We will show in the present chapter how an uninformed party, from King Solomon to the buyer of a second-hand car, can extract hidden information from an informed party. The magic word in this context is contracting. The uninformed party offers the informed party a deal—a contract—which the informed party must either accept or reject. By accepting or rejecting, the informed party discloses their hidden information. This is, roughly speaking, how the economic lie detector works.

In Chapter 1 we introduced Lady Kunigund who tried to make use of the economic lie detector. She threw her glove into an arena full of wild beasts and asked her lover, Knight Delorges, to fetch it and thus prove his love. The knight would have to show his colors: He would not fetch the glove unless his love for the lady was stronger than his fear of the savage animals. The lady's technique is known among economists as *screening*. In a screening game the *uninformed* party moves first by making an offer to the informed party. The lady played that part of the game well. What escaped her was that by asking him to fetch the glove she sent a signal about her own love: She gave away what a cold-hearted shrew she was. Unintentionally, she used an important technique of transmitting hidden information called *signaling*. In a signaling game the *informed* party moves first, with the intention of conveying some hidden information to an uninformed counterparty.

Both screening and signaling are important strategies in economic life. Health insurers screen customers by offering contracts with different combinations of premia and deductibles. Students try to obtain a diploma in order to help them signal their hidden abilities to potential employers. In this book we will focus on screening as our leading example of optimal contracting. Those interested in signaling will find a review of the literature in Riley (2001) and an intuitive introduction in Problem 14.7.

As reported in Chapter 13, Alice and Bob want to go on vacation, but are afraid of buying a car of unknown quality. They consider Fred's car which could be either reliable or unreliable. Fred, the owner, would sell a reliable car for at least $7,500 and an unreliable car for at least $5,000. Alice and Bob would gladly pay $10,000 for a reliable car, but they are not ready to pay anything for an unreliable car. What annoys Alice and Bob is that, obviously, everybody could be made happy. Fred could get up to $10,000 for a car he would sell for as little as $7,500, or he could keep a vehicle worth (to him) $5,000 for which Alice and Bob would not pay a cent anyway. Everybody would be fine—if Fred could only be trusted to be honest about the quality of his car!

How could they get Fred to disclose the true quality of his car?

We showed in Chapter 13 that a simple price offer would not work. It is not possible to kill two birds (price and quality) with one stone. If the buyers offer $10,000 they may end up with an unreliable car; if they offer less than $7,500 they either get no car or they may acquire a lemon.

Alice and Bob need a contract with a second dimension in addition to price. For example, they can offer Fred two different prices: One they will pay if the car runs smoothly, another if the car breaks down. Let us illustrate this with an example.

Assume for simplicity's sake that the car is equally likely to be reliable or unreliable. A reliable car never breaks down, while an unreliable car breaks down with a probability of 50 per cent. We also assume that both parties maximize their expected payoff; they are, in other words, risk-neutral. Under these assumptions, Alice and Bob can now formulate a contract that achieves their goal.

Bob thinks he has the solution: "We offer Fred $10,000 if he promises to pay the money back if the car breaks down on our vacation. To us a reliable car is worth $10,000—more than to him. If Fred's car really is reliable, he should be more than happy to accept. So we are sure to go on vacation if his car is good."

Alice agrees, but cautions: "I think your idea is fine, but a bit too generous. Fred might accept it, even if his car were unreliable. Although it may look mean, we should offer less than $10,000, and once we think about lowering the price, why don't we barter him down to $7,500? He is the source of uncertainty, after all! $7,500 is what he would be happy to accept if the car is good. But if the car is bad he has a fifty-fifty chance of paying the price back and he will reject an offer of $7,500. That makes sure we will not go on vacation in a lemon!"

Obviously there are several possible contracts that lead Fred to sell a reliable car and to keep a bad car. Their situation is represented by Figure 14.1. Alice and Bob have to maneuver within two constraints.

1 Fred should *accept* the offer if his car is reliable. In Figure 14.1 this constraint is represented by r, a vertical line through Fred's reservation price for a reliable car ($7,500). All points *to the right* of this line satisfy the constraint and lead Fred to sell a reliable car.
2 Fred should *reject* the offer if his car is unreliable. In Figure 14.1 this constraint is represented by u; all points on this line give Fred an expected return for an

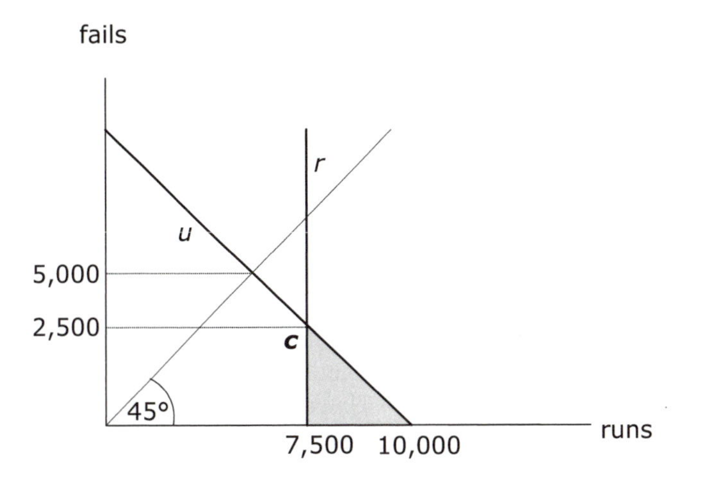

Figure 14.1 A contractual pair of payments must satisfy *r* and violate *u* to make the owner sell a reliable car and keep an unreliable car.

unreliable car of \$5,000. Line *u* has slope -1 (reflecting an unreliable car's fifty-fifty chance of failure) and runs through points (\$5,000, \$5,000) and (\$10,000, \$0). All points *below* this line violate the constraint and lead Fred to keep an unreliable car.

To acquire a reliable car while avoiding an unreliable car, Alice and Bob can offer any pair of payments represented by a point within the shaded area of Figure 14.1.

A rational buyer will try to get the car (if it is reliable) at the lowest possible cost. This means that Alice and Bob will offer a point on the (vertical) left border of the shaded area. Otherwise they would have paid too much if the car does in fact run smoothly. As they know that a reliable car will not break down (and an unreliable car would not be sold at the terms offered) they are indifferent at any points on *r* between point **c** and the horizontal axis. Fred, who shares their belief and is risk-neutral too, is equally indifferent.

It is common to assume that a party which is indifferent to several options chooses the option preferred by the counterparty. This would be the contract represented by point **c**.[96] Consequently we will call **c** the *optimal contract*. The payments specified in the contract are \$7,500 (car runs smoothly) and \$2,500 (car breaks down).

There are different real world interpretations of the optimal contract. We can see the contract as a promise of contingent payments of \$7,500 (car runs) and \$2,500 (car breaks down), respectively. Alternatively, we could say that Fred receives \$2,500 plus a bonus of \$5,000 if the car does not break down. Alternatively, we can say that Alice and Bob buy the car for \$7,500, but that Fred is liable for a guarantee of \$5,000 should the car break down.

The point of the optimal contract is the differential payments. These do the trick: their expected value is high enough for the owner of a good car, but too low to be attractive to the owner of an unreliable car. Contingent contracting thus bridges the information gap between the informed and the uninformed party. It may seem that the originally informed party, in our case Fred, regrets that his information advantage is eliminated. Yet, the contrary is the case: In the absence of a contingent contract he could not have sold his car. If Alice and Bob thought it equally likely that Fred's car was reliable or unreliable, a deal would have failed due to the lemons problem. Any non-contingent price offer below \$7,500 would only have produced an unreliable car. An offer of \$7,500 or above would have produced a car worth \$5,000 to the buyers in expected terms. Fred thus must be happy that contracting extracts his hidden information.

It may seem from our simple example that contingent contracts do away with the evil of information asymmetries. Unfortunately this is not completely the case. Contingent contracts in general improve upon simple non-contingent contracts. But they may not always achieve a Pareto optimum as would be attainable under full information. Contracts under asymmetric information may have to impose more risk on the informed party than arrangements under symmetric information. The utility loss from his "excess risk" is the deadweight cost of asymmetric information. Optimal contracts are therefore not necessarily in the Pareto sense. They are usually optimal from the view of the stipulating party (in the above example Alice and Bob) who can make take-it-or-leave-it offers to a counterparty.

Contracting normally extends over several stages: In a screening game one party offers, the other party accepts. In a signalling game one party makes itself observably attractive to an offer, the other offers, etc. In real life, parties often may not straightforwardly accept or reject an offer, but make a counter-offer. Finally, contracts, once established, are enforced. The contract itself thus is just the centerpiece of a more encompassing *"mechanism"*. Economists use the term mechanism to capture all relevant components of contracting. A typical example of a mechanism is an auction. First, the rules are laid down; bidders are invited to bid and the bids are then collected (in increasing or decreasing order). Sometimes, auctions are even structured into two or more stages. It is important to think of mechanisms in their entirety. Nevertheless, for simplicity's sake, we will mostly use contracts in the narrowest sense of the term.

Box 14.2 Econspeak versus Lawsperanto

The notion of a contract is central to both information economics and law. This makes it interesting to distinguish between the respective definitions of "contract" in the two disciplines.

In law a contract is defined as:

> an agreement with specific terms between two or more persons or entities in which there is a promise to do something in return for a valuable benefit known as consideration.
>
> (http://legal-dictionary.thefreedictionary.com/contract)

In economics a contract is defined as:

> an anticipated mapping of observable states of nature into payoffs.
>
> (Allen and Gale 1999, p. 1251)

The two definitions could hardly sound more different. Yet, they are complementary, rather than contradictory. The legal definition focuses on the promised *exchange* of something in return for something else (the "consideration"), while the economic definition stresses the *contingent* nature of such promises. But essentially, lawyers and economists mean the same thing. B promises to repair A's car for $500 by next Monday and for an additional express premium of $150 already by Friday evening, provided the required spare part arrives before lunchtime. A agrees to the offer and the express option. Lawyers would say, A promises the money for having his car fixed (the consideration). Economists, on the other hand, would say that A and B anticipate that the states of nature (e.g. the spare part arriving on time plus the car being fixed by Friday evening) are mapped into payoffs ($650 in this case) from A to B.

What if B finishes by Friday evening, but A refuses to pay? In legal terms this would be a breach of contract; at B's request, the contract would be enforced by law. In economic terms, there still is a mapping of states into payoffs. The state of nature "B finishes on time, A refuses to pay" leads to a payoff from A to B (the smaller of $650 or A's total assets). This payoff is defined by insolvency law, rather than by the contract alone.

Both economists and lawyers look at a contract as a part of a wider sequence of events. From a legal point of view, a contract comes into existence by (i) an offer and (ii) an acceptance of that offer which results in a meeting of the minds; the (iii) promises to perform are supported by (iv) the legal infrastructure to enforce the contract. Similarly, for economists, contracts are a part of wider "mechanisms", including the offer, acceptance, and the possibility of enforcement.

Lawyers and economists may talk very differently, but their respective languages are mutually comprehensible.

14 C Theory: Optimal contracts

14 C a Separating the wheat from the chaff

A preliminary clarification

To begin with, the topic of this chapter is mechanism design. This is a wider concept than optimal contracts: A mechanism includes the whole procedures of offer, acceptance, and—if needed—enforcement of a contract. A typical mechanism is an auction, consisting of a set of rules for bidding, for the allocation of the item on sale, and on payments due. It leads to a contract between the auctioneer and, normally, the highest bidder. The contract mainly specifies that the successful bidder (i) gets the item and (ii) has to pay the amount implicitly promised in the successful bid.

Even though "mechanism design" is the appropriate term, we will use the less ambitious label "optimal contracts". We will mostly focus on the use of contracts to screen individuals for their private information. We will be more concerned with contractual promises than with the wider mechanism in the background. In the following sections we will demonstrate how an insurer can screen individuals with different risk characteristics. The emphasis is on the (simpler) case of a monopolistic insurer. The case of competition among insurers will be treated briefly in a digression. Our model is strongly based on the seminal contribution by Rothschild and Stiglitz (1976) (written for the competitive case, though).

The agents

A population consists of a large number of individuals who face a health risk. In the event "good", an individual remains healthy; in the event "bad", the individual falls ill. The population consists of two types of individuals: A fraction β of the population are low-risk individuals who fall sick with probability p_L and remain healthy with probability $(1 - p_L)$. A fraction $(1 - \beta)$ are high-risk individuals with probabilities of falling sick and remaining healthy of $p_H > p_L$ and $(1 - p_H)$, respectively. Individual risks are independent (no epidemic, for example).

Individuals have utility function $u(c)$, with c denoting consumption (which in the present context is equal to terminal wealth). Individuals are risk-averse, i.e. $u' \geq 0$ and $u'' < 0$. We further assume that $u(0) = 0$, $u'(0) = \infty$, and $u'(\infty) = 0$.

Being healthy is equivalent to a payoff of e_g, falling ill is equivalent to a payoff of $e_b < e_g$. We call the pair of payoffs $\bar{\mathbf{c}} = \{e_g, e_b\}$ of an uninsured individual the individual's *endowment*. The endowment is the outside option for individuals who do not buy insurance.

A monopolistic insurer offers insurance policies against the event of falling ill on a take-it-or-leave-it basis. The insurer offers contractual payments in exchange for individuals' endowment. For clarity's sake, we assume that the insurer is risk-neutral, even though this is unnecessary, given the fact that with a large number of insured individuals the insurer can pool risks perfectly.

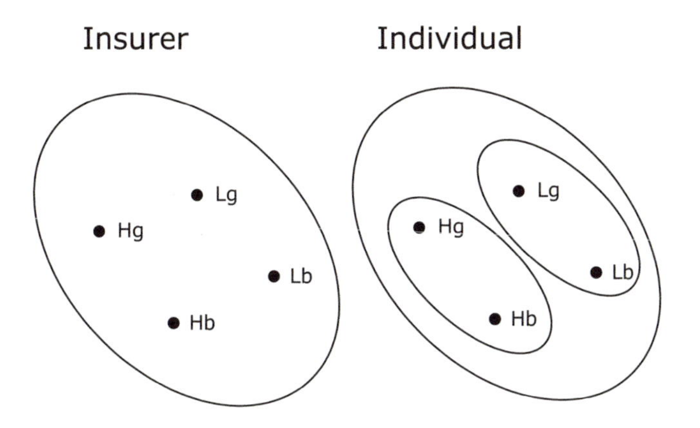

Figure 14.2 *Ex ante*, the insurer cannot distinguish either states (g, b) nor risk types (L, H), while the individual knows his type (L, H).

Information

Individuals privately know their own risk type before insurance contracts are offered. The insurer can observe final outcomes, i.e. whether an individual falls ill or not, but not individuals' types. The parties' *ex ante* information is summarized in Figure 14.2. We use the representation as information structures, or partitions, introduced in Chapter 3. Per individual there are four possible terminal states of nature, as shown in Figure 14.2: An individual may face a high or low risk (H, L) and may remain healthy or fall ill (g, b). Individuals know their type (H, L) but not future outcomes (g, b). The insurer cannot distinguish between any of the four terminal states (Hg, Hb, Lg, Lb). Everything except individual types is common knowledge, including probabilities and the structure of the game.

Contracts

Ex post, once individuals have or have not developed the disease, information is represented by Figure 14.3. Two subsets of the state space, the events (g, b)—health or sickness—, have become public knowledge, but not all of their four elements. Contracts can be written on these *observable states of nature*, but not on unobservable states. Under a contract specifying payments in unobservable states, the contracting parties might never agree on which state really prevails, and a third party, such as a judge, would also be unable to decide. The set of observable states of nature is called the *contract space*. A contract can be written as $\mathbf{c} = \{c_g, c_b\}$, subscripts denoting the two observable states. An insurance contract is a promise by the insurer specifying a pair of (gross) payments c_g or c_b (one for each of the two observable terminal states of nature) to the individual in exchange for the individual's endowment in the respective state, $\{e_g, e_b\}$.

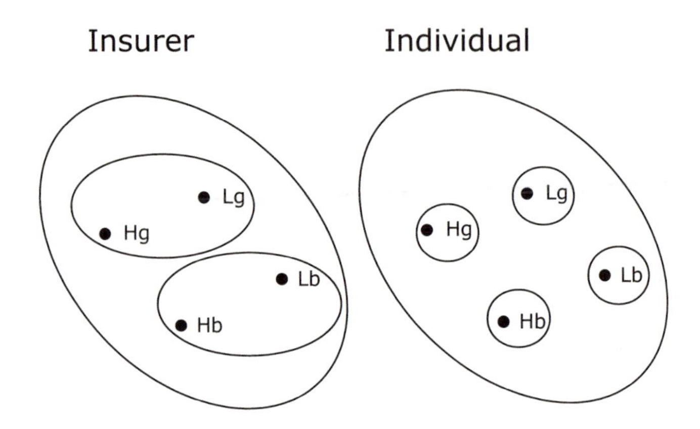

Figure 14.3 *Ex post*, the insurer can distinguish states (g, b) but not risk types (L, H), while the individual knows both.

14 C b The symmetric information case

We follow the structure already used to describe the insurance destruction effect (Chapter 13 D) and first look at the symmetric information case, where either nobody can observe risk types (no information) or everybody can (public information).

No information

If neither the individuals nor the insurer can observe individual risk, all individuals are equal from an *ex ante* point of view. The insurer is faced with the following average probability of an individual falling ill:

$$\bar{p} = \beta p_L + (1 - \beta) p_H,$$

A monopolistic insurer maximizes profits under the constraint that individuals accept the insurance contract. The insurer solves:

$$\max_{c_b, c_g} \left[\bar{p} \left(e_b - c_b \right) + (1 - \bar{p}) \left(e_g - c_g \right) \right].$$

which is equivalent to minimizing expenditures:

$$\min_{c_b, c_g} \left[\bar{p} c_b + (1 - \bar{p}) \, c_g \right],$$

subject to individuals' *participation constraint*:

$$\bar{p} u \left(c_b \right) + (1 - \bar{p}) \, u \left(c_g \right) \geq \bar{p} u \left(e_b \right) + (1 - \bar{p}) \, u \left(e_g \right).$$

This constraint holds with equality, otherwise the insurer could reduce one of the payments c without losing the customers. Furthermore, as the insurer and the insured hold the same beliefs regarding the probability of an individual falling ill, $\bar{p}$, the risk-neutral insurer assumes all the risk and effectively insures individuals by offering them a risk-free policy with $c_g = c_b = c^*$, c^* being the solution to:

$$u(c^*) = \bar{p}u\,(e_b) + (1 - \bar{p})\,u\left(e_g\right).$$

Due to the concavity of $u(.)$, the policy is profitable to the insurer: $c^* \leq \bar{p}e_b + (1 - \bar{p})\,e_g$. It also implements a Pareto optimum from the *ex ante* point of view of individuals who do not (yet) know their risk type.

Public information

When individuals' risk types are observable, the insurer can offer separate contracts to each of the two types. The optimal contracts $\mathbf{c}^{*L} = \{c^{*L}_g, c^{*L}_b\}$ and $\mathbf{c}^{*H} = \{c^{*H}_g, c^{*H}_b\}$ have four payments $\mathbf{c}$ solving the insurers' *objective function* which we again write as an expenditure minimization problem:

$$\min_{\mathbf{c}} \beta\left[p_L c^L_b + (1 - p_L)\,c^L_g\right] + (1 - \beta)\left[p_H c^H_b + (1 - p_H)\,c^H_g\right] \qquad (14.1)$$

subject to both types' *participation constraints*:

$$p_L u\left(c^L_b\right) + (1 - p_L)\,u\left(c^L_g\right) \geq p_L u\,(e_b) + (1 - p_L)\,u\left(e_g\right),$$

$$p_H u\left(c^H_b\right) + (1 - p_H)\,u\left(c^H_g\right) \geq p_H u\,(e_b) + (1 - p_H)\,u\left(e_g\right).$$

These constraints are both binding, otherwise the monopolistic insurer would pay more than required to bring both risk types on board.

The solution is illustrated by Figure 14.4. Both contracts, $\mathbf{c}^{*L}$ and $\mathbf{c}^{*H}$, lie on the intersections of the 45°-line and the respective participation constraints, lines PC_L and PC_H. Participation constraints are indifference curves that go through individuals' endowment (point E). At the intersection with the 45°-line an indifference curve has a slope reflecting the odds of falling ill or remaining healthy. It is easy to see why: On the 45°-line individuals have equal amounts of wealth in both states; they are thus risk-neutral between marginal changes in one or the other direction. A risk-neutral individual would trade p dollars in case of good health against $(1 - p)$ dollars in case of bad health. The tangents l and h thus have slope $-(1 - p)/p$. From the insurers' point of view, they are fair-odds lines, or iso-profit lines, for contracts with the respective types. Both tangents indicate a profit maximum, as the insurer cannot reduce payments without losing one or the other group of individuals. Their intersection, point E', is closer to the origin than point E, indicating that the insurer as a monopolist makes a profit.

The optimal contracts also implement a Pareto optimum: Risk-averse individuals are fully insured by the risk-neutral insurer. Note, however, that this is a Pareto

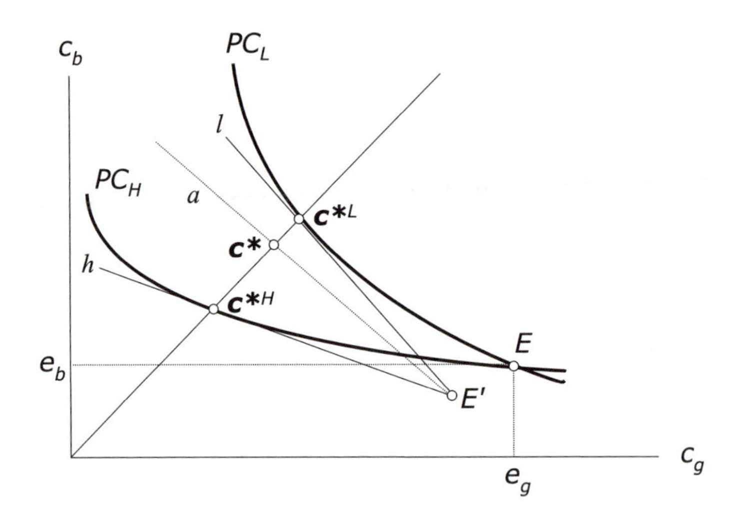

Figure 14.4 Under symmetric information contracts optimal for a monopolistic insurer, $\mathbf{c}^{*H}$ and $\mathbf{c}^{*L}$, are defined by (i) high- and low-risk types' participation constraints and (ii) full insurance (the 45°-line).

optimum only from an *ad interim* point of view of individuals who already know their risk type (but not yet the outcome of whether they will fall ill or not). Individuals are insured against falling ill *according to* their type. Individuals are not insured against being one type or the other. This is the insurance destruction effect of public information which we discussed in Section 13 D. *Ex ante*, individuals would be better off if nobody had any information about types. The insurer would offer the pooling contract **c** represented in Figure 14.4 by the intersection of the 45°-line with line a, the iso-profit line for the average risk of the population, $\bar{p}$.

14 C c Monopolistic screening: A single contract

We now turn to hidden information. Individuals know their types, but cannot be distinguished by the insurer. Under hidden information the insurer cannot prevent high-risk individuals from buying a contract designed for low-risk individuals (or vice versa). Individuals only comply with the insurer's plan if it is attractive for them to do so. The insurer has the choice between two basic strategies: offering either a pooling or a separating contract.

Assume the insurer issues a pooling contract. In principle, this might be a contract either for low-risk or for high-risk individuals. Yet, only the high-risk contract is feasible. It is easy to attract only *high-risk* individuals. As Figure 14.4 shows, the insurer can offer contract $\mathbf{c}^{*H}$, with the insurer's optimal contract for high-risk individuals derived above. This contract lies on PC_H and is accepted by high-risk individuals, but it is far below PC_L and thus unattractive to low-risk individuals.

It is impossible, however, to attract only *low-risk* individuals. The insurer would have to offer a contract represented in Figure 14.4 by a point above PC_L, but below PC_H, i.e. between the solid curves to the right of point E. Such a contract would not actually insure individuals, as it is more risky than their endowment E. Nor would such a contract be profitable to the insurer. The profit maximizing contract which would satisfy both constraints would be point E itself (yielding zero profit). Due to the information asymmetry, the insurer cannot generate positive profit from insuring only low-risk individuals. The optimal single contract therefore is $\mathbf{c}^{*H}$, the high-risk policy.

14 C d Monopolistic screening: A dual contract

The insurer may do better by offering a dual contract, i.e. a menu of two contracts from which individuals may choose. Under hidden information they will choose the contract that is more attractive for their risk type. The insurer faces an additional constraint: Not only should individuals prefer a contract to no contract (participation constraints), they should also buy the "right" contract that is designed for their type. The conditions under which they do so are called *incentive constraints*, or self-selection constraints. These have to be satisfied in the insurer's optimization strategy.

The insurer's problem

The insurer's *objective function* is the same as in the case of public information (14.1) above. But in addition to individuals' *participation constraints*:

$$p_L u\left(c_b^L\right) + (1 - p_L)\, u\left(c_g^L\right) \geq p_L u\left(e_b\right) + (1 - p_L)\, u\left(e_g\right) \tag{14.2}$$

$$p_H u\left(c_b^H\right) + (1 - p_H)\, u\left(c_g^H\right) \geq p_H u\left(e_b\right) + (1 - p_H)\, u\left(e_g\right) \tag{14.3}$$

the insurer also has to respect two *incentive constraints*:

$$p_L u\left(c_b^L\right) + (1 - p_L)\, u\left(c_g^L\right) \geq p_L u\left(c_b^H\right) + (1 - p_L)\, u\left(c_g^H\right) \tag{14.4}$$

$$p_H u\left(c_b^H\right) + (1 - p_H)\, u\left(c_g^H\right) \geq p_H u\left(c_b^L\right) + (1 - p_H)\, u\left(c_g^L\right) \tag{14.5}$$

These ensure that it does not pay for individuals misrepresenting their types. Individuals must not find it attractive to play the game with the contract designed for the other type.

Stepwise solution

The optimal solution is presented in Figure 14.5. It consists of a full insurance contract for high-risk individuals, $\mathbf{c}^H$, and a risky (incomplete insurance) contract

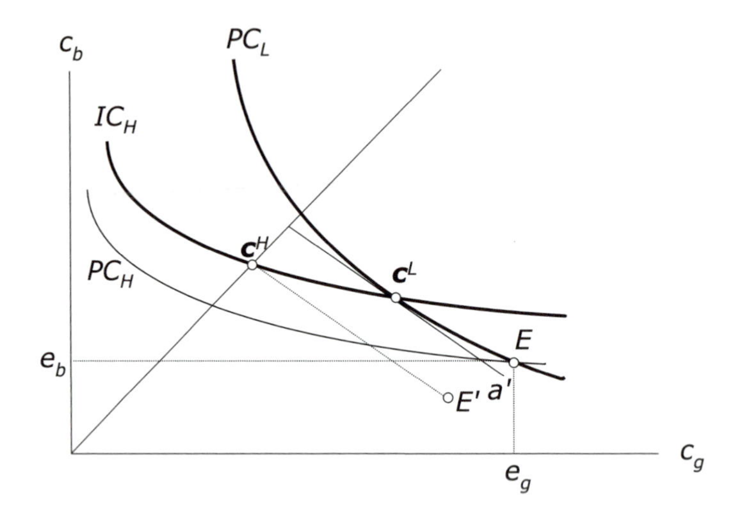

Figure 14.5 Under asymmetric information contracts optimal for a monopolistic insurer, c^H and c^L, are defined by (i) low-risk types' participation constraint and (ii) high-risk types' incentive constraint.

for low-risk individuals, $\mathbf{c}^L$. The solution is found through the following steps:

1 **Low-risk individuals** are *not* restricted by their *incentive* constraint (14.4). They need no extra reward for not buying a contract targeted at high-risk individuals. Any contract designed for high-risk individuals is unattractive to low-risk individuals. In Figure 14.5 PC_L lies above PC_H (except in the irrelevant area of "negative" insurance to the right of E).

2 As a consequence, PC_L, the *participation constraint* for low-risk individuals (14.2) binds, i.e. it holds with equality. The insurer would like to make the payments promised to low-risk individuals, c_b^L and c_g^L, as small as possible. The lower they are, the higher the insurers' profit. Starting at any point above PC_L, the insurer can lower these payments until the contract hits the participation constraint PC_L.

3 **High-risk individuals** are restricted by their *incentive constraint* (14.5). High-risk individuals are tempted to buy contracts designed for low-risk individuals. For high-risk individuals, all contracts satisfying the low-risk individuals' participation constraint (to the left of E) are attractive. High-risk individuals thus need an incentive not to sign the contract for low-risk individuals. Assume the insurer targets some contract $\mathbf{c}^L$ at low-risk individuals. Contracts high-risk individuals prefer to $\mathbf{c}^L$ then lie on or below IC_H, the high risk individuals' indifference curve running through $\mathbf{c}^L$. IC_H is therefore called the high-risk individuals' incentive constraint (14.5).

4 The *participation constraint* for high-risk individuals (14.3) is slack. As their incentive constraint (14.5) binds, their participation constraint (14.3) could

only bind if $c_b^L < e_b$ and $c_g^L > e_g$, which is not possible. The insurer would lose money by *buying* insurance from risk-averse individuals.

5 **The insurer** prefers the contract on the highest *iso-profit line* of all pairs of contracts that satisfy the above constraints. As both types buy insurance, this is the line a' reflecting the average risk of the population $\bar{p}$.[97] The insurer therefore transfers as little risk as possible to high-risk individuals (who have a strong demand for insurance), which implies a full insurance contract $c_b^H = c_g^H$, represented by point $\mathbf{c}^H$ on the 45°-line. At the other end, the low-risk individuals are burdened with a level of risk that maximizes the insurer's profit. It is given by point $\mathbf{c}^L$ on the iso-profit line, tangent to the low risks' participation constraint.

6 **To summarize:** The above logic leaves us with four equations for four unknowns: The binding PC_L and IC_H, plus $c_b^H = c_g^H$ and finally the equality of the slopes of a' and PC_L. These yield the payments under the optimal contracts represented in Figure 14.5.

Discussion of results

Under the optimal contract menu, high-risk individuals receive full insurance, just as in the benchmark case where risk types are observable. This is because the contract for high-risk individuals does not need to imply risk as a repellent for low-risk individuals. Yet, unlike in the case with observable types, high-risk individuals receive an information rent: Their contract is on a higher indifference curve (the incentive constraint) than in the symmetric information case (where it is on the participation constraint). The incentive premium is necessary to prevent high-risk individuals from buying the contract designed for low-risk individuals, who cannot obtain full insurance. The intuition of this paragraph carries forward to a world with more than two risk types. Among an arbitrary number of types the individual with the highest risk receives full insurance and an incentive premium, while the type with the lowest risk receives partial insurance and no incentive premium.

The optimal solution to the dual contract menu depends on the fractions of the different types in the population. When the fraction of low-risk individuals, β, goes towards unity, $\mathbf{c}^L$ and $\mathbf{c}^H$ converge towards the intersection of PC_H and the 45°-line. High-risk individuals can hide behind the good risks; they receive a large incentive premium, as the insurer cares less about the few bad risks than about the majority of good risks. Conversely, when the fraction of low-risk individuals, β, falls, line b becomes flatter and $\mathbf{c}^L$ drifts towards the endowment point E. At the limit, low-risk individuals receive no insurance at all, but are left with their endowment, while high-risk individuals are still fully insured, but hampered by their participation constraint. There is thus a lower bound to the fraction of low-risk individuals at which insurance for both types ceases to be profitable (when $\mathbf{c}^L$ would move beyond E).

The type-specific contracts $\mathbf{c}^L$ and $\mathbf{c}^H$ represent contracts often observed in the real world. Health insurers, for example, often offer contracts with a low deductible (zero in the case of $\mathbf{c}^H$) catering to bad risks, and contracts with a higher deductible attracting the better risks. Separating unobservable types is possible whenever the indifference curves exhibit the property represented in Figures 14.4 and 14.5. An indifference curve of a type 1 individual crosses an indifference curve of a type 2 individual once and only once. This is the so-called *single-crossing property* (see Wolfstetter, 1999; Bolton and Dewatripont, 2005).

Non-linear pricing

In the above model, high-risk types receive full insurance, while low-risk types receive only partial insurance. Such "rationing" is important to prevent the high-risk individuals from buying insurance at the low rates calculated for low-risk individuals. The simultaneous offer of different price-quantity bundles does the job of extracting information from the informed party. This technique is often referred to as "non-linear pricing". Non-linear pricing is defined rather generally by Riley (2001, p. 432) as "the design of an incentive scheme by an imperfectly informed monopolist". Examples include quantity discounts, frequent-flyer bonuses and membership schemes where members pay a fixed entrance fee and enjoy reductions on bought items, unlike non-members. Entry fees to amusement parks often are a kind of "one-day-membership" scheme. Since the price paid per item tends to vary with quantity, such schemes are also called non-linear pricing, a standard technique in optimal contracting. For a general introduction, see Bolton and Dewatripont (2005, Ch. 2). In terms of monopolistic price discrimination, non-linear pricing is discussed in Wolfstetter (1999, Ch. 1).

14 C e Digression: Competitive screening

The above discussion focused on so-called monopolistic screening. Here we will informally try to provide some intuition for what happens if the insurer is in competition with at least one other insurer. For the original analysis by Rothschild and Stiglitz (1976), we refer to Wolfstetter (1999) or Bolton and Dewatripont (2005).

Under competitive conditions an insurer is set at the zero-profit line. Now, it is the insurer who is restricted by a participation constraint. The gains from insurance go to the insured. Figure 14.6 illustrates the benchmark case in which the insurers can observe types. A competitive insurer offers contracts $\mathbf{c}^{*L}$ and $\mathbf{c}^{*H}$ to low- and high-risk individuals, respectively. Both types enjoy full insurance. As in the monopolistic case, observability of types leads to a Pareto-optimal outcome. The difference is that here, under competitive conditions, the insured attain a utility level above their respective participation constraints. With a monopolistic insurer, the insured would need an endowment of E'' to attain the same utility level. The insurer, by contrast, is set at iso-profit lines that run through E, implying zero profit.

Under asymmetric information, when risk types are not observable, the insurer again has the choice between offering a single contract and a dual contract.

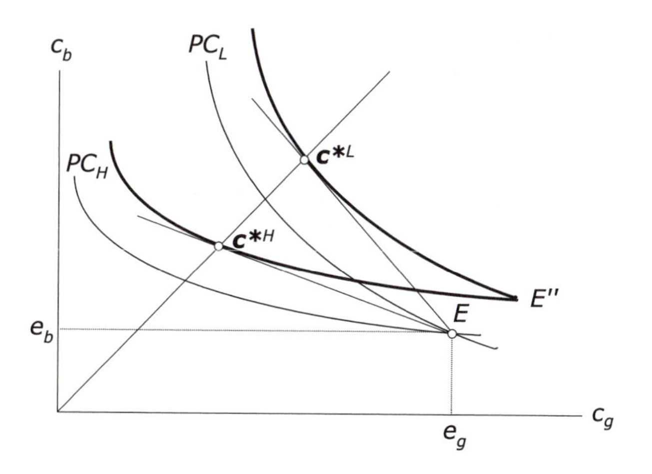

Figure 14.6 Under symmetric information a competitive insurer offers contracts $\mathbf{c}^{*H}$ and $\mathbf{c}^{*L}$, defined by (i) the respective fair odds lines for high- and low-risk types' and (ii) full insurance (the 45°-line).

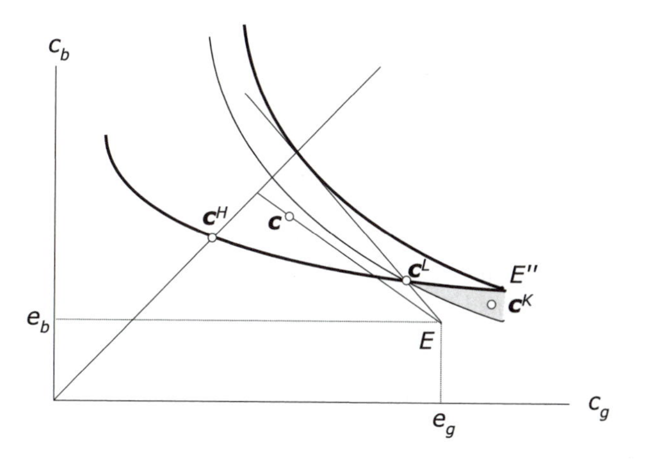

Figure 14.7 Under asymmetric information a competitive insurer cannot offer a pooling contract **c**. The optimal separating contract offers full insurance to high-risk types at high rates ($\mathbf{c}^H$) and partial insurance to high-risk types at low rates ($\mathbf{c}^L$). It is vulnerable to the "predatory" contract $\mathbf{c}^K$, which, in turn, is not viable in the absence of the separating contract.

Figure 14.7 shows that a single contract attracting both types—a so-called *pooling contract*—is not feasible. Such a contract is represented by point **c**, which lies on a zero-profit line. Obviously, contract **c** would attract high-risk individuals, but not the low risks. But then, a flatter iso-profit line applies, one that lies above *E*, which means that the contract loses money.

The insurer can try *separating contracts*, offering a menu with $\mathbf{c}^L$ and $\mathbf{c}^H$. As in the monopolistic case, it is the high-risk type who is tempted to buy the contract designed for low-risk individuals. Therefore, $\mathbf{c}^L$ cannot lie above the high-risk indifference curve through $\mathbf{c}^H$. Under competitive conditions an insurer is compelled to set $\mathbf{c}^L$, such that it is situated on a fair-odds line or iso-profit line for insuring low-risk individuals.

The dual contracts are represented in Figure 14.7. High-risk individuals obtain a contract $\mathbf{c}^H$, identical to the one they obtain when risk types are observable. Unlike in the monopolistic case, the insurer cannot afford to pay an incentive premium. The contract for low-risk individuals $\mathbf{c}^L$ thus must contain a different repellent for high-risk individuals. This repellent is the incomplete insurance that comes with $\mathbf{c}^L$. For high-risk individuals, the degree of insurance under $\mathbf{c}^L$ is unsatisfactory. In the competitive case, low-risk individuals do worse when types are not observable. Contract $\mathbf{c}^L$ lies on a lower indifference curve than the contract they obtain when types are observable (Figure 14.6).

The problem with the competitive case is that an equilibrium may not exist at all. Assume an insurer offers the optimal pair of contracts $\mathbf{c}^L$ and $\mathbf{c}^H$. Now, another insurer enters the market with contract $\mathbf{c}^K$ (to the right of E). Contract $\mathbf{c}^K$ is targeted at low-risk individuals only. In the presence of the contract pair $\mathbf{c}^L$ and $\mathbf{c}^H$, $\mathbf{c}^K$ will indeed attract the low-risk individuals, and only them (high-risk individuals prefer $\mathbf{c}^H$.) The insurer offering the separating contracts $\mathbf{c}^L$ and $\mathbf{c}^H$ will end up with only the bad risks and will lose money. Yet, once this insurer has left the market, contract $\mathbf{c}^K$ will attract high-risk types as well, as it still betters their endowment E. Insuring both types, contract $\mathbf{c}^K$ will now lose money (it lies above a fair-odds line through E). Once the contract is withdrawn, the market is free again for an insurer offering a separating contracts ..., and so on. In a competitive insurance market, an equilibrium therefore may not exist. For a more detailed discussion of this finding by Rothschild and Stiglitz (1976) and for references to further work, see Wolfstetter (1999) and Bolton and Dewatripont (2005).

14 D Application: Price–quality discrimination

Alice and Bob are out for dinner. Both would like a glass of wine. On the list they find two wines served by the glass, a *Table Wine* and a *Cabernet*. Although the former is less expensive, the latter seems to offer more quality per dollar. Alice goes for the *Cabernet*. Bob calls for the head waiter, and complains: “Today, I’m not much in the mood for a sophisticated wine, but when I look at the price of your *Table Wine* I may just as well drink water.” “I’m very sorry, Sir,” the waiter replies, “We used to have a decently priced and, I must admit, better *Merlot*, but we replaced it with the *Table Wine*. The problem was that nobody ordered the *Cabernet*.”

An uninformed party can extract hidden information from an informed party if contracts have a second dimension in addition to price. In our insurance example the second dimension was quantity. Yet, quantity is not the only possible choice for a second dimension besides price. A seller may just as well use *quality* as a screening device. We will try to illustrate a price-quality screening with what we will call the "*Cabernet–Merlot* model".

14 D a The Cabernet–Merlot model

The model (though not its name) is borrowed from Salanié (1997, Ch. 2.2). We try to present the intuition of the model graphically and with a minimum amount of formal apparatus. For a more stringent presentation, see Salanié (1997, pp. 18–26).

Assumptions

At the center of the model is a wine seller deciding about the qualities he should sell and at what prices. The seller faces a customer who drinks one bottle within the relevant period. The utility of a customer from a bottle of quality q bought for the amount t is:

$$\theta q - t$$

where $\theta > 0$ is an index for taste. A customer who buys nothing ($q = 0$, $t = 0$) derives a utility level of zero.

The customer may either be "coarse" (θ_L) or "sophisticated" ($\theta_H > \theta_L$). A fraction β of the population is coarse, fraction $(1 - \beta)$ is sophisticated. A sophisticated customer is keener on quality than a coarse customer. The difference in utility:

$$(\theta_H q - t) - (\theta_L q - t) = (\theta_H - \theta_L)q$$

increases with q. More generally, a higher type (an agent with higher θ) is willing to pay more for a given increase in q than a lower type (with a lower θ). This reflects the *Spence–Mirrlees condition* (See Salanié 1997, p. 31).

The seller, as a monopolist, can produce (grow) any quality level $0 < q < \infty$ at cost:

$$G = G(q).$$

We assume that $G(q)$ is strictly convex with $G'(0) = 0$ and $G'(\infty) = \infty$. The seller's utility is the difference between receipts and cost, or $t - G(q)$.

14 D b Observable types

If the seller can tell whether a customer is coarse or sophisticated, he can formulate two separate offers, one for coarse customers only, the other exclusively for sophisticated customers. For both types i, the seller separately maximizes:

$$\max_{q_i,t_i} (t_i - G(q_i)).$$

subject to condition that the customer buys the respective wine (participation constraint):

$$\theta_i q_i - t_i \geq 0.$$

The solution is illustrated in Figure 14.8. The solid lines are participation constraints for a sophisticated customer (PC_H) and a coarse customer (PC_L), respectively. Since the utility of rejecting the sellers' offer is zero, both participation constraints run through the origin. Their vertical distance increases in quality, as assumed above (Spence–Mirrlees condition). The two upward bending curves are iso-profit lines of the seller. Their convexity reflects the increasing marginal cost of quality. While the customer's utility increases towards south-east, the seller's profit increases towards north-west.

The two tangent points, $\mathbf{c}^{*L}$ and $\mathbf{c}^{*H}$, reflect the two first-best contracts. We characterize the contracts with *superscripts* rather than subscripts in order to remind that they are only *designed for* but may be rejected by the respective type, in principle. In fact, they are accepted by the right type, as contracts satisfy the respective customers' participation constraint.

Sophisticated customers obtain contract $\mathbf{c}^{*H}$; they drink quality q_H^* which we refer to as *Cabernet*. Coarse customers receive $\mathbf{c}^{*L}$. Not surprisingly, they buy a somewhat lower quality ($q_L^* < q_H^*$), but at a significantly better price $t_L^* < t_H^*$. We will refer to their choice as *Merlot*.

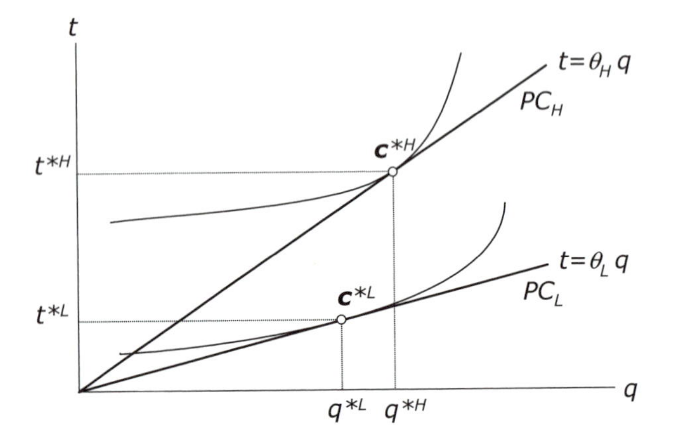

Figure 14.8 Optimal contracts, $\mathbf{c}^{*L}$ and $\mathbf{c}^{*H}$ are represented by the tangent points between the seller's (upward-sloping) iso-profit lines and the participation constraints for a customer with sophisticated taste (PC_H) or with coarse taste (PC_L), respectively.

14 D c Hidden types

The incompatibility of first-best contracts

When the seller cannot observe types, he cannot sell the contract pair given in Figure 14.8. A coarse customer would buy *Merlot* ($\mathbf{c}^{*L}$). A sophisticated customer would find that *Cabernet* ($\mathbf{c}^{*H}$), though acceptable *per se*, is clearly less attractive than *Merlot*. In order to make a sophisticated customer prefer *Cabernet*, the seller would have to make it cheaper by an amount equal to at least Δ, as indicated in Figure 14.9. The maximum a sophisticated customer would pay for *Cabernet* in the presence of *Merlot* is indicated by contract $\mathbf{c}'^H$. Contract $\mathbf{c}'^H$ lies on IC_H, an incentive constraint, i.e. on a parallel to PC_H running through $\mathbf{c}^{*L}$.

Mutually compatible screening contracts

The contracts $\mathbf{c}^{*L}$ and $\mathbf{c}'^H$ are compatible with each other, in the sense that both are bought by a customer of each type. However, they are not the optimal pair in the sense of maximizing the seller's profit. In Figure 14.9 it can be seen that a small reduction in the quality of the cheaper wine from q_L^* would increase the seller's total profit. The profit on the contract for a coarse customer would only fall by a marginal amount (the contract point moving almost along the iso-profit line). Yet, shifting point $\mathbf{c}^L$, and with it line IC_H, to the left would permit the seller to increase the price of the more sophisticated wine above t'_H. This would significantly increase the profit from the contract for a sophisticated customer, since $\mathbf{c}^H$ would move directly "up hill" in the seller's iso-profit landscape. A small reduction in q_L thus has a positive net impact on the seller's profit.

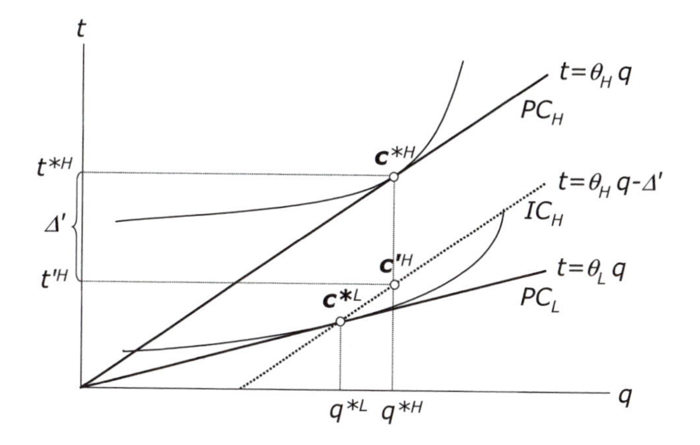

Figure 14.9 Under hidden information, contracts $\mathbf{c}^{*L}$ and $\mathbf{c}^{*H}$ cannot coexist. The highest possible price for the more expensive wine lies on an incentive constraint for a sophisticated customer IC_H running through $\mathbf{c}^{*L}$.

It is important to note that the optimal quality of the expensive wine remains constant when the seller changes the quality of the cheaper wine. At any given q, all iso-profit lines $t - G(q)$ have the same slope. Their tangent points with incentive constraints (parallel lines to IC_H) therefore all lie on the vertical line through q_H. The optimal quality to sell to a sophisticated customer remains at its first-best level q_H^*. Despite lowering the quality of the cheaper wine, the seller thus continues to offer *Cabernet*.

Optimal contracts

When the seller, starting from q_L^*, reduces the quality of the cheaper wine even further, the loss in profit from a coarse customer becomes more important relative to the concomitant increase in profit from selling the sophisticated wine more expensively. At some point, both effects cancel each other out. That point represents the optimum for the seller. It is illustrated in Figure 14.10 by q_L and the respective incentive constraint. The seller offers (i) a cheap *Table Wine* of low quality q_L and (ii) *Cabernet* of unchanged quality q_H, albeit at a lower price than under observable types. The former is targeted at coarse customers, the latter at sophisticated customers. The two optimal contracts, $\mathbf{c}^L, \mathbf{c}^H$, are defined by (i) the participation constraint for a coarse customer, and (ii) the incentive constraint, IC_H, which prevents a sophisticated customer from drinking the cheap wine. In the optimum any movement of IC_H, either to the left or to the right, would reduce the seller's profit.

Of course, there may not be an optimum at a positive quality $q_L > 0$. This is particularly likely if the probability β of a customer being coarse is low. In that case the seller does not offer any wine for a coarse palate. Instead he sells the better wine under the original conditions, $\mathbf{c}^H$, illustrated in Figure 14.8. A coarse

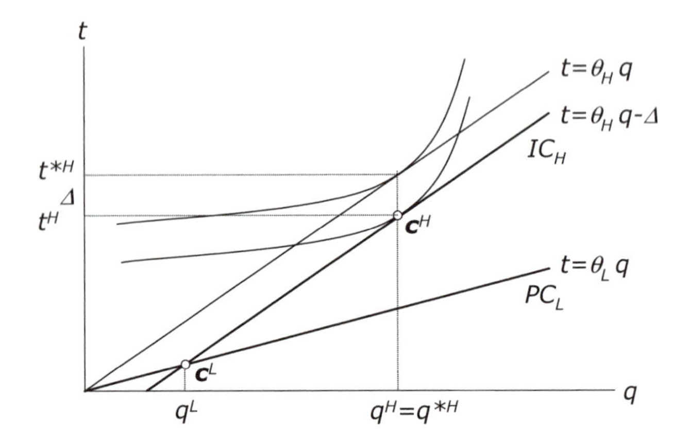

Figure 14.10 Optimal contracts under hidden information compared to the first-best imply (i) a lower quality for a coarse customer and (ii) a lower price for a sophisticated customer.

customer would have to drink water, while a sophisticated customer would receive *Cabernet*, albeit at a price that would make her indifferent to drinking water.

Digression: The seller's problem formally

Let us finally state formally the seller's problem under hidden information. The seller solves:

$$\max_{q_L, t_L, q_H, t_H} \{\beta\,[t_L - G(q_L)] + (1-\beta)\,[t_H - G(q_H)]\}.$$

subject to two participation and two incentive constraints:

$$\begin{aligned} \theta_L q_L - t_L &\geq 0 \\ \theta_H q_H - t_H &\geq 0 \\ \theta_L q_L - t_L &\geq \theta_L q_H - t_H \\ \theta_H q_H - t_H &\geq \theta_H q_L - t_L \end{aligned}$$

As should have become clear from the above figures, the solution is defined by the first and the last of these constraints: The participation constraint for a coarse customer and the incentive constraint for a sophisticated customer bind, i.e. they hold with equality. Therefore, we can substitute both constraints, $\theta_H q_H - t_H = \theta_H q_L - t_L$ and $\theta_L q_L = t_L$, in the seller's maximization problem.

In addition, we know that the optimal higher quality, q_H, is equal to its first-best level. The expression $q_H\theta_H - G(q_H)$ (the profit from selling the superior wine $\mathbf{c}^H$ to an observably sophisticated customer) thus is a constant in the maximization problem. Therefore, we can write the problem as:

$$\max_{q_L}\{\beta\,[q_L\theta_L - G(q_L)] + (1-\beta)\,[-q_L(\theta_H - \theta_L) + q_H\theta_H - G(q_H)]\}.$$

The first-order condition reads:

$$\beta\left[\theta_L - \frac{\partial G(q_L)}{\partial q_L}\right] - (1-\beta)(\theta_H - \theta_L) = 0,$$

or:

$$\frac{\partial G(q_L)}{\partial q_L} = \theta_L - \frac{\beta}{(1-\beta)}(\theta_H - \theta_L) < \theta_L,$$

from which it follows that $q_L < q_L^*$. A coarse customer gets a wine which is inferior to what he would get if types were observable (*Table Wine* instead of *Merlot*).

Characteristics of the optimum

The optimum exhibits five properties which also hold in more general models (Salanié 1997, pp. 25–26):

- A sophisticated customer obtains an *efficient* allocation (*Cabernet*).
- A coarse customer obtains an *inefficient* allocation (*Table Wine* instead of *Merlot*).
- A sophisticated customer is indifferent to her contract and to that of the next lower type.
- A sophisticated customer obtains an *information rent* $\Delta = t_H^* - t_H$.
- A coarse customer obtains zero surplus.

The sophisticated customer corresponds to the high-risk individuals in the insurance model. Both have a strong demand for the seller's product (insurance and wine respectively). What may look confusing is the fact that wine connoisseurs seem rather fortunate, while individuals with a fragile health are rather jinxed. Yet, previously in Chapter 13 we encountered an example where the fortunate group (professors) were the low-risk type in one context (life insurance) and the high-risk type in another (annuities). The situation was reversed for the less fortunate group (construction workers).

Optimal contracts under hidden information with more than two types are characterized by (i) information rents increasing from zero for the lowest customer type to the highest value for the highest type, and (ii) distortion (efficiency loss) increasing from zero for the highest customer type up to a maximum for the lowest type. The fact that the highest type gets an undistorted allocation (here: *Cabernet*) is often referred to as "no distortion at the top". As mentioned in Bolton and Dewatripont (2005, Ch. 2), it plays a role in many contexts, such as credit rationing and optimal income taxation.

The two effects—an information rent for the sophisticated customer and the distortion for the coarse customer—are closely linked, as Bolton and Dewatripont (2005, p. 56) point out. The seller would like to reduce the sophisticated customer's information rent. Yet charging a higher price for the superior good makes the inferior good more attractive. In order to prevent a sophisticated customer mimicking a coarse customer, the seller has to lower—i.e. to distort downwards —the quality of the good for the coarse customer.

14 E Application: Subordinated debt

14 E a Subordinated debt and bank supervision

Financial markets have developed a whole gamut of so-called hybrid securities, covering the middle ground between senior or secured debt and equity. Examples are junior or subordinated debt, convertible debt or preferred equity.

Among these, subordinated debt has come to play an important role in the discussion on the optimal supervision of banks. The starting point is a deficit in corporate governance that characterizes banks. As discussed in Section 6 F, banks typically borrow from a large number of depositors. These lack the ability and/or the incentive to collect information about the bank's solvency (Dewatripont and Tirole 1994, Section 13). This deficit could be made up by delegating supervision (i) to a specialized agency, or (ii) to holders of subordinated debt.

Subordinated debt is the lowest ranking type of debt in the hierarchy of claims. If a debtor falls bankrupt, secured and senior claims are paid back first. Junior or subordinated debt is only repaid after all other liabilities, including deposits, have been honored. Putting it the other way round, subordinated debt is the first shock absorber after equity.

Unlike the value equity, however, the value of subordinated debt does not reflect any upward potential of the debtor's assets. Therefore it does not increase with an increase in those asset's risk.[98] In addition, subordinated debt, compared to bank depositors, is held by professional investors who are not protected by deposit insurance.

For these reasons, subordinated debt has been proposed as:

- an instrument of market discipline;
- a source of information.

Subordinated debt provides some *market discipline* as a bad or risky bank either cannot raise subordinated debt or has to pay a high risk premium (for an example, see Figure 7.3). Subordinated debt is a *source of information*, since the yield has to compensate the holders of such debt for expected losses. It is often believed that the subordinated debt spread (the difference in yield compared to a treasury security with comparable characteristics like maturity, etc.) would directly measure the default risk of a bank (as perceived by the sophisticated investors who hold subordinated debt).

A number of proposals have been made that would require banks to issue subordinated debt instruments (Board of Governors and Department of the Treasury, Appendix A). Supervisors may also benefit from the information content of subordinated debt prices, as discussed in Chapter 7 E. Here we do not evaluate any such proposals. Our aim is to understand why subordinated debt is issued in the first place. In other words, we try to describe subordinated debt as an optimal contract.

14 E b A model of subordinated debt

Optimists and pessimists

The basis of our explanation of subordinated debt is the unobservable diversity of opinion. Relatively pessimistic investors buy senior debt; optimistic investors buy subordinated, or junior debt. The model we use is a simplified version of the model in Birchler (2000).

A risk-neutral bank wants to finance a project (think of the project as the bank's aggregate portfolio). The bank has no liquid funds; to invest in the project it has to borrow from a number of equally risk-neutral investors. Investors have one dollar each, and we normalize their number to one. Investors can either buy a risk-free asset (a treasury bill) with a safe-per-dollar gross return r, or they may lend to the bank. The bank invests borrowed funds into some project. The project is risky; when the bank has to repay its debt, it may either be solvent or fail.

Investors have private beliefs about the probability p with which the bank will remain solvent. A fraction β of investors is optimistic, fraction $(1 - \beta)$ is pessimistic. Optimists believe that the bank remains solvent with high probability p_H; pessimists believe it remains solvent with low probability $p_L < p_H$. The bank holds the average belief $\bar{p} = \beta p_H + (1 - \beta)p_L$.

The bank's project yields a gross, per-dollar return of $e_g > r$ if the outcome is good (bank solvent) and of e_b if the outcome is bad (bank fails). The project is risky in the sense that in the bad case it does not repay the investment, let alone the return on a risk-free asset or of a successful project:

$$e_b < 1 < r < e_g.$$

Yet, the net value of the project is strictly positive even from a pessimistic point of view:

$$p_L e_g + (1 - p_L)\, e_b - 1 > 0.$$

Contracts

The bank as a monopolist offers investors take-it-or-leave-it contracts. Contractual payments may depend on observable states of nature, i.e. on terminal solvency or failure. Contracts cannot be written based upon investors' unobservable beliefs. They thus have the form $\mathbf{c} = \{c_g, c_b\}$, with subscripts to payments referring to the two possible states.

We assume that it is profitable for the bank to borrow from both types. It may do so with a pooling contract or it may try to sell different, separating contracts to each of the two groups. In either case, the bank has to observe a number of constraints, as we will shortly see.

One set of constraints relevant in the present context comes from the legal environment: Due to limited liability, investors cannot promise net payments to the bank, nor can the bank promise more than the value of its assets in the respective state. One consequence of limited liability is that the bank (which has no capital of its own) would always want to borrow, even if it held pessimistic beliefs.

Observable types

Again we shall first look at the benchmark case of public information. What contracts would the bank offer if it were able to distinguish types? We presume readers are now sufficiently familiar with writing down maximization problems and we shall go directly to the graphical solution as given in Figure 14.11.

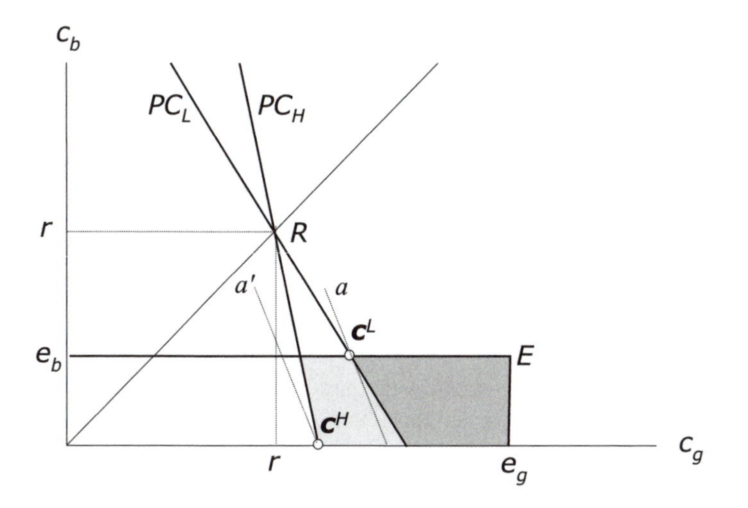

Figure 14.11 Optimal contracts sold to pessimistic investors, $\mathbf{c}^L$, and to optimistic investors, $\mathbf{c}^H$, are defined by (i) the two participation constraints and (ii) wealth constraints $c_b \leq e_b$ and $c_b \geq 0$, respectively.

In Figure 14.11, a contract is represented by a point in c_g-c_b–space. Limited liability of both the bank and investors means that a contract has to lie within the rectangle between the origin and the endowment point E. Payments by the bank cannot be negative, nor may they exceed the gross project return. These are *wealth* or *feasibility constraints*.

Furthermore, contracts have to "beat" the risk-free asset which is represented by point R. To do so, a contract must lie on or above a line running through R, called the investor's *participation constraint*. Participation constraints have slope $-p/(1-p)$, i.e. investors' rate of substitution between income available in the two outcomes. For optimists, the participation constraint is PC_H, for pessimists it is PC_L. PC_H is steeper than PC_L, because optimists believe in a higher solvency probability than pessimists. Compared to pessimists, they are ready to give up a smaller amount of income in the good state (bank is solvent) for one more dollar of return in the bad state (bank fails).

Finally, the bank's iso-profit lines have a slope between those of investors' participation constraints, since the bank holds the average belief $\bar{p}$. An iso-profit line closer to the origin means higher profits to the bank (smaller payments to investors).

Having all the elements of a solution ready, we first look at each group in isolation. Let us start with *pessimists*. A contract that is feasible and that will be accepted by pessimists lies in the darkly shaded area in Figure 14.11. The maximum profit the bank can achieve is indicated by the iso-profit line that intersects with the area in $\mathbf{c}^L$. That point represents the optimal contract offered to pessimists. It is defined by the pessimists' participation constraint PC_L and by the bank's limited liability in case of failure. The bank insures pessimists as good as it can by leaving them all its assets in case of failure.

A contract for *optimists* must lie on or to the right of PC_H, which includes the lightly shaded area in Figure 14.11. The maximum profit the bank attains within this area corresponds to iso-profit line a'. It intersects with the permissible area in point $\mathbf{c}^H$. Therefore, $\mathbf{c}^H$ represents the optimal contract for optimists. It is defined by optimists' participation constraint PC_H and by investors' limited liability in case of failure, $c_b \geq 0$. Under this contract investors insure the bank by claiming no income if the bank fails.

The subordination effect

Points $\mathbf{c}^L$ and $\mathbf{c}^H$ indicate contracts that are optimal for the bank if it looks at each group of individuals in isolation, i.e. *as if* the other group did not exist. The contracts do not take into account that the bank borrows from both groups *simultaneously*. When the bank sells contracts to both types of investors, it can exploit the fact that optimists under $\mathbf{c}^H$ receive no income if the bank fails. The bank can promise pessimists a higher bankruptcy dividend, by leaving them all assets available in case of failure. This includes the assets originally financed by optimists. As optimists have fraction β in the population, the bank can promise pessimists a bankruptcy dividend of e_b/β, rather than just e_b. We refer to this softening of the wealth constraint for pessimists due to optimists' readiness to forgo failure income as the "subordination effect".

The subordination effect (for $\beta = 1/2$) is illustrated in Figure 14.12. The (binding) wealth constraint on the bankruptcy dividend for pessimists is lifted

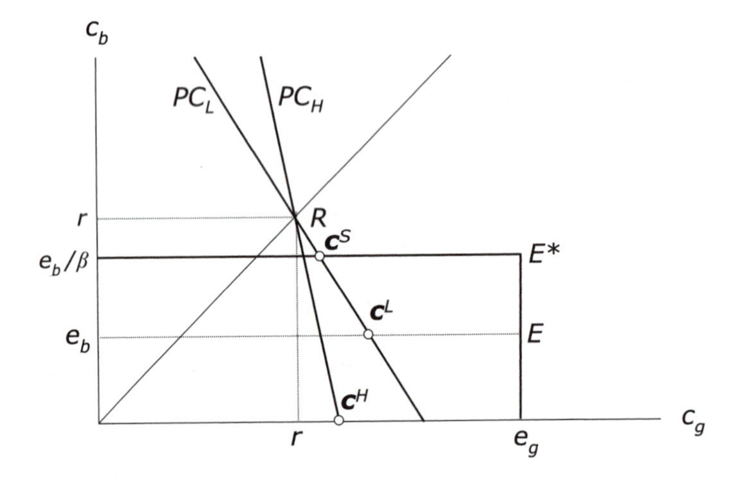

Figure 14.12 Due to the subordination of the contract sold to optimistic investors $\mathbf{c}^H$, the bank can promise pessimistic investors a higher bankruptcy dividend ($e_b/\beta > e_b$). Pessimistic investors get contract $\mathbf{c}^S$ rather than $\mathbf{c}^L$. The figure is drawn for $\beta = 1/2$.

to $c_b = e_b/\beta$. Together with the participation constraint PC_L, it defines $\mathbf{c}^S$, the optimal contract sold to pessimists *in the presence of optimist lenders*. To indicate its priority over optimists' claims we call $\mathbf{c}^S$ the *senior* contract.

Unobservable types

When the bank cannot distinguish types, a contract pair like $\mathbf{c}^S$ (or $\mathbf{c}^L$) and $\mathbf{c}^H$ in Figure 14.12 cannot coexist. As $\mathbf{c}^S$ lies above optimists' participation constraint PC_H it represents a higher utility level for optimists than $\mathbf{c}^H$. Optimists therefore would claim to be pessimistic and buy $\mathbf{c}^L$, the contract designed for pessimists, rather than their own, $\mathbf{c}^H$.

The bank therefore has to observe a further constraint, namely an *incentive constraint* preventing optimists from claiming to be pessimists. In Figure 14.13 this constraint is represented by line IC_H. That line runs through $\mathbf{c}^S$: An optimist is indifferent between the contract designed for pessimists, $\mathbf{c}^S$, and all other points on IC_H (we assume indifferent investors prefer "their" contract). To satisfy the incentive constraint the bank has to offer optimists slightly more than under the old contract $\mathbf{c}^H$. In Figure 14.13 the optimal contract sold to optimists corresponds to point $\mathbf{c}^J$. We call the contract $\mathbf{c}^J$ a *junior* or subordinated contract.

The contract pair $\mathbf{c}^S$ and $\mathbf{c}^J$ can coexist, even if the bank cannot distinguish between different types of investor. Pessimists buy senior debt, while optimists buy junior debt. Optimists, like the sophisticated wine lovers in Section 14 D, receive an information rent. In Figure 14.13 this rent has size Δ. It becomes smaller, the safer the senior contract is. The closer the bankruptcy dividend to

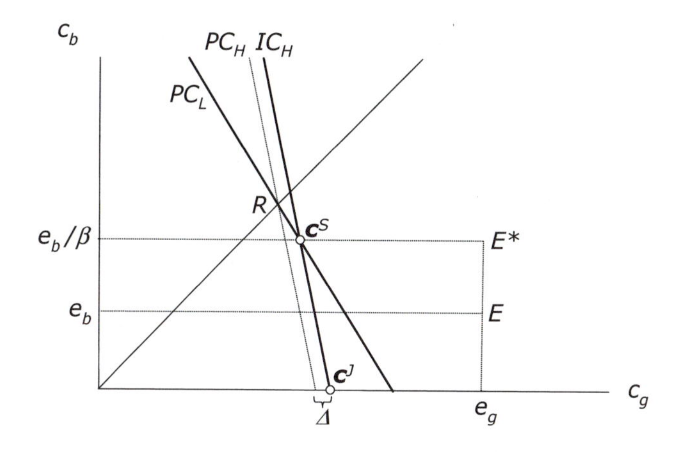

Figure 14.13 Pessimists buy a senior contract, $\mathbf{c}^S$, optimists a junior contract, $\mathbf{c}^J$. The two contracts are defined by (i) pessimists' participation constraint, (ii) optimists' incentive constraint, and (iii) wealth constraints $c_b \leq e_b/\beta$ and $c_b \geq 0$, respectively. The difference between e_b/β and e_b reflects the subordination effect.

pessimists moves towards the riskless gross return r, the smaller the information rent that is required to make optimists accept the junior contract. The yield on junior or subordinated debt therefore does not necessarily reflect the bankruptcy probability of the bank. Partly it depends on beliefs of its holders, but partly it reflects beliefs held by pessimistic investors who do not even hold such debt. The latter effect may help to explain swings in banks' subordinated debt yield that would otherwise look erratic (Birchler and Hancock, 2004).

14 F Conclusions and further reading

We tried to show how optimal contracts or, more generally speaking, optimal mechanisms, can bridge the information gap between informed and uninformed parties. An optimal mechanism can provide the uninformed party with an economic "lie detector" to extract the informed party's secret. To the informed party, optimal contracts provide a form of credible communication, freeing the uninformed party from the "rational paranoia" called for in the situation where the cat is sold in a bag.

Compared to the extensive literature on optimal contracts, both theoretical and empirical, we have not managed to present even the tip of the iceberg. You could say, we have shown the equivalent of an ice cube! More complete, and often also more demanding textbooks are Bolton and Dewatripont (2005), Mas-Colell, Whinston and Green (1995, Ch. 14) and Wolfstetter (1999).

The present chapter focused on *screening* (the uninformed party moving first). As supplementary reading we would recommend Wolfstetter (1999, Chs. 9.5–9.6). By contrast, *signaling* (the informed party moving first) was somewhat neglected. We tried to convey the idea in Problem 14.7; a systematic treatment can be found in Wolfstetter (1999, Ch. 10) and an overview of the literature in Riley (2001).

The main technique in finding optimal contracts is optimization under inequality constraints (represented by the Kuhn-Tucker conditions). We have not explicitly introduced this method here; an excellent reference is Wolfstetter (1999, IV, B).

Finally, we should mention two issues spared for later chapters. One is the hidden action side of contracts: Contracts are not only used to separate the wheat from the chaff; they are also written to motivate parties to take the right action. This topic will be discussed in Chapter 16. The other issue is a very powerful device for finding optimal contracts under hidden information. It is called the revelation principle and will be treated in the following Chapter 15.

Checklist: Concepts introduced in this chapter

- Objective function
- Participation and incentive constraints
- Limited liability and wealth constraints
- Single (pooling) versus dual (separating) contracts
- Deadweight cost of contracting under hidden information
- Monopolistic versus competitive screening
- Insurance contracts

- Non-linear pricing
- Price-quality-discrimination
- Screening versus signaling
- Subordinated debt

14 G Problem sets: Deal or no deal?

For solutions see: www.alicebob.info.

Problem 14.1 Monopoly versus competition
Take the two-type model of Section 14 C. Assume that neither individuals nor insurers have any information about individuals' risk types (except their publicly known frequencies). What pooling contract is offered by:

(a) a competitive insurer?
(b) a monopolistic insurer?

Problem 14.2 Unobservable demand
This problem is based on Wolfstetter (1999, Ch. 1). A monopolist faces two customers, one of which is of type $\theta_1 = 1$, the other of type $\theta_2 = 2$. Customers' utility functions are $u_i = x - \frac{1}{2\theta_i}x^2 - T_i$, where a high θ indicates that the customer has stronger preferences for the good. T_i is the total amount an individual pays to the monopolist in exchange for goods. The monopolist maximizes profits; her unit costs are constant and equalized to 0.

(a) What is the *marginal willingness to pay* of each customer?
(b) What quantities does the monopolist sell to each type, if she can distinguish between the types? What is her profit?
(c) Consider the case, where the monopolist cannot distinguish between the types. Derive the demand functions in the case of linear pricing ($T_i = px$). What are each buyer's surpluses? Calculate the joint demand. What price does the profit maximizing monopolist set? What is her profit?
(d) Consider the case, where the monopolist offers two different price/quantity combinations (contracts) $[T_i, x_i]$. Set up her maximization problem subject to the participation and incentive constraints. Are all of the constraints binding? What are the two optimal contracts? Calculate the monopolist's and consumers' surpluses.
(e) Compare and comment on the different outcomes from above.

Problem 14.3 Subordinated debt
In Section 14 E we introduced a model of subordinated debt. For that model, write down the bank's maximization problem (highlighting the binding constraints) for:

(a) public information (observable types)
(b) hidden information (unobservable types)

Problem 14.4 Beta Julia

You are offered a 1950 Beta Julia, a rare veteran car model you have been dreaming of for a long time. You know that the model only exists in two qualities. The good quality is twice as frequent as the bad quality. The value of the qualities to you is: $3,000 for the good quality and $1,500 for the bad quality. For the seller the values are between $2,000 and $2,500 (good quality) and between $1,000 and $2,000 bad quality. You are very risk-averse; the seller is risk-neutral. You are supposed to make a take-it-or-leave-it offer.

(a) What price would you offer if the car's quality were observable?
(b) What price do you offer for a car with unobservable quality?
(c) How much would you pay for a test revealing the quality of the car?
(d) What contract would you offer if the quality of a bad car can be verified *ex post* with 60 per cent probability?

Problem 14.5 Gill Bates

You work in a bank specializing on wealth management. One day, a guy calls you, claiming you know him from college. You vaguely remember him but you could not say whether he is trustworthy or not. He claims to be a close acquaintance of Gill Bates, the richest woman in the world. He could persuade her to become your customer, and, yes, he could use some cash.

You assume that your chances to do business with Gill Bates without any help are 5 per cent. With your colleague's help they would increase to 50 per cent—provided he told you the truth about his contact with the billionaire.

You assume that your colleague has a utility function $u = \sqrt{w}$, where w is the amount of cash you pay him. If he really is an acquaintance of Gill Bates and builds a contact for you he foregoes 20 utility units. If he is just bluffing, he only foregoes 11 utility units. If he rejects your offer, his utility level remains unchanged. You are risk-neutral, but you do not want to be bluffed by this guy.

(a) What would you offer if you could verify the person's claim?
(b) Is there a contract the person would only accept if he really is an acquaintance of Gill Bates?
(c) What contract do you offer (if you cannot verify the person's claim)?
(d) What would you pay for the ability to verify the person's claim?

Problem 14.6 The new headquarters

As chairman of Highfly.com you want to build new corporate headquarters. You have the choice among two risk-neutral contractors. *Ex ante*, these are of equal quality. However, once you have made an offer, the contractors privately learn about their individual cost to build your headquarters which may be either $240 billion ("high") or $160 billion ("low"). The costs of the two contractors are independent; each contractor has a fifty-fifty chance to have either high or low cost. A contractor builds the headquarters if your offer covers at least the cost.

You are risk-neutral. The value of the new headquarters to Highfly.com is $360 million.

(a) What are your expected costs of the new headquarters if you can observe the contractors' constructing costs?
(b) What is your expected net benefit if you can observe contracting costs?
(c) What contract do you offer if you cannot observe construction costs?
(d) What is your loss of expected benefit due to the non-observability of construction costs?
(e) What is the aggregate loss (to all parties) from the non-observability of construction costs?
(f) What contract do you offer if the value of the new headquarters to Highfly.com is $540 million (and construction costs are not observable)?
(g) What is the aggregate loss from the non-observability of construction costs in case (e)?

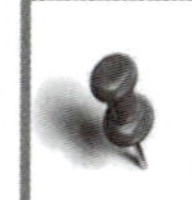

Bob would like to convince Alice that he is a cool guy. His first idea is to rent a Ferrari for the next weekend. However, Alice may not really appreciate it. A better idea is to spend his spare time as a babysitter, even though it's poorly paid. The question is: How many hours need Bob spend babysitting in order to convince Alice of his coolness?

Problem 14.7 Baby signals

In Chapter 1 Lady Kunigund inadvertently signaled the absence of love for Knight Delorges. In many economic contexts, signaling is an important technique to bridge the "information gap" between an informed and an uninformed party. Unlike the Lady, agents send signals consciously and deliberately, with the clear purpose of gaining credibility.

In a signaling game, the *informed* party moves first, unlike in the screening game used in the main text. Signaling is an activity which is less costly to a party offering a high quality product than to a party offering a low quality product. Guarantees, brand names (the value of which would be damaged by selling a bad quality), education—these can all serve as signals.

Signaling is not part of contracting, strictly. The signaling party just expects his or her payoff to depend on the level of the signaling activity. For example, a student expects that wages increase in the education level. Such expectations are the contractual background for the signaling activity, so to say. Signaling is considerably more tricky than screening, since the payoff a signaling party may expect depends on the beliefs the uninformed party (the employer, in our example) holds after observing the signaling activity. Instead of going into the complexities, let us tackle an example. For more elaborate presentations of signaling, see Hirshleifer and Riley (1992, Sec. 11.3) (from where the example is inspired), Salanié (1997, Ch. 4) and Riley (2001).

Let us assume, that Alice's appreciation, a, of a person's (not directly observable) coolness varies on a scale from 0 (totally uncool) to 1 (cool). Bob's utility is the difference between Alice's appreciation of him and his cost of winning the appreciation. In the present context, this is the cost of babysitting, c, which depends on the hours spent and on the babysitter's type. Bob's utility is:

$$U = a - \frac{1}{\theta} b,$$

where b is the number of hours spent with babies, and θ is Bob's type. We assume that θ is in a range from θ_0 (uncool) to θ_1 (cool).

(a) Assume Bob is cool, having $\theta = 1$. Draw a set of Bob's indifference curves in a b-a-graph. Which indifference curve is Bob's participation ($b \geq 0$) constraint (assuming that $a(0) = 0$)? [Exceptionally, for those who get stuck, we offer some help in Figure 14.14.]
(b) What should Alice believe regarding Bob's coolness if she observes him babysitting for θ hours?
(c) Is it possible to say what Alice should believe if she observes Bob babysitting for $\theta_0 < b < \theta_1$ hours?

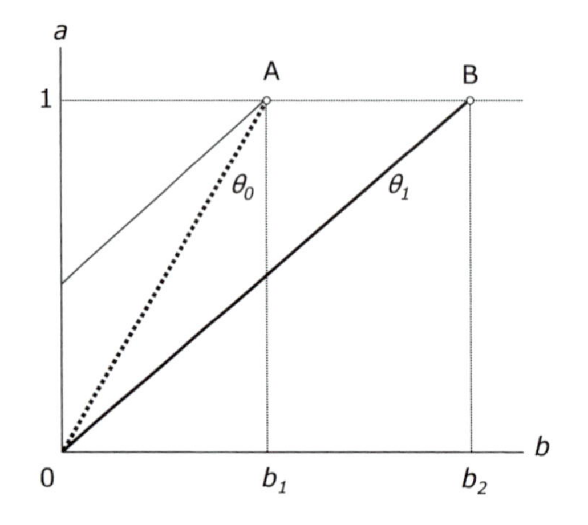

Figure 14.14 Curves of indifference between reward (a) and signaling activity (b) are flatter for high ability individuals (solid lines) than for low ability individuals (dashed line). The maximum signaling level acceptable to a high ability individual is b_2. The minimum required to signal high ability in the presence of on type of low ability individuals is b_1.

(d) Assume that there is a continuum of competitors for Alice's appreciation with types $\theta \in [\theta_0, \theta_1]$. Bob has the highest θ, θ_1.

(d1) What is the number of hours, b^*, Bob will babysit?
(d2) and what will be Alice's appreciation of him?

(e) Assume that there are only two competitors for Alice's appreciation, Bob and Simon. Alice does not know which of the two has type $\theta_0 = 1/2$ and which has type $\theta_1 = 1$.

(e1) What is the maximum number of hours type θ_0 will babysit for an appreciation of $a = 1$?
(e2) What is the optimal number of hours type θ_1 will babysit (given the competitor has $\theta_0 = 1/2$)?
(e3) What is the optimal number of hours type θ_0 will babysit (given the competitor has θ_1)?
(e5) What utility levels do the two competitors achieve if types are observable?
(e6) What utility levels do the two competitors achieve if types are not observable?

(f) Would better wages for babysitters or nicer babies increase or decrease Bob's utility loss from asymmetric information?

15 The revelation principle

15 A Introduction

The famous Muslim theologian Al Ghazali (AD 1058–1111) wrote: "Know that a lie is not wrong in itself. If a lie is the only way of obtaining a good result, it is permissible. We must lie when truth leads to unpleasant results." Today, Al Ghazzali's pragmatic view is repeated again and again in education manuals: Children cannot but lie if they fear harsh punishment for admitting the truth. In this respect, *homo œconomicus* is merely an overgrown child; having an incentive to lie (s)he will. For economists the main sinner is not the liar but the person who creates an incentive for others to lie.

A stunning result of game theory says that there is no reason to create such an incentive. Nobody can gain by making others lie. This is the quintessence of the so-called "revelation principle". Though an abstract mathematical concept, the revelation principle matters in practice. It explains, for example, why borrowers often ask lenders to repay a fixed sum, rather than a percentage of their returns, as we will discuss in our first Application, devoted to the debt contract. The revelation principle also explains why in an auction the seller should not expect to get more than what the second highest bidder would be ready to pay.

The revelation principle is also relevant for important economic policy issues. Mookherjee (2006, p. 360), for example, links the revelation principle to the debate about centralization or decentralization: "Under some additional assumptions, this Principle establishes that centralized control cannot be dominated by any delegation arrangement", as "the outcome of any decentralized organization can be mimicked by a centralized organization in which the responsibility of each agent is merely to communicate their information to a central authority and await instructions on what to do." If a central body could do just as well, why then did Von Hayek (1945) put so much stress on the market's role of aggregating decentralized information (see Chapter 7)? He must have felt that the assumptions behind the revelation principle are often violated in reality (Mookherjee 2006, p. 360). Some accounting scams that occurred in the early years of the twenty-first century suggest that it may indeed be difficult to set incentives for truthful reporting in a large centralized organization. This will be the topic of our second Application, devoted to the failure of Enron.

In the present chapter we will introduce the revelation principle in a non-formal way. In particular, we will try to support the reader's intuition with a number of examples. We will try to show that the revelation principle is an extremely useful tool, both for understanding existing financial contracts and for formulating new contracts (or, more generally, mechanisms). We will also try to indicate the limits of the revelation principle.

15 B Main ideas: Many lies, one truth

The author of a widely used microeconomics textbook begins his explanation of the revelation principle by stating: "The revelation principle is one of those things that is obvious once you understand it but somewhat cumbersome to explain" (Kreps 1990, p. 691). Let us nevertheless try to explain it. As indicated in the Introduction, we will follow an intuitive approach, avoiding the use of formal language as much as possible.

The very point of information asymmetries is that the better informed person may lie to the lesser informed person. In many cases he even has an incentive to do so. The seller of a used car tends to say nothing about the difficulties starting the engine in winter. As a result, a potential buyer will try to anticipate the seller's potential for lying. As different contracts lead to different lies, and different lies require different contracts, the number of candidates for an optimal contract can become very large. Writing an optimal contract looks like a tricky business indeed.

Then an idea comes along which makes life much more simple: the *revelation principle*. The revelation principle says that the party formulating a contract can, without any loss of generality, restrict attention dramatically. Whatever can be achieved, can be achieved by a contract under which the counterparty does not mind telling the truth. More technically, the revelation principle says that any allocation that can be generated by some mechanism can be generated by a direct truthful mechanism. "Any allocation" means any monetary and non-monetary payoffs, including their respective probabilities. In a direct, truthful mechanism the uninformed person first asks the informed person to (directly) declare the (true) value of the secret parameter. Payoffs are then shared according to a scheme that does not punish the informed party for being truthful.

The revelation principle is very practical for two reasons. First, it dramatically simplifies the search for optimal contracts. We only have to consider the subset of all contracts that lead to truth-telling. As there is only one truth about the value of an economic parameter but an infinite number of potential lies, this is indeed a considerable simplification. Second, the revelation principle shows which outcomes are achievable and which are not. Any outcome that cannot be achieved by a truthful mechanism cannot be achieved at all. The uninformed party cannot derive from the informed party that which it derives from a truth-telling mechanism. Implementing a complicated scheme under which the informed party does its utmost to resort to a complicated web of lies thus serves no purpose, unless the informed party is too confused to behave rationally!

15 C Theory: The revelation principle

15 C a Do not punish for truth

A starting example

Alice hates it when Bob devours her homemade cookies, while never complimenting on her cooking. This time she will make him admit that he likes a cookie. He'll even pay for it! "Here's a cookie, Bob. Maybe you'd like to have it. You might even get it. Just tell me how much you would pay for it. But don't think you can get it for next to nothing. I'll put a minimum price in this envelope. If you bid less, I'll keep the cookie. Clear? So, make your offer!"

We have to assume, of course, that Bob does not know Alice's secret minimum price. Let him believe it lies in the interval [0,V]. A negative minimum price would not match Alice's self-esteem as a chef. Nor would she believe that his valuation of her cookies exceeds V. With anything in the interval appearing equally likely, Bob thinks of Alice's minimum price as drawn randomly from a Uniform distribution in the interval [0,V].

Bob's true valuation of Alice's cookies is either zero (when he thinks he is on a diet) or V (when he has a craving). Let us assume, Bob mentally tosses a coin, making both outcomes equally likely. If the coin says "no" (0) Bob would bid nothing. But if the coin says "yes" (V), as happens to be the case, what should he bid? He ponders two extremes: A bid of zero would maximize his net benefit from getting the cookie, but certainly not beat Alice's minimum price. A bid of V, on the other hand, would maximize the odds of beating the minimum price (without paying more than the cookie is worth), but would wipe out any net benefit. Before he gives himself a headache, Bob might decide to keep the golden mean and to bid $V/2$.

Bob's optimization problem

Indeed, with valuation V, Bob should bid $V/2$. To demonstrate this, we shall express Bob's optimization problem in words. Bob maximizes his expected benefit W, which is the probability that his bid B exceeds Alice's secret reservation price R, times the net surplus value of the cookie ($V - B$):

$$\max_B W = \Pr(B \geq R)(V - B). \qquad (15.1)$$

Bob knows that, with Alice's reservation price R drawn from a Uniform distribution $U \sim [0, V]$, the probability of beating R is:

$$\Pr(B \geq R) = B/V.$$

Substituting for $\Pr(B \geq R)$ in (15.1) yields

$$\max_B W = \frac{B}{V}(V - B) = B - \frac{B^2}{V}.$$

Both extreme offers, $B = 0$ and $B = V$, yield a zero net benefit, but any value of B in between leads to a positive expected benefit. The optimal bid B^* can be found by computing the first-order condition with respect to B:

$$\frac{\partial W}{\partial B} = 1 - \frac{2B^*}{V} = 0 \Rightarrow B^* = \frac{V}{2}.$$

Implications

It is therefore rational for Bob to bid half his valuation. What does this result imply? Recall that he never gets the cookie in half of the cases when his valuation is low. When his valuation is high, he offers B^* and gets it with a probability of 50 per cent. Taken together, he gets the cookie in 25 per cent of all cases. Whenever he gets the cookie he pays $V/2$. Alice's expected revenue (Bob's expected expenditure) thus is $V/8$. Note that due to informational asymmetry, Bob sometimes does not get the cookie although his valuation exceeds Alice's secret price. Most importantly, though, Bob *lies* about his true valuation. He bids less than what he would be ready to pay.

The example continued

Alice suspects Bob is lying, but Bob replies: "*You* made me lie; you tried to get as much out of me as possible!" Alice reflects and comes up with a different set of rules. "Bid again," she says, "I'll make a concession. You only pay my secret price, even if your bid is much higher—just to show that I don't want to get 'as much out of you as possible', as you claim.

These are the new rules:

- If you bid less than my secret price, I keep the cookie.
- If you bid at least my secret price I toss a coin: 'heads' I sell, 'tails' I don't.
- If I sell, you pay my secret price, even if your bid is higher. Ready?"

Bob may briefly reflect. With valuation 0, he had better make no bid. With a valuation V he might still bid $V/2$ as before. However, Alice's secret price might be $0.7V$, say. This is an amount he would be happy to pay, but his bid of $V/2$ would be rejected. All things considered, there seems to be only one way to qualify for the coin-tossing option and to get the cookie. Bob summons up his courage and bids V.

With Bob's new strategy, Alice sells her cookie in 25 per cent of all cases—whenever Bob has a high valuation V *and* the coin shows 'heads'; two events that have 50 per cent probability each. As $E(R) = V/2$, Alice's expected net revenue is $V/8$ $(= 0.25 * V/2)$, just like in the first game. The final outcome (probability of the transaction, and expected revenue) is exactly the same as in the first game. The main difference is that Bob now comes out best by truthfully declaring his valuation. Instead of asking for a bid, Alice could just as well have invited him to declare his valuation directly. (Note that Bob never overstates his valuation by declaring it to be $2V$, even if he thinks Alice's secret price may be above V. Overstating the valuation means that his only risk would be overpaying for the cookie).

Implication: Truth-telling

Alice has replaced the first game by another game which leads to the same outcome, but has two remarkable properties:

1 Alice directly asks Bob to declare his valuation;
2 Bob has an incentive to declare truthfully.

The revelation principle is nothing more than a generalization of this example. It says that any mechanism which leads to lying can be replaced with another direct and truthful mechanism, which in turn leads to the same outcome. "Mechanism" means the game, starting from Alice's offer and ending with a virtual authority-enforcing agreement if necessary. "Direct" refers to the invitation to declare the unobservable variable; "truthful" refers to the incentive to declare truthfully.

This is a fairly vague definition of the revelation principle. In the literature there are a number of more formal definitions (Gibbons, 1992; Kreps, 1990; Mas-Colell, Whinston and Green, 1995; Laffont and Martimort, 2002; Bolton and Dewatripont, 2005), some followed by a proof of the principle. There are remarkable differences between individual treatments, for several reasons. First, the revelation principle, starting with special cases, was generalized over time. Second, it can be introduced in different settings (public procurement, monopolist pricing, auctions). Third, every textbook has its own style. Each reader may thus find a personal favorite among the various formulations. We find that the approach in Laffont and Martimort (2002) strikes a good balance between being intuitive and formal. We thus try to present it here in a somewhat simplified form.[99] The setting is that of a principal (Alice) writing a contract for an agent (Bob), who knows his unobservable type (valuation for cookies).

A direct revelation mechanism

First, we need a few definitions:

- The agent is of a privately known type $\theta \in \Theta$.
- Contract parties can agree on allocations A.

- The message space M is the set of possible messages m the agent can give.
- A mechanism is a message space plus a mapping g from the message space M into allocations A.
- A *direct* revelation mechanism is a mapping from Θ to A.
- A direct revelation mechanism is *truthful*, if truth-telling maximizes the agent's utility, i.e. if

$$u(a(m=\theta)) \geq u(a(m' \neq \theta)).$$

In the language of the above example: Bob knows his type θ. He and Alice can agree on all allocations for which they have sufficient cookies and money. The message space consists of all possible declarations Bob can make, e.g. on his valuation (measured by some real number). The mechanism is the message space plus the game Alice sets up to move from messages to allocations of cookies and money. A direct revelation mechanism is any game in which the allocation directly depends on Bob's message about his type, and an example of a truthful mechanism is the second game Alice designs.

The revelation principle

The revelation principle now says that:

> *Any allocation rule $a(\theta)$ obtained with a mechanism $(M, g(.))$ can also be implemented with a truthful direct revelation mechanism.*

Again in terms of the example: The outcome of the first game, in which Bob had to pay what he bid, or of any other game that makes Bob lie, can be implemented with a truthful mechanism in which Bob fares best if he declares his true type.

The revelation principle is extremely useful as it dramatically simplifies the contracting problem. Whatever outcome contract parties want to achieve can be achieved with a direct, truthful mechanism. As there are many lies but only one truth, this simplifies the contracting problem dramatically. Parties can, without loss of generality, concentrate on truthful mechanisms.

15 C b Assumptions and exceptions

A person filling out her personal income tax declaration just after having learnt about the revelation principle may be a bit puzzled. "Why don't they give me more of an incentive to declare truthfully? And why do they let others get away with even bigger lies?" This illustrates that the revelation principle rests on certain assumptions. Putting it simply, the revelation principle is applicable whenever truth-telling mechanisms can be implemented, but does not apply when there are barriers to truthful implementation.

The three big Cs

There are three important revelation principle violations found in practice. Arya, Glover and Sunder (2003) call these “the big Cs”: communication, commitment and contracting. We will discuss these briefly here, but we will meet them again in Application 15 F.

Communication

First, the implementation of truthful contracts is inhibited if parties cannot sufficiently *communicate* the relevant truth. In a principal–agent setting, the agent (the manager of a firm, say) may want to communicate both current earnings and expected permanent income; accounting standards, however, may only permit the declaration of one single figure as the firm’s “true” earnings. This may prevent the manager from making a truthful declaration. Another example is an auction with risk-averse bidders. If risk-averse bidders can only bid their valuation for the object to be sold, i.e. a single figure, they cannot truthfully reveal their preferences. Revelation of risk-averse preferences would require a statement about their convexity.[100]

Commitment

Furthermore, the principal’s inability to *commit* may prevent parties from employing truthful mechanisms. A manager may obtain a long-term contract in order to have an incentive for truthful reporting of interim performance. However, the firm’s owners may be unable to commit to not breaking the contract, and let the manager quit “by mutual agreement”. Or, imagine a job applicant who tells the employer: “I will accept the job for $5,000 a month. Just tell me what you would have been ready to pay.” This is a (weakly) truthful mechanism in a one-shot game with commitment. However, as the applicant may break his promise (or quit after a few months) once he learns that he is worth $10,000 to the firm, the mechanism is not implementable.

Contracting

Finally, parties may be unable to *contract* truthfully due to legal or political obstacles. A tax code, to take the example we began with, could in fact be written as a truthful mechanism. Only the result would be quite different from anything consistent with at least some notion of distributive justice. In addition, one might need a different tax code for every individual. Similarly, competition law often prevents monopolists from maximizing their monopoly rent by giving customers the opportunity to fully reveal their individual preferences. But restrictions on contracting are not always enshrined in the law. Even private parties, facing certain contracting and transaction costs, often refrain from “re-inventing the wheel” and resort to standardized contracts. These are very common in labor, rental and financial markets (insurance contracts, debt contract).

The above examples show that it may not be possible to use direct truthful mechanisms in practice. This makes the revelation principle less valuable in some situations than it potentially could be. But even where it is not directly applied, it remains useful. When we observe a mechanism that does not induce truth-telling by the agent, the revelation principle reminds us to ask: "What particular circumstances prevent contract parties using a truthful mechanism?" Arya, Glover and Sunder (2003), for example, wondered why firms give their managers an incentive to rig the books (see Section 15 F).

15 C c False implications

We have tried to explain what the revelation principle means. Let us also list a few things it does *not* mean.

"A truthful direct mechanism is optimal."

This is a wrong inversion of a true statement. An optimal contract can always be implemented as a truthful direct mechanism. But *not all truthful mechanisms are optimal.*

Just think of the Alice and Bob example. Alice could say: "You can have a cookie for free, but please tell me what you would have paid for it." Bob then has no reason to lie, but Alice certainly does not maximize her revenue. Indeed, the mechanism illustrated in the second game is truthful, but not optimal. Alice forgoes some revenue by tossing a coin, when Bob makes an offer she would accept in principle. The rule that she tosses a coin was only used to make the outcome of the second game identical to that of the first game. One of the most famous truth-telling mechanisms, the second price auction, is not optimal *per se* (see below); it is only optimal if the seller also sets an optimal reservation price. To recap, of the numerous mechanisms that do lead to truth-telling, many are not optimal. But each mechanism that leads to lying, whether optimal or not, has a "brother" in the world of truthful mechanisms, a brother who leads to the same outcome.

"Truth-telling solves the problem of asymmetric information."

Wrong again! In a transaction between an informed and an uninformed party information asymmetry normally implies some welfare loss. This loss, the so-called agency cost, can be minimized but not eliminated. Truth-telling *per se* does nothing to minimize agency cost; truthful mechanisms are just one relatively simple way of implementing a particular allocation. Under a truthful mechanism, agency cost comes in the form of the suboptimal risk allocation required as an incentive to tell the truth. While the truth comes for free if the mechanism is compared to any non-truthful mechanism, it comes at a cost even if the best truthful allocation (i.e. the second-best) is compared to the allocation possible under symmetric information (first-best).

"In contractual affairs one should always tell the truth."

While this may be true on moral grounds or within very long-term bilateral relationships, it is obviously wrong in a one-time transaction like the example of Alice selling Bob her home-made cookies. Truth-telling is attractive to the agent under some special class of contracts (so-called truth-telling contracts), while lying is attractive under all others. In any case, truth-telling is not an attractive option to the *principal*. The seller in an auction (unlike perhaps the bidders) has no incentive to reveal his true reservation price (see below).

A similarly wrong statement would be: "Never accept a contract that makes you lie." This may be morally correct but it is definitely not profit-maximizing. Contracts that make lying attractive to the informed party may be among the most profitable.

15 D Application: The debt contract

15 D a Why is there debt?

Debt is a standard contract . . .

Debt contracts are among the most common financial contracts. Under a debt contract the borrower promises a fixed repayment to the lender. Whenever the borrower can, he pays back that fixed amount when the debt becomes due (abstracting from interim interest payments). Only in the event of the borrower defaulting will the lender acquire whatever assets the former has.

. . . yet puzzling

Many people do not realize how puzzling the prevalence of debt contracts actually is. In fact, the debt contract implies very one-sided risk-sharing. For a broad range of results (terminal wealth of the borrower), the lender obtains a fixed repayment, while the borrower keeps the residual income which is a risky claim. The borrower "insures" the lender in this range. Only in the range where the borrower is unable to pay the full face value of the debt, does a certain degree of risk-sharing arise. Risk-sharing under a debt contract might be optimal if the lender is more risk-averse than the borrower. There are no indications, however, that this should generally be the case. To the contrary, lenders are often professional lenders, like banks, which should be able to diversify a major part of their risk and thus should be almost risk-neutral. Borrowers, though, are often individuals who are commonly assumed to be risk-averse. Why then would parties agree on an arrangement like a debt contract, which places most risk on the shoulders of the more risk-averse party? To put it simply: Why is there debt? (Lacker, 1991).

Model assumptions

The original analysis of the debt contract can be found in Diamond (1984).[101] In the Diamond model a risk-averse entrepreneur can invest in a project which yields an uncertain return $\tilde{y} \in [0, \overline{Y}]$. The realization Y is only observable by the entrepreneur, not by third parties. The entrepreneur has no liquid assets and has to borrow from a risk-neutral investor in order to run the project. The investor (the principal) formulates a contract as a take-it-or-leave-it option for the entrepreneur (the agent). The contract is enforced by a higher authority (who cannot observe the entrepreneur's return either). The authority can enforce financial payments or execute a contractual penalty on the borrower for breach of contract (for example, sending the entrepreneur to jail for a number of days).

The time line

The time line of events is given in Figure 15.1. In $t = 0$, the investor (the principal) offers a contract to the entrepreneur (the agent). The distribution of returns from the agent's project, $\tilde{y}$, and the fact that the project return is the agent's private information are both common knowledge as of $t = 0$. In $t = 1$, the agent accepts or rejects the contract. Where the contract is accepted, the first part of the contract is executed, i.e. the principal pays the agreed amount to the agent. In $t = 2$, the project return is realized and becomes known to the agent. The second part of the contract is executed, i.e. the agent pays back whatever was agreed with the principal, or incurs a contractual penalty. The fact that Y, the realization of $\tilde{y}$, is observed by the entrepreneur after contracting is sometimes referred to as *post-contractual* hidden information. The time line of the story differs somewhat from the time line in the case of *pre-contractual* hidden information, as discussed above and illustrated in Figure (13.1).

Contracting and the revelation principle

The principal's task in $t = 1$ is to formulate the optimal contract. A contract is a mapping of observable states of nature into agreed payments. The principal would like to contract on the possible realizations Y of the random variable $\tilde{y}$. Unfortunately, Y is not observable. However, the principal can find an observable substitute for Y: a *declaration* of Y by the agent. A declaration *is* observable.

The problem is that the declaration may not be truthful. Asked about realized returns the entrepreneur may answer with an understatement. At this point the

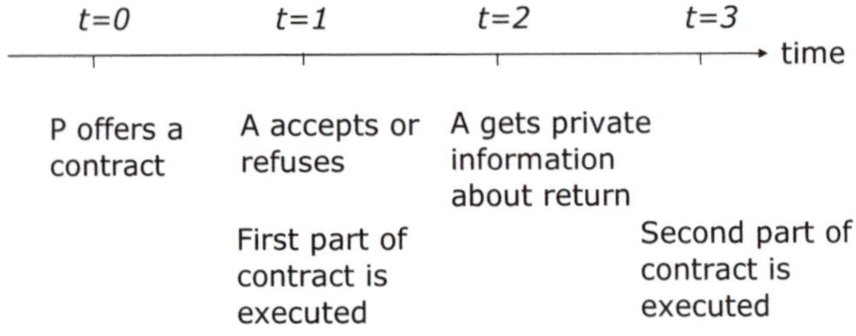

t=0 t=1 t=2 t=3 time

P offers a contract

A accepts or refuses

A gets private information about return

First part of contract is executed

Second part of contract is executed

Figure 15.1 The time line in a post contractual hidden information model.

revelation principle comes into play. Without loss of generality, the principal can restrict the search for an optimal contract to contracts that induce the entrepreneur to tell the truth.

Let us assume that the principal formulates such a truthful contract. A rational agent's declaration of Y is then identical to the true value of Y. By writing contracts on the truthful declaration, parties can indirectly contract on the unobservable parameter Y, as it stands for the—observable—declaration of Y. But we must never forget that this only holds under contracts that give the agent no incentive to lie.

Contracts in Y–z-space

A (truth-telling) contract $z(Y)$ is a mapping of return Y into the agent's repayment z. Graphically, a contract is any line in (Y–z)-space, as illustrated by Figure 15.2.

The optimal contract maximizes the principal's wealth subject to the constraint that it is (i) attractive to the agent (participation constraint), (ii) that it leads to truth-telling by the agent (incentive or truth-telling constraint), and (iii) that it is feasible (feasibility or wealth constraints).

The optimal contract

The optimal contract is illustrated in Figure 15.3. The steps to find it are the following:

1. The agent is invited to declare the true realization of return Y (revelation principle).
2. The agent's promised repayment, D, must be flat in Y (incentive constraint); this gives the agent no incentive to lie about the outcome.

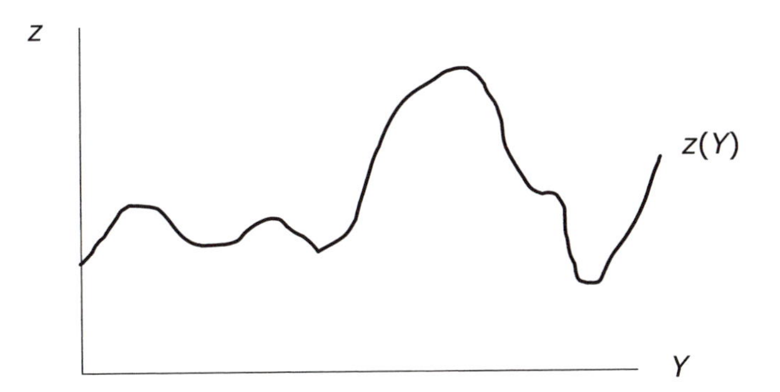

Figure 15.2 A contract can be any mapping of states (Y) into agreed payments (z).

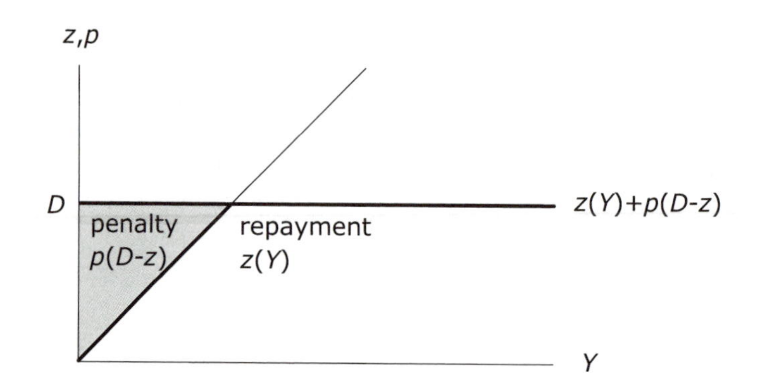

Figure 15.3 The debt contract.

3. The agent's monetary payment cannot be negative or exceed Y, i.e. $0 \leq z \leq Y$. In graphical terms, the function $z(Y)$ must be situated between the Y-axis and the 45°-line (feasibility constraints).
4. The promised payment is the sum of a monetary payment, z, and a contractual penalty p applicable if z falls short of D.
5. A positive difference between the promised (incentive-compatible) payment D and its feasible monetary component $\max(D, Y)$ must be compensated with a penalty.
6. Parties agree that the agent pays money whenever possible. The principal values money more highly than a penalty to the agent.
7. The expected (dis)utility of repayment D (cash and/or penalty) must not exceed the expected utility of the project return to the agent (participation constraint); nor must the expected cash payment fall short of the present value of the credit (non-negative profit to investor).

Truth-telling versus risk-sharing

This arrangement requires some further comment. First, the optimal contract is a debt contract, featuring a fixed repayment whenever possible, and a transfer of the borrower's assets to the lender plus a penalty to the borrower if the latter defaults. This implies suboptimal risk-sharing. In a first-best world, a risk-neutral investor would assume all risks of a risk-averse borrower. Truthful reporting of the private information requires a flat repayment scheme which imposes the most residual risk on the borrower.

The role of collateral

Second, the need for a penalty of some sort shows that "there is no credit without collateral". Even when no explicit collateral is agreed upon, the borrower's residual assets plus a loss of reputation or of future profits, serve as *de facto* collateral. The primary purpose of collateral in this model is not to give the lender access to some additional funds should the borrower not pay; rather, the role of collateral is to give the borrower an incentive to repay. As mentioned before, the collateral good is less valuable to the lender than to the borrower. At the limit, the collateral may be completely useless for the lender but painful to lose for the borrower (days in jail). When the collateral good is worth more to the borrower than to the lender, the positive probability of collateral transfer implies a *deadweight cost of the contract*. Note that a compassionate lender, suffering from seeing the borrower punished, is not credible. With such a lender, a borrower might try to renege in order to obtain better terms from the lender *ex post*.

Lack of collateral can block economic development, even in an environment where profitable projects abound. This is typically the situation faced by many potential borrower-entrepreneurs in poorer regions of the world. In the absence of collateral, intense screening of credit applicants and monitoring of borrowers is required. The information cost involved, which often leads to a monopoly of the local moneylender, may make credit prohibitively expensive. This is where the idea of microfinance comes in (see Box 15.1).

Limits to borrowing and lending

Finally, the terms of credit (D) and the implied interest rate are determined by the borrower's participation constraint and by the investor's profit maximization. A higher nominal value D increases the probability of default and the expected cost of losing the collateral for the borrower. There is an upper limit to the repayment the lender can ask for. Similarly, the amount of credit the lender can grant depends on the collateral the borrower will give and on the expected cost of losing the collateral (Lacker, 1991).

Box 15.1 Microfinance

During the famine of 1974 in Bangladesh, Dr. Muhammad Yunus, an economics professor at Chittagong University, lent twenty-seven dollars to a group of forty-two villagers. Convinced of the development impact of small loans, he founded the Grameen Bank in 1976 (licensed under a special law in 1983). The bank offered "microcredit" to members of small groups, particularly women. Its loans were intended to assist borrowers in starting a small business, making bamboo furniture or buying a dairy goat, thereby offering them a way out of poverty.

The developing world has a long history of similar microfinance schemes. The common denominator of microcredit or microfinance is the "joint liability" concept. A group of individuals form an association to apply for loans. Loans to individuals within the group are approved by its other members;

the group is jointly responsible for repayment. Often, members of a group receive credit on a rotating basis. In the absence of institutions like the Grameen Bank, the only source of credit for many people is a pawnshop or a moneylender who often charge staggeringly high interest rates, up to several hundred per cent per annum. There is also anecdotal evidence that some professional lenders beat up clients who fail to pay on time.

Since its modest beginnings, the microfinance movement has been very successful by most standards. The number of microfinance banks and projects have grown impressively over the last decades. The same is true for the volume of outstanding credit. Small loans have been granted to 65 million poor people without collateral around the world. Default rates on micro-loans are low (1–3 per cent). Critics have doubts about the low default rates reported by the lending institutions. They also find fault with the accounting standards observed by many of these non-profit organizations. Such qualms notwithstanding, microfinance is a source of tremendous hope for development organizations—and for the poor in developing countries. The United Nations proclaimed the year 2005 as the International Year of Microcredit. In October 2006, Dr. Yunus and the Grameen Bank were awarded the Nobel Peace Prize.

Sources: Armendariz de Aghion and Morduch (2005); http://en.wikipedia.org/wiki/Microfinance; http://www.mixmarket.org/; *The Economist*, "Microfinance, The hidden wealth of the poor", 3 November 2005; *The New Yorker*, "Millions for Millions" (by Connie Bruck), 30 October 2006.

15 D b The Merchant of Venice

William Shakespeare's *The Merchant of Venice* reads like an illustration of many of the issues discussed above. Most importantly, the credit relationship between Shylock on the one hand, and Antonio or Bassanio on the other offers several lessons on financial contracting.

The female star of the piece is Portia, a young woman hard to beat in terms of beauty, intelligence and wealth. There are many contenders for her hand in marriage. Yet, she is bound by her late father's will: She must marry the candidate who chooses the correct casket out of three—each being made of a different metal (gold, silver, and lead) and each bearing a different inscription. This is an elegant screening game, and fortunately both—the mechanism designed by the father and Portia's heart—will finally select the same candidate, the handsome but financially broke Bassanio.

In order to court Portia, Bassanio has to borrow heavily. Already weighed down by debt but determined to marry Portia, he seeks help from his friend Antonio, a well-to-do merchant. As it happens, Antonio's wealth is tied up in several undertakings at sea. They have to ask Shylock, the usurer, to help them out. Shylock is reluctant to lend; first, he has to borrow a part of the required 3000 ducats himself; second, he is well informed about Antonio's wealth being diversified but nevertheless at risk on the high seas; but most importantly, as a member of the oppressed Jewish community, Shylock has repeatedly been humiliated by Antonio. "Why should I lend to my enemy?" he asks. Here is Antonio's reply (Act 1, Scene III):

> If thou wilt lend this money, lend it not
> As to thy friends; for when did friendship take
> A breed for barren metal of his friend?
> But lend it rather to thine enemy,
> Who, if he break, thou mayst with better face
> Exact the penalty.

Shylock finally agrees, but asks Antonio for an unusual pledge:

> This kindness will I show.
> Go with me to a notary, seal me there
> Your single bond; and, in a merry sport,
> If you repay me not on such a day,
> In such a place, such sum or sums as are
> Express'd in the condition, let the forfeit
> Be nominated for an equal pound
> Of your fair flesh, to be cut off and taken
> In what part of your body pleaseth me.

Antonio accepts without hesitation as he expects his ships to return long before the three months after which he must repay have passed. Yet, two unlikely things happen. Bassanio, dressed up on borrowed money, wins Portia's heart and, of the final three candidates, is the only one to choose the right casket. Antonio's ships, though, are lost one after the other. When the bond matures he cannot repay. Shylock insists on the contractual penalty, clinging to a very literal interpretation of the bond. Bassanio, sent by his wealthy new wife, offers to repay twice the amount since repayment is two days overdue. Yet, Shylock insists on exacting the penalty. In the Duke's court, Portia by an ingenious intrigue saves Antonio's flesh and life. Disguised in men's clothes and acting as the court's external expert, she argues even more to the letter of the text than Shylock. She points out that Shylock may indeed cut out Antonio's heart, but that he has no right to spill a single drop of blood (no blood being mentioned in the contract).

The play ends with a final screening game: Portia (still disguised as a legal expert) begs her unsuspicious husband for a sign of gratitude, to wit, a ring he had sworn to Portia he would keep forever. Poor Bassanio has the choice between betraying his wife (who gave him the ring) and affronting the lawyer (who just saved his friend's life). He decides against the party he believes is absent.

The play is clearly full of examples from game and contract theory. The main financial transaction is the credit from Shylock, secured by Antonio's flesh. Both

Shylock and Antonio are part of a chain of credit, as Antonio borrows on behalf of his friend and Shylock has to refinance part of the credit with a third party. Both are "financial intermediaries". The contract is a debt contract collateralized by the threat of a penalty, very much in the sense of Diamond (1984). The threat would not be credible, if Shylock felt just a touch of mercy. But, as Antonio had told him, lending to an enemy removes the credibility problem. Shylock perverts the notion of the deadweight cost of a debt contract. Normally this cost consists in the positive probability that the debtor will be punished without any (*ex post*) benefit to the lender. Shylock, however, very much regrets that the penalty could not be enforced.

There are some further contractual issues: Why do the parties agree on a debt contract rather than on equity financing? Antonio could have offered a share of the profits from his shipping expeditions. In this case he would have avoided the risk of bankruptcy; and he might even have borrowed among his own community, rather than go to Shylock who, being Jewish, was not bound by usury laws. The debt contract model suggests that Antonio's business profits were not sufficiently observable to third parties and therefore not contractible. The example also illustrates that in reality contracts are never absolutely complete. There is always at least a margin of incompleteness which calls for interpretation by the court, if contested. There is a wealth of literature on law and economics which deals with issues such as: How should a judge interpret a contract if she assumes that parties *ex ante* wanted to maximize their joint welfare? Readers interested in law and economics are referred to Cooter and Ulen (2004).

15 E Application: Auctions

15 E a The second price auction

Auctions as information extracting mechanisms

Auctions are a good illustration of the revelation principle. The deeper economic purpose of an auction is to allocate a good to the person with the highest valuation. The seller, however, is normally interested only in maximizing the proceeds from the auction. As the highest bid tends to be made by the bidder with the highest valuation, both goals coincide, thus making auctions fairly efficient mechanisms. Auctions extract private information (individual valuations) from the bidders. This is possible since the seller can commit to sell the object to the highest bidder. An auction without such a commitment would not work. The seller could use the information content of the bids and then start a new procedure. Bidders, anticipating the seller's trick, would not submit meaningful bids in the first place.

Bidding is endogenous

While it is clear who should acquire the auctioned object (the bidder with the highest valuation), it is less obvious what price he should pay. The main choices are: (i) His own bid and (ii) the second highest bid made in the auction. At first sight it might seem that (i) is always preferable to (ii), the highest bid exceeding the second highest bid by definition. However, this is not true. Bidders are conscious beings who react to auction rules. The very same bidder thus rationally submits different bids in a "first price auction" where the winning bidder has to pay his own bid, and in a "second price auction" where the winning bidder pays the second highest bid.

The first price auction

It is easy to see that in a *first price auction* bidders do not bid their true valuations of an object. For simplicity's sake, let us assume that bidders know their subjective valuation of the object, i.e. that we talk of a *private value* auction, rather than of a *common value* auction where all bidders are uncertain about the objective value of an object (like an oil field or an air wave spectrum right). A bidder bidding his true valuation in a first price auction is certain to make zero profit. Either someone else bids more or the bidder acquires the object exactly for what it is worth to him. The profit of bidder i is $u_i = max(v_i - b_i, 0)$, where v_i are valuations, and b_i are individual bids. Obviously, if $b_i = v_i$, bidder i makes zero profit. Therefore, underbidding one's valuation increases the chance of not winning the auction but at the same time leads to positive profit, which in turn is conditional on winning. Under some simplifying assumptions (symmetrical bidders, i.e. individual valuations are independently drawn from an identical distribution, risk-neutrality) all bidders underbid their valuation by the same percentage. If valuations v_i are uniformly distributed, a risk-neutral bidder i's optimal bid is $b_i(v_i) = v_i(n-1)/n$. Note the two limiting cases: If $n = 1$, $b_i = 0$, i.e. a monopolistic bidder acquires the object for free. If n becomes larger, the optimal bid gets arbitrarily close to the true valuation. This is the "atomistic competition" assumption which underpins the idea of Walrasian general equilibrium.

The second price auction

It is also easy to see why in a *second price auction* bidders optimally bid their true valuation. As a bidder never pays his own bid, there is no reason to lie about one's valuation of the object. On the contrary, in fact, as underbidding would lead to the risk of losing the object to a bidder with a lower valuation, while overbidding might lead to a loss but never to a profit. The second price auction thus is a truthful mechanism.

From the revelation principle it follows that for any first price auction (which is not a truthful mechanism) there is a second price auction leading to the same expected revenue for the seller. To phrase it differently, the information rent the highest bidder obtains in the second price auction (the difference between the highest and the second highest valuation) can in no way be appropriated by the seller. He might try a first price auction. However, as mentioned above, bidders underbid their valuations. In fact, the highest bidder's optimal bid is exactly equal to the expected second highest valuation, i.e. to the revenue in the second price auction!

Revenue equivalence

First and second price auctions are revenue-equivalent (in expected terms). It is equally irrelevant (under the assumptions made) whether auctions are sealed or open. The revenue equivalence of these four basic auction types has been shown by William Vickrey as early as 1961. Later it was shown that any mechanism that (i) allocates a good to the bidder with the highest valuation and (ii) gives the bidder with the lowest valuation zero expected utility, leads to the same revenue for the seller. For the necessary assumptions (e.g., risk neutrality of bidders) and an elegant and highly intuitive proof, we recommend Brunnermeier (2001).

Revenue equivalence and optimality

The fact that the four basic auction types are revenue-equivalent under the required assumptions does not mean that they are also optimal. Not every second price auction, for instance, is automatically revenue maximizing. The reason is that the specification "second price auction" is incomplete; there are many types of second price auctions.

One degree of freedom in designing a second price auction is the seller's reserve price. Assume that the seller values the object to be auctioned at r. It can be shown that the seller maximizes her revenue by setting a reserve price $b_0 > r$. The seller thus acts as a further, virtual bidder. And she does not bid truthfully, but in the optimum overbids her valuation r. Setting a reserve price $b_0 > r$ increases the seller's revenue in all cases where b_0 exceeds the second highest valuation (but not the highest). At the same time, there is a positive probability that the good is not sold even though the highest bidder's valuation exceeds the seller's. For more details, see Wolfstetter (1999).

Note that the two games Alice used to sell cookies to Bob were, in fact, degenerate versions of both a first and a second price auction. Neither was optimal though, as the random draw of Alice's reservation price in the first game (the first price auction) and the equivalent randomization of accepting Bob's offer in the second game (the corresponding second price auction) did not maximize Alice's revenue.

"What really matters in auction design"

In practice, there are many more degrees of freedom in auction design. The above title is from Paul Klemperer (2002). He and others have shown that slight changes in assumptions can have important consequences, and that minuscule details of auction design can dramatically reduce the seller's prospects. The proceeds from government auctions of indexspectrum licence UMTS spectrum rights around the year 2000, for example, led to revenues (Euro per head of population) of 630 in the UK, 615 in Germany, and 20 in Switzerland. It is not only important to know that the four basic types are revenue-equivalent under simplifying assumptions; it is also important to remember that auction design greatly matters in reality.

15 E b Goethe's second price auction

Author versus publisher

Johann Wolfgang von Goethe (1749–1832) was not only the leading German author of his time (some would say the greatest ever), but a cunning economist as well. Many economists would probably first think of the scene on paper money in *Faust II*. However, as Moldovanu and Tietzel (1998) show, Goethe also provided practical proof of his economic intuition. Having finished his manuscript *Hermann und Dorothea*, he was not satisfied with the usual fees to authors. In a letter Goethe complains: "Let me . . . name the main evil. It is this: the publisher always knows the profit to himself and his family, whereas the author is totally in the dark" (p. 856; all citations from Moldovanu and Tietzel, 1998). Trying to assess his value, Goethe set up a sophisticated mechanism. In a letter, dated 16 January 1797, he writes:

> Concerning the royalty we will proceed as follows: I will hand over to Counsel Böttiger a sealed note which contains my demand, and I wait for what Mr. Vieweg [the publisher] will suggest to offer for my work. If his offer is lower than my demand, then I take my note back, unopened, and the negotiation is broken. If, however, his offer is higher, then I will not ask for more than what is written in the note to be opened by Mr. Böttiger.
>
> (p. 855)

A second price auction

This scheme is in fact a second price auction with two bidders: Vieweg, the potential publisher, and Goethe (via his secret reservation price). If Vieweg bids the reservation price, he seals the deal and pays the reservation price, i.e. the second highest bid. If he bids below the reservation price, Goethe "wins" the auction, i.e. he keeps the manuscript and foregoes Vieweg's offer which now is the second highest bid. It is easy to verify that Vieweg's best strategy is to reveal his true valuation for the manuscript. With *underbidding* he would have run the risk of not acquiring a manuscript which he would have preferred to obtain at the reservation price. This risk is not compensated with a lower expected price to pay (as in a first price auction). With *overbidding* he would have run the risk of buying at a reservation price above his own valuation. Bidding the true valuation thus is an equilibrium.

Of course, Goethe could have achieved the same result with a much simpler mechanism, an explicit take-it-or-leave-it offer. The resulting allocation would have been identical. However, Goethe would never have learned his value, as the publisher Vieweg would simply have answered "yes" or "no". This is an important difference, as Goethe was curious about his market value. It should be noted that Goethe's secret reservation price was 1000 Talers, about three or for times the usual fee for similar volumes by other popular authors.

The outcome

What was the outcome of the game? Moldovanu and Tietzel (1998) report that the Counsel Böttiger wrote to Vieweg:

> The sealed note with the Golden Wolf is really in my office. Now tell me what can and will you pay? Given what I approximately know . . ., let me just add one thing: you cannot bid under 200 Friedrichs d'or.
>
> (p. 857)

Indeed 200 Friedrichs d'or were exactly 1000 Talers. This is what Vieweg offered and the unsuspicious Goethe accepted. Moldovanu and Tietzel (1998) point out that Böttiger's treason did not directly affect Goethe's revenue, but it frustrated Goethe's attempt to learn his true value. Almost twenty years later, Goethe did better. For his collected works he invited several bidders, ending up with 60,000 Talers for around 40 volumes. Moldovanu and Tietzel compute that Goethe thus kept a much larger share of the cake.

15 F Application: Why Enron should not have happened

Accounting scandals

The early years of the 21st century witnessed a number of corporate scandals affecting (formerly) prestigious firms such as Enron, WorldCom and Arthur Andersen. A common characteristic of these incidents was (to put it mildly) inadequate financial reporting. The public, politicians, and the press were appalled: "Earnings management is a scourge in this country . . . We need to put a stop to earnings management," an accounting expert said.[102] One consequence was legislation to ensure greater transparency. In the US public disgust at accounting practices led to enactment of the Sarbanes–Oxley Act, which set (among other things) stricter rules for corporate accounting and reporting.

Are managers and accountants professional liars?

Like legislators, most press reports took it for granted that in the absence of external regulations, corporate managers have an incentive to lie. Some economists (e.g., Arya, Glover and Sunder 2003), though, pointed out that lying managers, from an economic point of view at least, are an anomaly. The revelation principle would tend to predict that firms (shareholders) use contracts with their managers, which provide the latter with an incentive to tell the truth rather than to rig the firm's books. How can the obvious incomplete use of the revelation principle in practice be explained?

Arya, Glover and Sunder (2003) review the literature to explain why managers are left with some discretion to manipulate performance measures, i.e. why shareholders may prefer to let managers manipulate performance reports. They start out from the observation that installing video cameras in offices might not represent an optimal degree of revelation of employees' activity. They conclude that "the optimal mix of rigidity and flexibility . . . is unlikely to be a corner solution" (p. 12).

More generally, explanations of managerial discretion in reporting build on a violation of the revelation principle. Possible suspects are to be found under the three big Cs: communication, commitment and contract. Arya, Glover and Sunder (2003) present several examples.

Communication problems

First, *communication* between the agent and the principal may be restricted by the requirement that the agent, i.e. the manager, may only report one figure for the firm's income. Assume that the firm has an income stream with both a permanent and a transient component. Selfless managers would want to report both components separately, as they are in a better position than shareholders to distinguish between permanent and transient income. If transient income cancels out over time, managers might be content only to report permanent income. "Manipulation" of earnings' figures would be in the shareholders' interest, as transient income might lead to a wrong valuation of the firm. Shareholders may also allow a self-interested manager to manipulate earnings if the ability to smooth earnings reflects the ability of the manager to forecast the future and thus to run the firm. A manager who has to report a big jump in earnings is fired.

Limits to commitment

Second, the inability of the principal to *commit* may prevent the use of the revelation principle. Arya, Glover and Sunder (2003) use the possibility of early termination of a project by the principal as the setting for their example. In a somewhat modified form, the story goes as follows. A hedge fund specializes in predicting the fundamental value of a given financial asset. It may have heavily invested in information acquisition in order to make a good guess of the fundamental value. The fund manager knows that the market price will eventually reach its fundamental value, but that presently the asset is, say, undervalued. The manager thus buys a position in the asset. This is called "risky arbitrage" (Brunnermeier, 2001).

Now, let us assume the hedge fund finances its investment partly with bank credit. The market price of the asset therefore may or may not revert to its fundamental value before the bank credits are due. Assume again the asset price continues to fall. If the hedge fund has to disclose its financial position, the bank may become nervous and may not renew the credit. Even if the bank's credit officer perfectly understands (and believes in) the fund's arbitrage strategy, the bank may be unable to extend credit: The bank itself borrows from many short-term investors who themselves may want to terminate their exposure. If these investors observe that an important hedge fund in which the bank has invested is in poor condition, they may withdraw their funds, thus forcing the bank to liquidate part or all of its assets. The bank then cannot continue to finance the hedge fund, even if it has made a contractual promise to roll-over its credits. As a consequence, the hedge fund runs into liquidity problems and has to liquidate part of its investment. Note that early liquidation is most likely when continuation would be most profitable (i.e. when the underlying asset is most undervalued). Expecting this possibility, the hedge fund might not invest in information on the asset's fundamental value in the first place. Transparency, in this case, may harm the informational efficiency of financial markets. It may thus be optimal to allow the hedge fund to remain "opaque" as to its financial situation. To be more precise, it lets fund managers value assets at their fundamental rather than market value.

There are other sources for inefficiency with full disclosure. Arya, Glover and Sunder (2003) discuss *posturing*, i.e. wasting resources to improve measured performance. We would add that budgets may be another, and typically ambiguous, example. On the one hand, budgets improve transparency and facilitate truthful revelation of performance by employees; on the other, they may invite a concentration of efforts on budgeted activities and the neglect of non-measurable contributions to firm value (e.g. by accelerating sales at the cost of jeopardizing future sales prospects).

Smiling models

To conclude, when there are limits to communication, commitment or contracting, the revelation principle does not help. In these cases, the principal may be unable or unwilling to use truthful mechanisms and may prefer to leave managers with an imperfect incentive for transparency. Asking for more transparency and disclosure in such cases may not always be beneficial in relation to exposing the "truth". Arya, Glover and Sunder (2003, p. 11) point out the analogy between accounting and photography. Financial reporting, they say, is not like taking a "true" picture of a landscape. It is rather like photographing a model: "The model smiles and poses for the camera even as the photographer changes camera angle and settings in reaction to the model. The state of a firm and its financial reports are reflexive in the sense of being dependent on each other."

15 G Conclusions and further reading

The revelation principle is an invaluable compass in the search for optimal contracts. It reduces the range of relevant candidates to direct truthful mechanisms. It also helps to understand mechanisms or contractual arrangements we find in reality, such as the second price auction or the debt contract. The wide use of such mechanisms is indirect evidence for the practical relevance of this principle.

The revelation principle has its natural limitations when truthful mechanisms cannot be implemented due to limitations on communication, commitment or contracting. Financial reporting was discussed as an area where the principal (shareholders) may not find it in their interest or within their power to implement mechanisms that make managers report truthfully.

We have offered a rather intuitive presentation of the revelation principle. Readers interested in a more rigorous presentation, in formal proofs, or simply in an alternative way of tackling the issue, will find these in a number of textbooks. As a starting point, we would recommend Laffont and Martimort (2002, Ch. 2, Sec. 2.9) who also offer a relatively accessible proof.

Box 15.2 Reporting workplace injuries

On 30 July 2001, the British Health and Safety Executive (HSE) issued a press release (http://www.hse.gov.uk/statistics/press/c01031.pdf) with the title:

> New health and safety statistics show fatalities up, reported non-fatal injuries down.

The seeming puzzle—fewer injuries, more casualties—led HSE chairman Bill Callaghan to doubt the figures:

> It is good to see that non-fatal injury rates continue to decline, but . . . I am concerned by the possibility that this may indicate . . . an increase in under-reporting.

Did employers take a cavalier attitude with respect to workers' safety and under-report casualties? The answer to the latter question seems to be yes. Yet, this does not imply that firms did not care about workers' safety. Quite the contrary! Workplace accidents and injuries are not only a source of suffering for employees but also represent a considerable loss to firms and to the economy. More and more firms therefore use safety incentive programs in order to prevent workplace accidents and injuries. Yet, such programs do not always have the intended consequences.

Safety incentive programs offer explicit rewards to employees for reductions in the number of accidents and incidents. In a report for the HSE Daniels and Marlow (2005) point out that employees who work under incentive programs tend not to report incidents, as they do not want to lose prospective rewards. Under-reporting seems to be typical of injuries that are not obvious to others (such as musculoskeletal disorders), and it is particularly severe in teams, where a reporting individual would spoil the team record. However, the authors also claim a "lack of scientific literature providing evidence of accident reporting accuracy by age, gender, and occupation" (Daniels and Marlow, 2005).

Due to the incentive to suppress information, and hence prevention, safety programs often fail to reduce the true number of accidents. The British Fire Service, for example, despite a strong safety culture, symbolized by the goal of zero reported injuries, had nearly 2,000 major and over-three-day injuries in the period 2002–03. For the same period, experts estimated that near misses were under-reported by a factor of 20 to 60. This suggests that the safety goals led to a reduced number of reported, rather than of total incidents.

Daniels and Marlow (2005) find that the negative effects of safety incentive programs on truthful reporting are strongest when incentives:

- are of a financial nature (as distinct from recognition, for instance),
- possess a high financial value,
- are of the "all-or-nothing" variety,
- are reinforced by group pressure.

Even without explicit incentive programs, employees tend to under-report for several reasons, such as the administrative burden or time pressure, fear of reprisals, fear of being considered careless, feeling that suffering is a kind of weakness, or employer disinterest. The problem of safety incentive programs illustrates that it may be very difficult to reward both the effort to avoid mistakes *and* the reporting of mistakes once made.

Checklist: Concepts introduced in this chapter

- Truthful mechanism
- Direct revelation mechanism
- The revelation principle
- Assumptions: Communication, commitment, contracting
- The debt contract
- The deadweight cost of collateral
- The second price auction
- Revenue equivalence
- Optimal reservation price

15 H Problem sets: Know your value

For solutions see: www.alicebob.info.

Problem 15.1 Your value

A large law firm is interested in inviting you to give a lecture on Contract Design. You are supposed to make a proposal as to a speaker's fee, but you are quite uncertain about what they are ready to pay. You assume that their valuation in dollars is in the range of 0 to $50,000; all valuations in that range being equally likely.

Your opportunity cost to give the lecture (the value of your best alternative, like leisure, studying or lecturing somewhere else) is \$10,000. What do you propose if you maximize your expected fee and, at the same time, try to show your expertise in contract theory?

Problem 15.2 Risk-averse bidders

With risk-neutral bidders the first and second price auctions are revenue-equivalent. The second price auction is also a truthful mechanism. Assume that bidders are risk-averse.

(a) Does a risk-averse bidder bid more or less in a first price auction than an otherwise identical risk-neutral bidder?
(b) Does revenue equivalence still hold?
(c) Does the revelation principle still hold? (Hint: what would bidders have to reveal?).

Problem 15.3 The optimal reservation price

Assume you are risk-neutral and you want to sell an object in an auction. There are two risk-neutral bidders with independent private valuations v_i for the object to be sold. You do not know these valuations, but you know that both are drawn from a Uniform distribution between zero and one, i.e.$v \sim U[0,1]$. To yourself the object is worth v.

(a) What is a truth-telling mechanism to sell the good?
(b) What is the optimal reservation price?
(c) What is your expected revenue?
(d) What would you pay for knowing one (both) bidders' valuations? (Hint: The expectation for the k-th highest value of n uniformly distributed valuations $v \sim U[0,1]$ is $(n+1-r)/(n+1)$, see Box 5.2. Note also that Problem 15.1 is the one bidder special case to this exercise.)

Problem 15.4 Efficiency or revenue?

This exercise is from Jehiel and Moldovanu (2003, p. 279). You want to sell two objects, A and B. There are two bidders. You cannot observe their valuations but you assume they are the following:

Bidder 1: $V_1^A = 10$, $V_1^B = 7$
Bidder 2: $V_2^A = 8$, $V_2^B = 12$

A second price auction is (weakly) value maximizing, i.e. no other procedure yields a higher revenue to the seller. Answer the following questions:

(a) What is your revenue in two separate single object auctions?
(b) What is your revenue if you auction objects A and B as a bundle?
(c) Which procedure (two single auctions or the bundle auction) maximizes your revenue?
(d) Which procedure maximizes social welfare?

16 Creating incentives

16 A Introduction

In 1919, the English writer D. H. Lawrence (1885–1930) travelled through Sardinia. As he later recalls in *Sea and Sardinia* he was most amazed by a railway trip:

> It is a queer railway. I would like to know who made it. It pelts up hill and down dale and round sudden bends in the most unconcerned fashion, not as proper big railways do, grunting inside deep cuttings and stinking their way through tunnels, but running up the hill like a panting small dog, and having a look round, and starting off in another direction, whisking us behind unconcernedly.

As Lawrence experienced first-hand, the railway line from Cagliari leading north towards the heart of the island is unusually bendy. The curves were in part a cheaper alternative to tunnels and bridges. Yet, to a certain extent, they have a different if equally economic explanation: The British company who built the line was paid *per mile*—per mile built, not per mile required. Not surprisingly, the curvy option too often prevailed against its straight alternative.

This is one of many practical examples which illustrate the impact of incentives. Incentives play a role whenever one person, called the principal, asks another person, called the agent, to do a particular job. Typically, the principal cannot survey or control all aspects of the agent's way of performing the task. Nor are the agent's interests identical to those of the principal. Without adequate measures, the agent thus may not act in the principal's best interest, a phenomenon known as "moral hazard".

The underlying problem is called "hidden action", as distinct from the "hidden information" problems dealt with in the previous chapters. The standard answer is the design of incentive contracts or mechanisms: Promotions, bonus payments, stock options, "employee of the month" awards and similar instruments are used to motivate workers; patents are granted to inventors or innovative firms; competitions are organized in order to get the best out of architects; or prisoners are released on parole, to name but a few examples.

Often, incentive schemes achieve their purpose: Salespeople try harder if they participate in their sales. Waiters smile in the expectation of tips. Assistant professors work night and day in the hope of publishing enough to be awarded tenure.

Sometimes, incentive schemes even work too well, creating excess pressure on the individuals involved. One example—also from the world of railways—is the fatal contract bid by Louis Favre for the construction of the Swiss Gotthard tunnel (see Box 16.1).

Box 16.1 The Gotthard tunnel

The 15 km-long railway tunnel through the Gotthard, a breakthrough link between Northern and Southern Europe, was built between 1872 and 1881. Construction was surveyed by the Swiss engineer Louis Favre from Geneva. Favre, competing with seven other contestants, had won the contract with a promise to build the tunnel within eight years and to pay heavy contractual penalties for delays. For each day of delay, a contractual penalty of CHF 5,000 was due (after six months: CHF 10,000); in addition, after one year, a construction bond of CHF 8 million would be forfeited (total construction cost was CHF 55.8 million).

Difficult geological conditions made construction a race against the clock. At peak times, 3,800 workers drudged in three eight-hour shifts a day. Temperatures above 35 centigrade, miserable working conditions and the time pressure only served to increase the number of accidents. The official count is 177 casualties (and many more injured), mainly due to water inrushes or accidents with the compressed-air driven lorries carrying excavated material out of the tunnel. In 1875 a strike by the workers was crushed by military force, killing four and wounding 13.

But the race went on. The builders used a new technology never tried in a tunnel before: “Nobel’s Safety Blasting Powder”, a new explosive patented by the Swedish chemist and engineer Alfred Nobel in 1867. This powder, later known as dynamite, laid the foundation to Nobel’s fortune and thus to the Nobel prize (established by the inventor’s will in 1985).

Yet, Louis Favre fought a losing battle. During the seventh year of construction, on 19 July 1879, he suffered a fatal heart attack on a visit inside the tunnel. The stress from the approaching deadline and the looming penalties proved too heavy. Seven months later, the miners cut through, the north and south sections of the tunnel meeting with a tiny difference of 33 cm. The first “passenger” to cross the tunnel was Favre himself: his picture in a box was handed through the hole a day after cut-through. In 1881 the tunnel was finished—in an impressive time for such an undertaking, but too late for Favre.

Source: Kovári and Fechtig (2005).

Incentive contracts do work—but not always as intended. Premia for scoring a goal may lead soccer players to shoot from a bad angle rather than pass to a better placed colleague. Premia for winning a game, by contrast, may prevent selfish play but lead to suboptimal effort in a team. The design of optimal contracts is therefore a tricky business.

The present chapter is a sister to Chapter 13. Both are driven by a basic information asymmetry. There, one party had hidden (unobservable) *information*. Here, one party can take a hidden (unobservable) *action*. While hidden information leads to adverse selection, hidden action leads to *moral hazard*, again a term from insurance literature.

There is a wealth of literature, theoretical and empirical, on incentives or the so-called principal–agent relationship. We will focus on the basic model, neglecting many of the possible refinements. Instead we will devote two Application sections to problems in which information plays a particularly important role: one on bank deposit insurance and moral hazard, the other on so-called credence goods.

16 B Main ideas: Delegation and moral hazard

Alice and Bob are almost ready to set out on vacation. The only problem is Syd, Alice's cockatiel. The bird needs fresh food and water once a day. And, like most cockatiels, Syd suffers from night-frights.[103] The lights of a passing car could make him panic. It is important that the curtains are tightly closed at night but, of course, open during the day. Alice could ask her neighbor, Luca, to care for the bird. She helped Luca when, during his vacation, a food parcel from his mother in Italy arrived. But Luca often works long hours at the university and sometimes throws parties, so he may forget the cockatiel.

Alice is in a dilemma. She could, of course, threaten to kill Luca if something happened to the bird. But under such terms he would hardly agree to care for it. Or she might offer some monetary reward if she finds Syd alive and well upon her return. What terms would be acceptable to Luca and yet attractive enough to ensure that he never forgets to pull those curtains?

As the saying goes: "If you want something done right, you have to do it yourself."[104] In reality, we cannot do everything ourselves. We have to delegate important tasks: Shareholders have their companies run by specially skilled managers; conflicting parties let the lawyers argue in their place; and patients rely on their doctors for both diagnosis and treatment.

Delegation is not only a form of division of labor, and thus a source of higher productivity, as Adam Smith pointed out, but also a source of a special kind of problem. The delegating person, normally called the "principal", cannot perfectly instruct and supervise the "agent", the person trusted with the delegated task.

The principal is unable to see whether the agent invests extra hours, engages in research, or is focused on the important aspects of the job. The agent, being on a long leash, often finds sufficient room to pursue some private goals. Employees spend time on private phone calls or surf on the internet during working hours. Managers try to build empires, not always to the benefit of their shareholders. Insured patients see their doctors too often and ask for expensive treatment. Taxi drivers choose unnecessarily longer routes.

Economists have thoroughly analyzed the theory of incentives or the "principal–agent relationship". A recent monograph from the vast literature is Laffont and Martimort (2002). The basic model has one principal facing one agent. The principal offers the agent a contract under which the agent performs the task delegated by the principal. The contract has to be sufficiently attractive for the agent to accept. In particular, the contract has to promise the agent enough expected return to compensate him for his effort. If the agent is risk-averse, the expected return must also compensate him for any risk imposed by the contract.

After accepting the contract, the agent performs his duty—more or less well. The agent's actions are not fully observable by the principal. The agent may therefore pursue his own goals—avoiding effort, neglecting risk borne by the principal, diverting the principal's resources, etc. In the simple model the agent's hidden action is reduced to a single dimension commonly called "effort". In the very basic version which we will present in the theory section, the agent can only choose between two effort levels: working and shirking.

The principal would like the agent to work, rather than to shirk. Telling him would not do, though: A shirking agent could still claim to have worked; the principal, unable to observe the agent's effort, could not prove otherwise. Therefore, the contracting parties cannot contract on effort.

However, they have a second best alternative: Instead of contracting on unobservable effort itself they may be able to contract on an observable variable correlated to effort: Final output or the success of the project managed by the agent. A working agent is more likely to finish the project successfully. Paying for success indirectly rewards effort—not deterministically, but stochastically, i.e. in expected terms.

Rewarding (paying more) for success or punishing (paying less) for failure has two consequences: It partly insures the principal and it de-insures the agent. Most contracts that are supposed to induce effort burden the agent with some risk. Labor contracts promise bonuses, health insurers use franchises, car insurers apply history-dependent premia—all with a view to inducing the agent to behave carefully, and all with the result of putting the agent partly at risk.

Yet, an agent who is free to accept or reject a contract, will not bear risk without adequate reward. If the agent is risk-averse, the principal has to pay a risk premium to make the agent bear the risk. This premium is a deadweight cost of the incentive contract: It is a cost to the principal but does not make the agent happier than he would have been without the risk. It is called the *agency cost* of the contract.

A principal may find that the agency cost required to stimulate effort by the agent is too high. The principal may be better off with a lower, but cheaper effort.

Agency cost thus comes in two forms: as a deadweight risk premium (above) or as a suboptimal (lower than first-best) effort.

A common example of a principal–agent problem is the employer–worker relationship. A hotly debated issue in this area is executive compensation, in other words the remuneration of company managers. Due to the separation of ownership and control, managers do not always act in the best interest of the owners of "their" firms (Jensen and Meckling, 1976). Despite great efforts by economists to understand compensation schemes, the role of observed remuneration packages remains controversial. For an introduction see Jensen, Murphy and Wruck (2004), Bebchuk and Fried (2003), Guay, Core and Larcker (2003).

Hidden action is not the only type of principal–agent problem. Another is what we may call "hidden appropriateness" of observed actions. The use of corporate aircraft, the time CEOs spend on golf courses, the occurrence of corporate mergers—these are all observable, at least in principle. The problem is that no outsider can judge (let alone prove) whether such actions are appropriate, i.e. in the principal's best interest. Did the CEO really spend time on the golf course because that is where the clients are or just because golf was more fun than going through accounting principles for the umpteenth time?

While some situations are characterized by hidden action and others by hidden appropriateness, some suffer from both. Once more, health care is a typical example. Medical doctors do not only have private knowledge on what a patient needs, they also provide a treatment, many elements of which (like surgery) cannot be observed by the patient. Such goods regarding which only the expert knows what the client needs as well as what the client receives are called *credence goods*. Credence goods create moral hazard in the form of over-treatment (doing more than necessary) or over-charging (charging for more than the treatment given).

16 C Theory: Incentive contracts

16 C a Assumptions: Actors, information, time line

Work or shirk?

In the basic principal–agent model, events have a time structure represented in Figure 16.1. In $t = 0$, a risk-neutral principal owns a project with an uncertain return X. The principal asks a risk-averse agent to run the project and offers the agent a (take-it-or-leave-it) contract. This implies that bargaining power is with the principal.

The agent in $t = 1$ accepts or refuses the contract. If the agent refuses, the relationship is terminated. If the agent accepts, the promised task is performed in $t = 2$. The agent has more than one way to perform this task. We will sum up all his potential actions in one dimension and refer to it as "effort". For simplicity's sake, we assume that there are two effort levels: a high effort ("work") or a low effort ("shirk"). Importantly, the effort chosen by the agent is *not observable* to the principal.

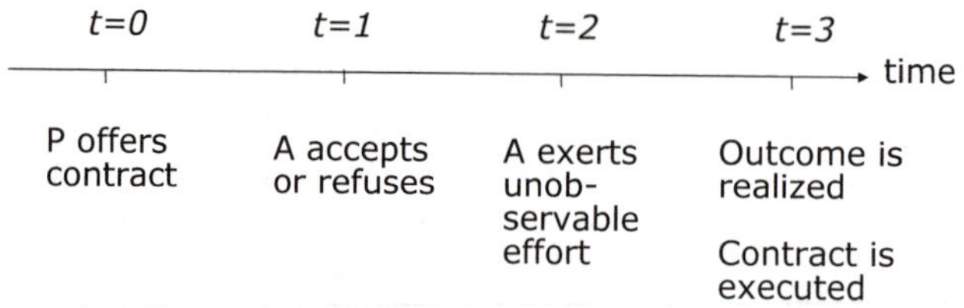

t=0	t=1	t=2	t=3
P offers contract	A accepts or refuses	A exerts unobservable effort	Outcome is realized Contract is executed

→ time

Figure 16.1 The sequence of events in a hidden action problem: A contract offer by the principal ($t = 0$) is accepted (or refused) by the agent ($t = 1$). Under the contract, the agent exerts unobservable effort ($t = 2$). The effort has a non-deterministic impact on the final outcome (payoff), which is shared among the parties according to the contract ($t = 3$).

In $t = 3$ the project yields some return. This return is likely to be greater if the agent has worked, rather than shirked. The return is shared among the principal and the agent, according to the contract signed in $t = 1$.

The agent dislikes effort; the disutility of a low and a high effort is e_1 or $e_2 > e_1$, respectively. The agent's total utility is the difference between utility from income $u(w)$ and disutility of effort e:

$$U = u(w) - e,$$

where $u(w)$ has positive but decreasing marginal utility ($u' > 0$, $u'' < 0$).

Both parties completely understand the game represented in Figure 16.1. All elements of the game are common knowledge, including the fact that the agent can take an unobservable action. Only the action itself remains private knowledge of the agent.

The impact of effort on expected return

The project has two possible outcomes X_H and $X_L < X_H$, the probabilities of which depend on chance as well as on the agent's effort. The agent has a choice of two effort levels, e_1 (shirk) and e_2 (work). If the agent works, the probability of X_H is $p > 1/2$ and the probability of X_L is $(1 - p)$. If the agent shirks, the reverse probabilities hold.

It is important that the outcome is not deterministically linked to the action. Otherwise the principal could tell the action from the outcome. Therefore, we assume that the link between the action and the outcome is stochastic. A higher effort increases the probability of a favorable outcome—but even after a negative outcome a shirking agent may still claim to have worked; or, the agent may have been unlucky despite hard work.

From $p > 1/2$ and $X_H > X_L$, it follows that a higher effort leads to higher expected return:

$$pX_H + (1 - p)X_L > (1 - p)X_H + pX_L.$$

The distribution of gross returns (not counting the cost of effort) is "better" after working than after shirking. In more precise terms, the former distribution first-order stochastically dominates the latter.[105]

The contract space

Contracts can be written for observable states of nature (Chapter 14). These are the final outcomes, X_H and X_L, but not effort levels e_1 and e_2. With two possible outcomes X_H and X_L, the contractual payment from the principal to the agent—the agent's "wage"—can take two values, w_H and w_L. A contract is a pair of payments $\mathbf{w} = \{w_H, w_L\}$, each for one of the two possible outcomes.

The game tree

The game tree for the principal–agent problem is given in Figure 16.2. The principal first posts a contract offer: this is a pair of payments (w_H, w_L), one for each of the two possible outcomes (X_H, X_L). Then the agent accepts or rejects that offer (forever). We assume that if the agent rejects the offer, the aggregate payoff to be shared among agent and principal is zero. If the agent accepts to run the project, he may either work or shirk. The "cloud" around the two decision nodes after "work" and "shirk" in Figure 16.2 indicates that the two nodes lie within the same information set (the same subset of the partition; see Chapter 3) of the principal: The principal cannot distinguish between the two states. Finally, nature "chooses" an outcome, the odds depending on the agent's effort.

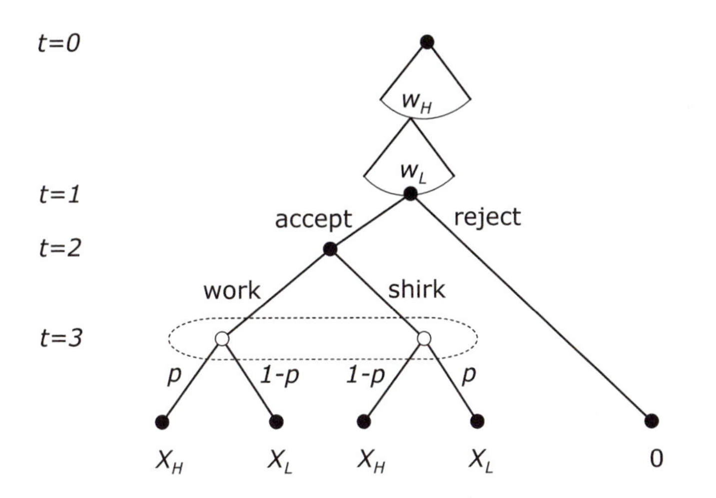

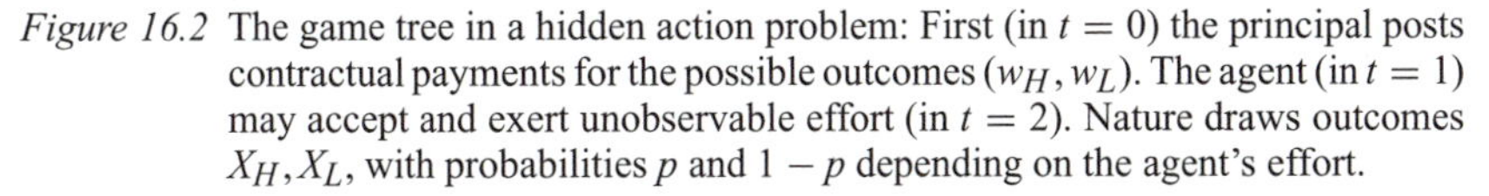

Figure 16.2 The game tree in a hidden action problem: First (in $t = 0$) the principal posts contractual payments for the possible outcomes (w_H, w_L). The agent (in $t = 1$) may accept and exert unobservable effort (in $t = 2$). Nature draws outcomes X_H,X_L, with probabilities p and $1 - p$ depending on the agent's effort.

16 C b Observable effort

The risk-neutral principal maximizes expected income. Solving the game in Figure 16.2 backwards, the principal first needs to find the optimal contract for each particular effort level; then a comparison of profits from the two optimal contracts reveals the overall optimal contract.

As a benchmark we first analyze the contract the principal would offer if the effort were *observable* (first-best). With observable effort, the principal can distinguish between the two nodes in $t = 3$ (see Figure 16.2). In that case she can directly write a contract on effort. The risk-neutral principal pays the agent a fixed wage, w, for the desired effort and nothing for any other effort level.

Inducing a low effort

If the principal wants the agent to accept the contract and to exert a *low effort*, the minimum uniform wage w_1^* solves:

$$u(w_1^*) = e_1$$

Inducing a high effort

Similarly, if the principal wants the agent to exert a *high effort*, the minimum uniform wage required to make the agent accept the contract, w_2^*, solves:

$$u(w_2^*) = e_2$$

In both cases, the optimal contract is a fixed wage offer with a utility just sufficient to compensate the agent for the desired effort. With a fixed wage, the agent is fully insured against the outcome. All risk falls on the principal who, being risk-neutral, also has a comparative advantage to bear it.

The optimal observable effort level

Which of the two effort levels is better for the principal? The answer is found by comparing profits from the two effort levels. Profits from a low effort are:

$$(1-p)X_H + pX_L - w_1^* \tag{16.1}$$

Profits from a high effort are:

$$pX_H + (1-p)X_L - w_2^* \tag{16.2}$$

Subtracting (16.1) from (16.2) yields the condition under which the principal prefers the agent to make a high, rather than a low effort (to work rather than to shirk):

$$(2p-1)(X_H - X_L) \geq w_2^* - w_1^*. \tag{16.3}$$

The intuition is straightforward: The left-hand side measures the increase in expected return from the higher effort, $2p - 1 = p - (1 - p)$ being the difference in the odds of the two outcomes achieved by the latter. The right-hand side is the difference in wages required to induce a high rather than a low effort.

16 C c *Unobservable effort*

The principal's problem

When effort is *not observable*, the principal cannot distinguish between the two nodes in $t = 3$ in Figure 16.2. With any contract offered the agent will not only choose whether to accept or reject but also whether to work or to shirk.

The risk-neutral principal maximizes the expected return (net wages). As in the first-best case, the principal initially has to find the optimal contracts for each of the possible effort levels; a comparison of profits then tells the principal which of the two optimal contracts is optimal overall.

Inducing a low effort

Again we start with the optimal contract based on a low effort, e_1. The optimal contractual payments $(w^*_{1,H}, w^*_{1,L})$ solve the principal's problem:

$$\max_{w_{1,H}, w_{1,L}} V = (1 - p)(X_H - w_{1,H}) + p(X_L - w_{1,L}), \tag{16.4}$$

subject to:

$$(1 - p)u(w_{1,H}) + pu(w_{1,L}) - e_1 \geq 0. \tag{16.5}$$

In words: The principal chooses contractual payments that maximize profits under the participation constraint (16.5) for a shirking agent. As will become clear below, the principal does not need to ensure that the agent provides a low rather than a high effort.

The solution is simple: The principal pays a uniform wage $w_1 = w_{1,H} = w_{1,L}$, where w_1 solves:

$$u(w^*_1) = e_1.$$

This is the cheapest way to satisfy (16.5); any difference between the two wages would cost the principal some expected return without adding to the risk-averse agent's utility. The solution is identical to the optimal shirking contract under observable effort. As the principal does not need to provide the agent with an incentive for shirking, effort is quasi-observable.

The solution is illustrated in Figure 16.3. Axes measure contractual payments, not in dollar amounts but in their utility equivalent. The horizontal (vertical) axis

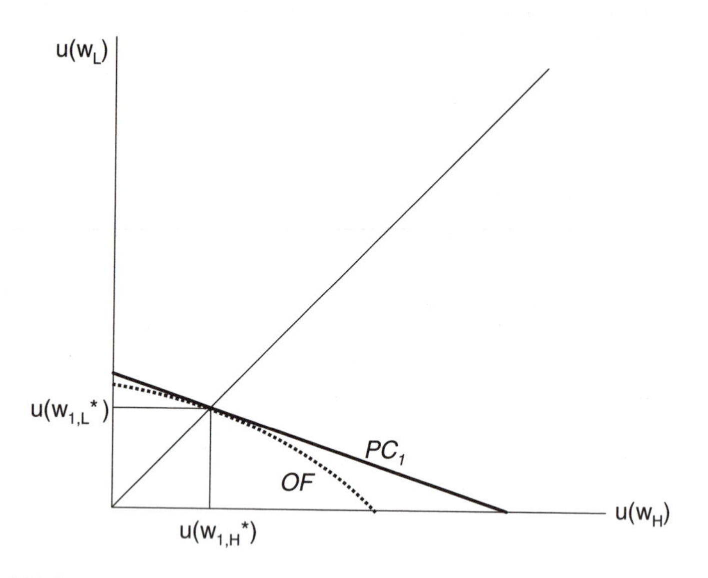

Figure 16.3 Payments to the agent under the optimal contract, $w^*_{1,H}, w^*_{1,L}$, for a low effort (e_1) are determined by the tangent point of the participation constraint for e_1 with the associated objective function. The figure is drawn for $p = 3/4$.

measures utility from payments in the case of a high (low) return X_H (X_L). The solid line PC_1 is the participation constraint for an agent who plans to shirk. It is a straight line because the figure is drawn in utility units; in dollar amounts the participation constraint would be convex (as it is an indifference curve for a risk-averse agent). Promising a lower wage in one state requires promising a higher wage in the other state; PC_1 thus has the negative slope $-(1-p)/p$. In Figure 16.3 it is drawn for a value of $p = 3/4$, implying a slope of $-1/3$.

The optimal contract for (unobservable or observable) low effort is given by the tangent point of the participation constraint PC_1 and the principal's objective function OF, representing the highest attainable iso-profit line. This tangent point lies on the 45°-line, because indifference curves (including the participation constraint) and objective functions have the same slope if $w_{1,H} = w_{1,L}$, as can be verified in (16.4) and (16.5). At that point, the agent is marginally risk-neutral.

Inducing a high effort

Finding the optimal contract that stimulates a high effort is slightly more difficult. The contract must satisfy one more constraint: It must not only be attractive to an agent who works (rather than shirks), but must also give the agent an incentive to work by promising a higher expected utility from working than from shirking.

The optimal contractual payments $(w^*_{2,H}, w^*_{2,L})$ solve the principal's problem:

$$\max_{w_{2,H}, w_{2,L}} V = p(X_H - w_{2,H}) + (1-p)(X_L - w_{2,L}), \tag{16.6}$$

subject to:

$$pu(w_{2,H}) + (1-p)u(w_{2,L}) - e_2 \geq 0 \tag{16.7}$$

$$pu(w_{2,H}) + (1-p)u(w_{2,L}) - e_2 - (1-p)u(w_{2,H}) - pu(w_{2,L}) + e_1 \geq 0 \tag{16.8}$$

where (16.8) can be simplified to:

$$u(w_{2,H}) - u(w_{2,L}) - \frac{e_2 - e_1}{(2p-1)} \geq 0 \tag{16.9}$$

The first constraint (16.7) is the participation constraint for an agent exerting a high effort. The second constraint is the incentive constraint, making sure that the agent does not prefer to shirk (it is assumed that an indifferent agent chooses the action preferred by the principal).

It may be helpful to look at the graphical solution first. In Figure 16.4, the participation constraint, PC_2, is steeper than its sister in Figure 16.3, because the probabilities of the two outcomes are reversed by a high effort. With $p = 3/4$, the slope is -3. It also lies more to the right, as the agent has to be compensated for his higher effort e_2. Still, both participation constraints (PC_2 and PC_1) intersect the 45°-line at the same point (the point representing the optimal contract for a low effort derived above). Admissible contracts lie on or above PC_2.

In addition, admissible contracts must lie on or to the right of IC, the incentive constraint for a high (rather than a low) effort. IC runs parallel to the 45°-line with a horizontal or vertical distance of $(e_2 - e_1)/(2p - 1)$, as follows from (16.9). Figure 16.4 also makes clear why no incentive constraint restricts the contract for inducing a low effort, as discussed above. In order to make the agent opt for a low effort, a contract has to lie on the left of IC. This obviously is not a restriction as it does not prevent full insurance of the agent by the principal, i.e. a contract on the 45°-line.

The optimal contract for a high effort, $\{w^*_{2,H}, w^*_{2,L}\}$, is given by the intersection of the participation and incentive constraints. It lies in the left corner of the admissible area (above PC_2 and to the right of IC). This point lies on the attainable objective function (iso-profit line) closest to the origin, i.e. on the line with the lowest wage payments and the highest profit to the principal.

As Figure 16.4 shows, both the participation and the incentive constraint bind, i.e. hold with equality. As long as the participation constraint does not bind, the principal can reduce at least one of the promised payments without losing the agent. As long as the incentive constraint does not bind, the principal can reduce the difference between $w^*_{2,H}, w^*_{2,L}$ without fear of the agent shirking. Formally,

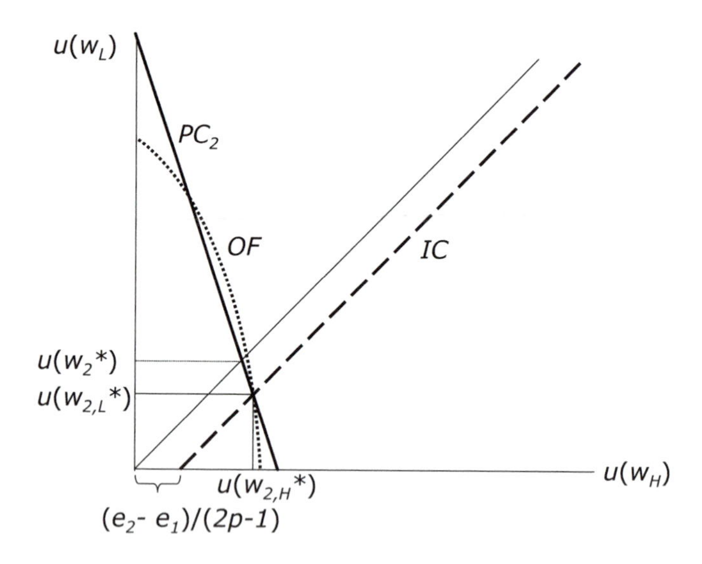

Figure 16.4 Payments (in utility equivalents) to the agent under the optimal contract, $w^*_{2,H}, w^*_{2,L}$, for a high effort (e_2) are determined by the intersection of the participation constraint for e_2 and the incentive constraint (for e_2 versus e_1). The figure is drawn for $p = 3/4$.

the optimal contract is defined by (16.7) and (16.8), both holding with equality. Contractual payments, $\{w_{2,H}, w_{2,L}\}$, therefore are the solutions to:

$$u(w^*_{2,H}) = \frac{pe_2 - (1-p)e_1}{(2p-1)} \tag{16.10}$$

$$u(w^*_{2,L}) = \frac{pe_1 - (1-p)e_2}{(2p-1)} \tag{16.11}$$

The optimal contract for a high effort is a contingent contract. It promises two different payments: A high wage for a high return and a low wage for a low return $(u(w^*_{2,H}), u(w^*_{2,L}))$. This is intuitively plausible; formally it follows from $p > 1/2$ and $e_2 > e_1$. The contract can be seen as the combination of a fixed wage component $(w^*_{1,L})$ and a "bonus" for a high return $(w^*_{2,H} - w^*_{2,L})$.

Compared to the contract which is optimal under observable effort, the contingent contract pays more after a high return and less after a low return. In Figure 16.4, the point $(u(w^*_{2,H}), u(w^*_{2,L}))$ lies to the right, below the intersection of PC_2 (or PC_1) and the 45°-line. In fact, the payment for a low return is even lower than under the contract for a low effort $(w^*_{2,L} < w^*_{1,L})$, as $(pe_1 - (1-p)e_2)/((2p-1)) < e_1$.

The monotone likelihood ratio property

It may seem obvious that the principal pays the agent more for a high return than for a low return. Yet, this is not always the case. The promised wage may not increase monotonically with the principal's return. An optimal compensation scheme pays more for those outcomes that are statistically more likely to occur after a high effort than after a low one. For example, an optimal incentive scheme for a portfolio manager may promise a high bonus for returns that are slightly above average, but not for very high returns. Somewhat above-average returns can be made more likely by the manager's effort, while very high results can only achieved by extremely good luck.

Whether the promised wage increases strictly in effort depends on the so-called *monotone likelihood ratio property* (see Mas-Colell, Whinston and Green, 1995, Ch. 14), Laffont (1993), or Laffont and Martimort (2002). The likelihood ratio LR is the ratio between the densities of the probability distributions of $\tilde{X}$ for e_1 and e_2, respectively:

$$LR(X) = \frac{f(X \mid e_2)}{f(X \mid e_1)}$$

If $LR(X)$ is increasing, the wage promised under the optimal contract also increases in X. In our example the likely ratio is monotonically increasing: X can only take two values for which $LR(X_L) = (1-p)/p = 1/3 < 3 = p/(1-p) = LR(X_H)$.[106]

Optimal unobservable effort

As in the case of observable effort, a comparison between profits from a working contract and a shirking contract tells the principal which effort level is optimal. When effort is *not observable*, the principal asks for a higher effort if:

$$(2p - 1)(X_H - X_L) \geq pw^*_{2,H} + (1-p)w^*_{2,L} - \bar{w}_1, \qquad (16.12)$$

as follows from (16.6) and (16.11).

16 C d Welfare comparisons

It might seem that the informational asymmetry—only the agent can observe the chosen level of effort—does not play a role. After all, the agent's effort choice is an open secret. The principal offering a contract knows perfectly well what effort the agent will choose under the proposed contract. So should we worry about an informational asymmetry?

Yes, we should. It turns out that the possibility of hidden action has a cost, even if the action can be perfectly forecast from promised contractual payments. The case is analogous to hidden information (Chapter 13), where information can be extracted from the agent, but only at a cost. In the context of hidden action, such costs can arise in two forms:

- Incomplete insurance of the agent;
- Inefficient choice of effort.

Incomplete insurance of the agent

Figure 16.4 illustrates that the incentive for the agent to exert a high effort requires that risk in the form of two different payments is imposed on the agent. As the agent is risk-averse, the uncertain payment $(w^*_{2,H}, w^*_{2,L})$ in expected terms must exceed the certain payment w^*_2 by a risk premium Δ:

$$\Delta = pw^*_{2,H} + (1-p)w^*_{2,L} - \bar{w}^*_2 \geq 0 \tag{16.13}$$

The risk premium, Δ, does not increase the agent's utility. The agent, being set to his participation constraint, is equally well off under observable effort (without risk premium) as he is under unobservable effort (with the risk premium). The risk premium simply compensates the agent for the additional risk.

Yet, for the principal Δ is a real cost which reduces expected profits. The reduction in profits due to the non-observability of effort can be seen in Figure 16.4: The optimal contract $(\{w^*_{2,H}, w^*_{2,L}\})$, compared to the contract which is optimal under observable effort, lies on a higher curve for the objective function OF (representing higher wage costs and lower profit). The risk premium therefore represents a deadweight loss of the contract, or *agency cost*.

Anticipation of effort and observation are not the same: The principal can anticipate the agent's effort, but only at the cost of imposing a (costly) risk on him. This is why an incentive contract is more expensive than the equivalent contract under observable action.

Inefficient choice of effort

The other form of *agency cost* is the inefficient choice of effort level. Under non-observable effort, the principal may choose to encourage a lower effort than under observable effort. The risk premium required to make the agent exert a hidden effort makes a higher effort more expensive than a lower effort. The principal may find, therefore, that the higher effort does not pay, although it would if effort were observable.

When effort is *observable* the principal asks for the higher effort if (16.3) holds. When effort is *not observable* the principal asks for a high effort if (16.12) holds. Effort choice is suboptimal if the principal would encourage high effort when the effort is observable, but prefers the lower effort when the effort is hidden, i.e. when (16.12) is violated. Combining condition (16.3) with the violated condition (16.12) and using the risk premium Δ from (16.13) yields the following condition for suboptimal effort choice:

$$w^*_2 - w^*_1 \leq (2p-1)(X_H - X_L) \leq w^*_2 - w^*_1 + \Delta$$

In words: The principal induces an inefficiently low effort if the additional expected return from a higher effort $((2p-1)(X_H - X_L))$ lies between the additional cost under observable and unobservable efforts. If we denote the difference in net expected profit by:

$$V_2 - V_1 = (2p - 1)(X_H - X_L) - (w_2^* - w_1^*),$$

the condition for suboptimal effort choice can be written as:

$$0 \leq V_2 - V_1 \leq \Delta. \tag{16.14}$$

The interpretation is straightforward: Effort choice is suboptimal if the net profit from a higher effort is positive but lower than the risk premium required to induce a higher effort by the agent without losing the agent.

More than two effort levels

The logic developed in this section also holds when there are more than two possible levels of unobservable effort. The lowest possible effort can always be induced with a fixed wage set to satisfy the agent's *participation constraint*. Any higher effort level is induced by an incentive contract defined by the respective participation constraint, as well as an *incentive constraint* to make the desired effort level (slightly) attractive relative to all lower effort levels.

The optimal effort level for the principal is not always the highest possible effort level. It does not pay to make agents work to death, to wit Louis Favre. The optimal effort level from the principal's perspective is always the result of balancing the incremental risk premium against the incremental expected project return which is linked to a higher effort level.

16 C e Further refinements of the basic model

Apart from a greater number of effort levels, there are many possible refinements to basic principal–agent models introduced in the preceding sections, such as intertemporal issues (repeated interactions, reputation), continuous action space, multidimensional action (effort and risk taking, multi-tasking), partly observable actions (signals), competition among principals, competition among agents, and teamwork.

There is a wealth of literature dealing with such refinements. A very readable introduction is Sappington (1991). For a more thorough and formal treatment of the main model of hidden action we would recommend Mas-Colell et al. (1995, Chs. 14, 23). An encompassing review of existing refinements is Laffont and Martimort (2002). The literature on incentives in firms is summarized in Prendergast (1999).

16 D Application: Bank deposit insurance and risk taking

16 D a Is deposit insurance a source of moral hazard?

Banks issue liquid claims, but hold partly illiquid assets. This makes them vulnerable to runs (see Chapter 9 D). Banks also have a large number of small depositors who have not much of an incentive nor the ability to supervise a bank (see Chapter 6). In order to protect banks against runs and to strengthen small depositors' confidence in banks they cannot assess, most countries have introduced some form of *deposit insurance*, see for example Laeven, Karacaovali and Demirgüç-Kunt (2005), and Dewatripont and Tirole (1994, Sec. 3.1.6).

One drawback of deposit insurance is that it may create perverse incentives. Insured depositors do not ask for an adequate risk premium on their deposits. Insofar as banks borrow from insured depositors, they need not care about their risks. As Freixas and Rochet (1997, Ch. 9.4) and the literature cited therein show, a compensating tax on risks via fair, risk-based pricing of deposit insurance is difficult to achieve, both conceptually and politically. Under most (if not all) existing deposit insurance systems, risk taking tends to be subsidized. This effect seems so obvious that the role of deposit insurance as a source of moral hazard has become almost commonplace.

In what follows we try to illustrate how deposit insurance can indeed lead a bank to take excessive risk. But, we will also present a simple model (Blum, 2002) challenging the conventional deposit insurance and moral hazard story. Somewhat surprisingly, deposit insurance may lead to a reduction in bank risk—exactly *because* it is underpriced!

A simple model

In the Blum (2002) model, a bank (the agent) borrows on its liability side from depositors (the principal) in order to finance its asset side (loans, etc.). The time structure follows that of Figure 16.1. First, depositors offer the bank a money deposit at a given interest rate, which the bank accepts. Next, the bank chooses the risk and the return on the pool of its assets. Finally, the bank is either successful or fails.

Blum (2002) analyzes the impact of three regimes:

- Known action: Depositors know what risk the bank will choose (or the bank can commit itself to a certain risk choice);
- Deposit insurance: Depositors are insured against loss of their deposits (and thus do not need to care about the bank's risk);
- Hidden action: The bank chooses the risk "behind the depositors' back".

It shows that deposit insurance has a different effect, depending on whether the bank's action is known by or hidden from depositors.

The risk-return trade-off

The bank's choice variable, asset risk, is continuous. This is a slightly more sophisticated case than the agent's simple choice between two discrete effort levels, as analyzed in the theory section (16 C). The bank's assets are summarized in one risky asset which has an uncertain gross return of either X (success) or nothing

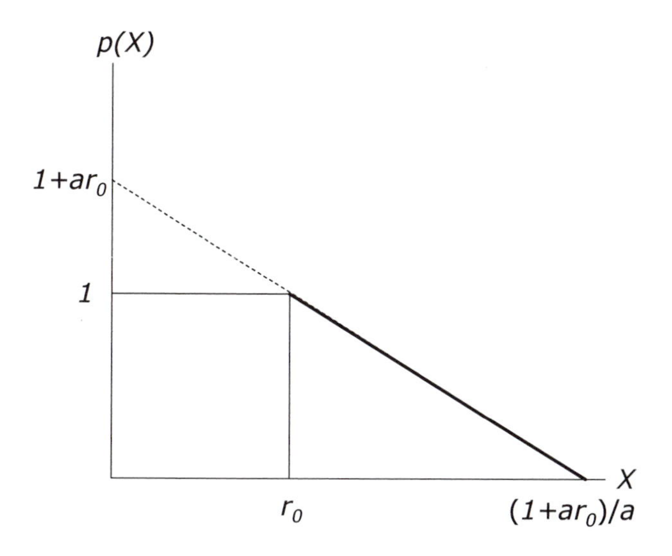

Figure 16.5 The probability of success p is a decreasing function of the project return X, $p(X) = 1 + ar_0 - aX$. In the relevant range, the relation is given by the decreasing solid line. If $X = r_0$, the project is safe ($p = 1$).

(failure). The bank chooses X and the related probability of success. The probability of success, p, (or of failure, $1 - p$) is a decreasing (increasing) function of asset return X:

$$p(X) = 1 + ar_0 - aX, \tag{16.15}$$

with a being a constant parameter ($a > 0$) and r_0 the (gross) risk-free rate of return. For simplicity's sake, we will later set $r_0 = 1$. But for the moment we will keep the notation r_0.

This risk-return trade-off (16.15) is represented by the solid line in Figure 16.5. The minimum value for X is equal to the risk-free rate of return r_0 and has a probability $p(r_0) = 1$. If X goes towards $(1 + ar_0)/a$ (which we assume to be an upper boundary), p goes to zero.

From the relation between risk and *potential* return X, Blum (2002) derives the relation between risk and *expected* return $p(X)X$. This relationship is illustrated in Figure 16.6. The inverse U-shaped line (equivalent to the area of a square under the straight line in Figure 16.5) shows that the expected return first increases and then decreases with X. The line for $p(X)X$ starts at point (r_0, r_0). The value of X, for which the expected return is maximized, is labelled X^*.

The first-best

A depositor who could hold the bank's assets directly would solve:

$$\max_{X} V_{FB} = p(X)X - r_0. \tag{16.16}$$

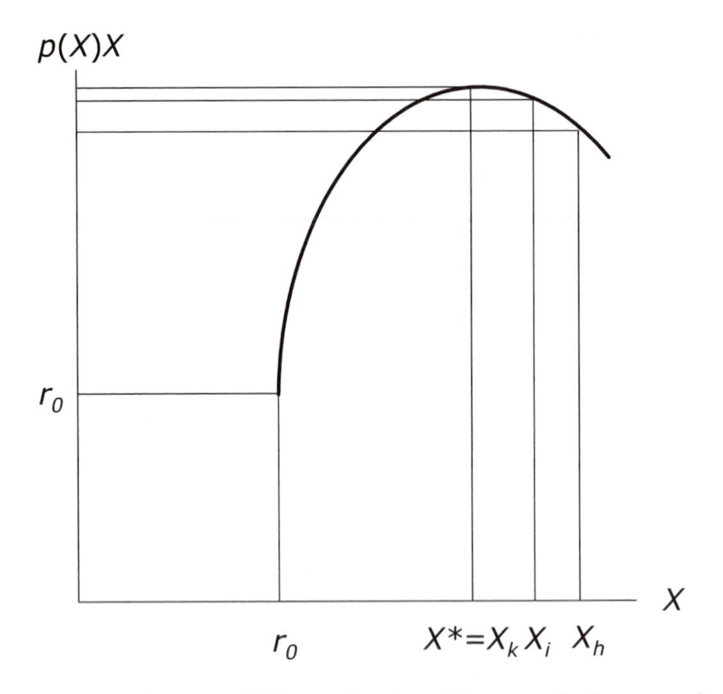

Figure 16.6 Expected return $p(X)X$ as a function of the return in the event of success X first increases and then decreases. The point (r_0, r_0) represents a risk-free project.

The optimal value for X, X^*, solves the necessary and sufficient first-order condition:

$$\frac{\partial p}{\partial X} X + p(X) = 0. \tag{16.17}$$

At X^* the expected asset return is maximized. The solution to equation (16.17) under the chosen specification (16.15) yields:

$$X^* = \frac{(1 + ar_0)}{2a},$$

and

$$p^* = \frac{(1 + ar_0)}{2},$$

Known risk choice: The private optimum

When the bank borrows from depositors it has to pay a (gross) interest rate which we denote by r. Due to limited liability, the bank repays r only in the event of success, if it remains solvent. The rate, therefore, has to compensate depositors for the risk of failure (in which case they receive nothing).

When depositors can observe the bank's risk choice before they lend, the bank solves:

$$\max_X V_{PO} = p(X)(X - r). \tag{16.18}$$

subject to the participation constraint for risk-neutral depositors $r \geq r_0/p(X)$.

The participation constraint binds (otherwise the bank would pay over the odds). Depositors get an expected return equal to the risk-free return: $pr = r_0$. Substituting r_0/p for r in (16.18) yields exactly the same problem as (16.16). For the bank, paying r only in the event of success is equivalent to always paying r_0. Observability of the risk choice fully internalizes the cost of risky debt. When depositors know the bank's risk choice, for example when the bank can commit to a certain risk, the bank chooses the first-best. Both the first-best and the privately optimal value for X are represented by $X^* = X_k$ (see Figure 16.6).

Deposit insurance

Next we assume that there is a deposit insurer who insures depositors against losing their claims in a bank failure. The insurer does not charge risk-adequate insurance premia. For simplicity's sake, we assume that deposit insurance is offered for free. Yet, the bank continues to pay depositors only in the event of success; in the event of failure the insurer steps in for the bank which enjoys limited liability. Therefore, insured depositors lend to the bank if it offers a rate r_d which is at least equal to the risk-free interest rate r_0.

The bank thus solves:

$$\max_X V_i = p(X)(X - r_0).$$

subject to the participation constraint

$$r_d \geq r_0$$

The first-order condition

$$\frac{\partial p}{\partial X} X + p(X) - \frac{\partial p}{\partial X} r_0 = 0,$$

yields optimal values for X and p of

$$X_i = \frac{(1 + ar_0) + ar_0}{2a} = X_{FB} + \frac{r_0}{2},$$

and

$$p_i = \frac{1}{2} = p_{FB} - \frac{ar_0}{2}.$$

Under deposit insurance the bank chooses higher X and lower p than in either the first-best or the (identical) case of observable risk choice. This is the moral hazard effect of deposit insurance. In Figure 16.6, the bank's choice under deposit insurance, represented by X_i, lies to the right of X^*.

Non-observable risk choice

Finally, we look at the case where depositors cannot observe the bank's risk choice before they have to decide about lending. The bank chooses asset risk secretly either before or after depositors lend. From a game theoretic point of view, both cases are equivalent. The key point is that depositors, when they agree on lending terms, do not know the risk the bank has chosen or will choose. Depositors and the bank play a *simultaneous* game, unlike the above case of known risk choice which leads to a *sequential* game.

In the simultaneous game, depositors try to anticipate the bank's decision. In fact, depositors *can* anticipate the bank's decision, knowing all the relevant parameters of the game. Depositors in particular know what combination of return and risk is optimal for the bank, given any repayment promised to depositors. In other words, depositors know the *bank's reaction function*. Conversely, the bank knows the *depositors' reaction function*. Combining the two reaction functions will yield the equilibrium risk and return.

The bank's reaction function

The bank promises depositors a return in the event of success of r_h (h for "hidden choice"). The bank's problem is to choose r_h, which solves:

$$\max_X V_h = p(X)(X - r_h).$$

The *bank's reaction function* $r^{(B)}(X)$ is the first-order condition:

$$\frac{\partial p}{\partial X}X + p(X) - \frac{\partial p}{\partial X}r_h = 0,$$

or

$$r_h^{(B)} = 2X - \frac{1 + ar_0}{a}. \tag{16.19}$$

Depositors' reaction function

A depositor lends to the bank if the expected return is not negative. The *depositor's reaction function* $r^{(D)}(X)$ is the participation constraint

$$r_h^{(D)} = \frac{r_0}{p(X)} = \frac{r_0}{1 + ar_0 - aX}. \tag{16.20}$$

The two reaction functions are illustrated in Figure 16.7. Both have a positive slope: The rate of return required by depositors and the level of risk set by the bank are strategic complements (see Section 9 B). The value for X chosen by the

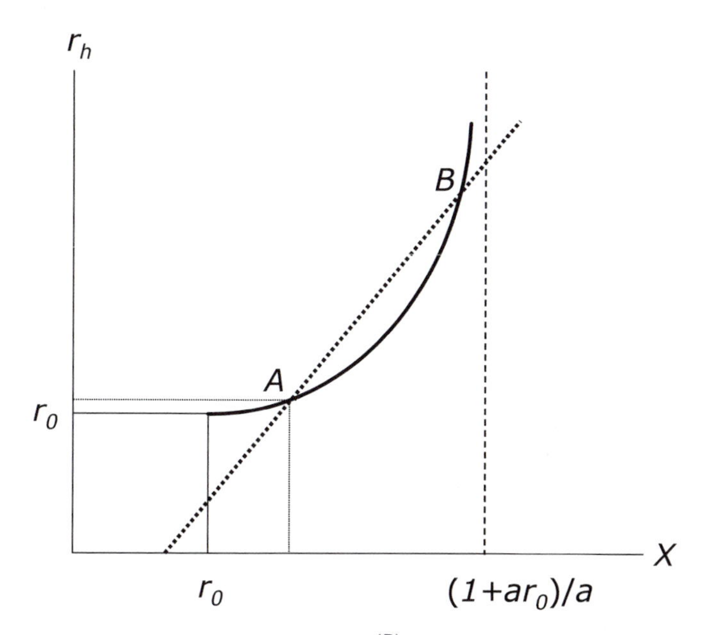

Figure 16.7 Reaction functions of depositors, $r_h^{(D)}(X)$ (solid line) and of the bank, $X(r_h^{(B)})$ (dotted line). The bank's choice of X increases linearly in r_h. Depositors demand a risk premium $r_h - r_0$ increasing in $1 - p(X)$ and thus in X; r_h asymptotically approaches $X = (1 + ar_0)/a$.

bank is a linear function (16.19) of the return required by depositors. The return required by depositors (16.20) is convex in X and goes to infinity when X goes to $(1 + ar_0)/a$.

Equilibrium, graphically . . .

Typically for strategic complements, there are multiple equilibria (see Chapter 9). The Nash equilibria (in pure strategies) are given by the intersections of the two reaction functions, as illustrated in Figure 16.7. The stable equilibrium is obtained at point A (B is an unstable equilibrium). However, an equilibrium may not exist, as the two reaction curves may not intersect; this occurs if $ar_0 > 0.17$ (Blum see 2002, p. 1435).

. . . and formally

Solving for equilibrium formally by setting $r^{(B)} = r^{(D)}$ yields a quadratic equation for X_h. The two solutions (equilibrium values) are:

$$X_h^{(1,2)} = \frac{3(1 + ar_0) \pm \sqrt{(1 + ar_0)^2 - 8ar_0}}{4a}.$$

The smaller solution, $X_h^{(1)}$, is the return chosen by the bank and anticipated by depositors in the stable equilibrium of point A. The probability of success at $X_h^{(1)}$ is:

$$p_h = \frac{(1 + ar_0)}{4} + \frac{\sqrt{(1 + ar_0)^2 - 8ar_0}}{4}.$$

In the range of ar_0 in which a meaningful equilibrium exists $(0 < ar_0 < 0.17)$, $p_h \leq 1/2$.[107] In that range, the expression under the square root cannot exceed unity (under parameter values $ar_0 < 0.17$ for which an equilibrium exists). Therefore:

$$p_h < p_i = \frac{1}{2}.$$

This is a surprising result: If depositors do not know the bank's risk choice, the bank chooses a *higher* risk in the *absence* of deposit insurance than under deposit insurance (when depositors do not care about the bank's risk). This calls for a systematic comparison of the risk levels the bank chooses under different scenarios.

16 D b Risk choice under different scenarios

The introduction of deposit insurance has an ambiguous effect, depending on whether depositors *know* the bank's risk choice or *anticipate* it. As can be seen in Figure 16.6, the following relations among the different risk levels chosen by the bank hold:

- If the bank's risk choice is *known* to depositors, deposit insurance leads to a socially inefficient choice of higher asset returns and higher risk ($p_i < p_k = p_{FB}$; $X_i > X_k$) (moral hazard).
- If the bank's risk choice is *not known* to depositors at the time of depositing money at the bank, depositors try to anticipate it. In that case, a bank with insured depositors chooses an even less efficient, high level of return and risk. Deposit insurance, by contrast, leads to a reduction of risk and increases efficiency ($p_h < p_i$; $X_h > X_i$).

The beneficial effect of deposit insurance when risk is unobservable contradicts conventional views on the impact of deposit insurance. Yet, the intuition is simple: When the bank's risk choice is known, depositors directly ask for a premium commensurate with the observed risk. When the bank's risk is not known, depositors anticipate that the bank may secretly take a higher risk. Consequently, they demand a higher premium, which in turn induces the bank to take more risk: Banks marginally prefer options that make it less likely that the higher premia have to be paid, with the higher bankruptcy risk $(1 - p)$ being compensated by higher return X. This leads banks with hidden risk to aim at higher returns when depositors are *not* insured.

The result resembles a phenomenon known as *credit rationing* described by Stiglitz and Weiss (1981). Higher interest rates charged by a bank crowd out the best borrowers. This adverse selection effect leads banks to limit the amount of credit and to keep interest rates moderate. In Blum (2002), the effect that higher interest rates lead to higher risk is caused by moral hazard, rather than by adverse selection. Otherwise the two problems are quite similar.[108]

16 E Application: Credence goods

16 E a Doctors, lawyers, mechanics, and other experts

Alice and Bob are on vacation. After a coffee break at a roadside diner, their car fails to start. In the end, the vehicle is towed to a garage in a nearby town. The garage owner is busy with another vehicle but promises to look at the problem early next morning. Alice and Bob spend an uneasy night. "It might be just a fuse," Alice says. "It could also be the whole ignition system, coil and all that," Bob adds, "that would be quite a bit of work." "Or, it may be the fuse, but the guy will go on to replace the ignition system," Alice snarls. "Even worse!" Bob exclaims: "It may be the fuse, he changes it, but charges us for the whole ignition system."

In many real-life situations we need the help of experts: mechanics, lawyers, medical doctors, advisors. These specialists have a double information advantage over their clients. Not only do they know how to *fix* a problem, they also know much better than their clients what the problem actually *is*. They sell the diagnosis as well as the related cure. This creates incentives to be dishonest (i) about the diagnosis and the cure that best serves the client, as well as about (ii) the cure that was actually applied, as distinct from the cure the client is charged for. The client, to put it bluntly, may get the wrong treatment, the wrong bill, or, in an extreme case, even both.

Goods that share these properties, like medical treatment, legal advice or repair services, are called *credence goods*. An overview on the economics of credence goods is given in Dulleck and Kerschbamer (2006). Assume a customer has a problem (a car engine that does not start) which may be "serious" (ignition system) or "minor" (fuse). The customer does not know whether the problem is serious or only minor. Only an expert (the mechanic) can diagnose the problem. The expert is also the person who can cure the problem. A serious problem requires expensive treatment characterized by its cost $\overline{c}$, a minor problem needs cheap treatment $\underline{c}$. While expensive treatment also cures the minor problem, cheap treatment is useless against the serious problem.

The customer's gross utility is either v (successful intervention) or 0 (unsuccessful or no intervention). Gross utility as a function of the problem and the

Table 16.1 The customer's gross utility from a credence good as a function of the treatment required ($\underline{c}, \overline{c}$) and the treatment received ($\underline{c}, \overline{c}$)

Customer's utility		Customer needs $\underline{c}$	$\overline{c}$
Customer gets	$\underline{c}$	v	0
	$\overline{c}$	v	v

intervention chosen is given in Table 16.1 (see Dulleck and Kerschbamer 2006, p. 11):

The expert's temptation

In the top left and bottom right cases in Table 16.1, the customer receives the appropriate treatment. In the upper right case, the customer receives *under-treatment*, in the bottom left case, the customer receives *over-treatment*. Under-treatment or over-treatment are a form of opportunistic behavior by the expert. The other is *over-charging*. The expert's scope for wrong treatment or over-charging depends on (i) the customer's type (probability of having a serious problem), (ii) the customer's options (the cost of an alternative diagnosis), (iii) the expert's liability for under-treatment, and (iv) whether the customer can verify that he has been over-charged.

Key criteria

Dulleck and Kerschbamer (2006) systematize models of credence goods according to the following four criteria:

- **Homogeneity:** All customers have the same probability of having the minor or the serious problem.
- **Commitment:** The client is committed to buy the repair service from the expert who provided the diagnosis. There are, in other words, strong economies of scale between diagnosis and repair. This can be modeled by a cost of diagnosis, d, identical for all possible experts, which is sufficiently "high" to prevent the client from duplication (asking for a second opinion).
- **Liability:** The expert cannot provide under-treatment (cheap treatment $\underline{c}$ when expensive treatment $\overline{c}$ is required).
- **Verifiability:** The expert cannot over-charge (charge for the expensive treatment $\overline{c}$ when cheap treatment $\underline{c}$ has been given).

These criteria do not mutually exclude each other. On the contrary, Dulleck and Kerschbamer (2006) show that a model is characterized by its assumptions regarding the above criteria.

16 E b A simple case: One customer, one expert

In what follows we will focus on the simple problem of one customer and one expert. Therefore, homogeneity and commitment hold automatically. The other two criteria or assumptions may or may not hold; we will look at the consequences of relaxing them.

The relevant game tree (a simplified version of the tree in Dulleck and Kerschbamer 2006, p. 14) is given in Figure 16.8. The game begins with the expert posting prices $\underline{p}$ and $\overline{p}$ for expensive and cheap treatment, respectively. Next, nature draws the type of customer problem: serious or minor. In line with the commitment assumption, we drop the subsequent move by the customer, choosing between the expert's repair service and asking a different expert for a second opinion. The following two moves are the expert's: First, the expert decides about the treatment the customer receives. Second, the expert writes the bill, telling the customer how much to pay.

A list of potential outcomes

Below the final nodes in Figure 16.8 we give the payoffs to the customer (C) and the expert (E) for the different outcomes. For easy reference, outcomes are numbered from (1) to (8). The individual outcomes have the following characteristics:

(1) Serious problem: adequate treatment, adequate charge
(2) Serious problem: adequate treatment, under-charge

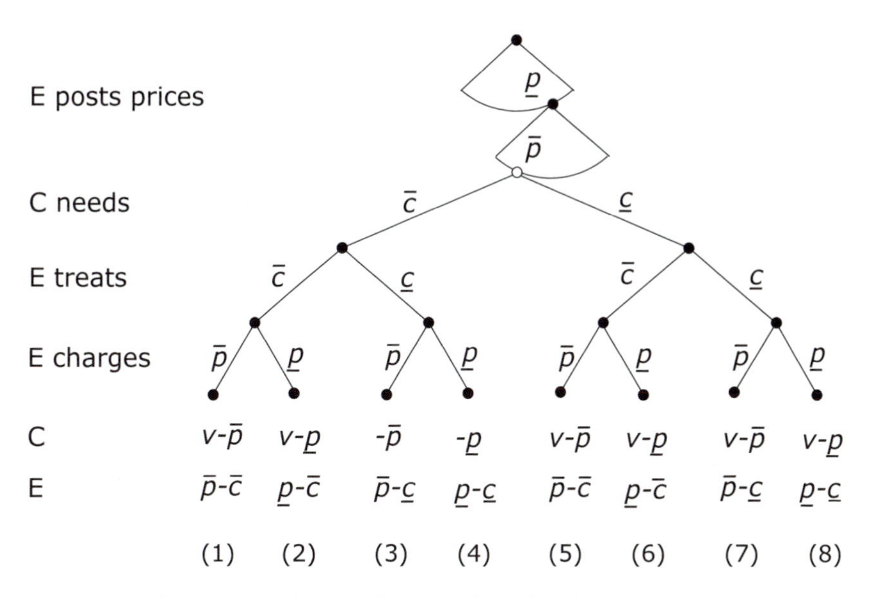

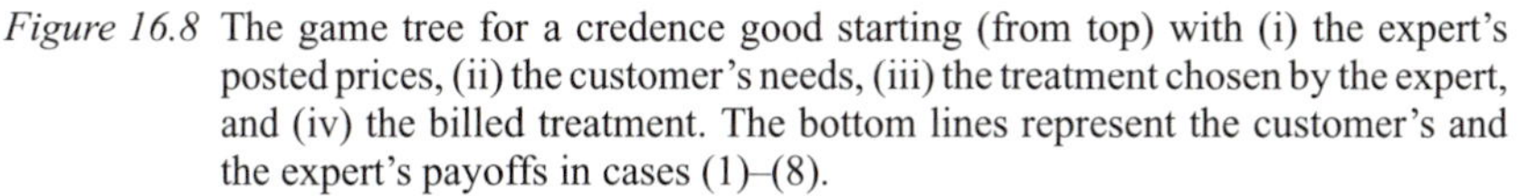

Figure 16.8 The game tree for a credence good starting (from top) with (i) the expert's posted prices, (ii) the customer's needs, (iii) the treatment chosen by the expert, and (iv) the billed treatment. The bottom lines represent the customer's and the expert's payoffs in cases (1)–(8).

(3) Serious problem: under-treatment, over-charge
(4) Serious problem: under-treatment, adequate (charge to under-treatment)
(5) Minor problem: over-treatment, adequate charge (to over-treatment)
(6) Minor problem: over-treatment, under-charge
(7) Minor problem: adequate treatment, over-charge
(8) Minor problem: adequate treatment, adequate charge

In the above list, there are two cases in which the customer receives the right treatment and pays the right price: (1) for the serious problem, and (8) for the minor problem. Both cases are consistent with the liability as well as the verifiability assumption. Two other cases, (2) and (6), can be ignored: these involve undercharging which obviously is never attractive for the expert. This leaves us with cases (3), (4), (5) and (7), in which the expert does not serve the customer's best interests.

Liability and verifiability . . .

We assume that experts maximize their own utility, not that of the customer. If indifferent, the expert chooses the option the customer would prefer. Let us look at the case where (in addition to homogeneity and commitment) both liability and verifiability hold. Under these assumptions, the only case the customer needs to be concerned about is (5), over-treatment of a minor problem. The customer gets adequate treatment (assuming that an indifferent expert chooses the option the customer would prefer) only if:

$$\overline{p} - \overline{c} \leq \underline{p} - \underline{c}. \tag{16.21}$$

. . . and the role of commitment

Here the role of the commitment criterion becomes clear. We did assume that the customer is "captured" by the expert (d, the cost of a second diagnosis is sufficiently high). Now, think of the opposite case where $d = 0$, and where the customer can opt for a free, second or third opinion. In this case an expert could only attract customers by posting prices that satisfy (16.21) with equality. The absolute markup would have to be identical for expensive and cheap treatment. In this case, customers would always receive the right treatment. By assumption (verifiability) they would also be charged correctly. And, ironically, they would never need the second opinion (unless we assume that experts may err). To sum up, under non-commitment, liability, and verifiability, the market solves the problem of the informational asymmetry between expert and customer.

Liability without verifiability

Under the liability assumption, the expert cannot provide cheap treatment where expensive treatment is required; this excludes cases (3) and (4). The lack of verifiability permits overcharging for cheap treatment where it is adequate (7), however. For the expert, (7) dominates all available alternatives. The customer with a minor problem will therefore receive the adequate treatment, albeit at a high price. An example may be a portfolio manager who claims to do a great deal of research for a customer, and charges for it, without actually doing so.

Verifiability without liability

If verifiability holds but liability is violated, a customer not only has to worry about the over-treatment of a minor problem (as in the case where both liability and verifiability hold). Now, conversely, a customer may get under-treatment for a serious problem (at a correct price, at least). The condition under which a serious problem will receive adequate treatment is 16.21, with the inequality sign reversed, i.e., $\overline{p} - \overline{c} \geq \underline{p} - \underline{c}$.

Neither liability nor verifiability

If neither verifiability nor liability hold, a customer ends up in the worst possible case: with a minor problem, the customer will simply be over-charged (7). With a serious problem, the customer is under-treated and over-charged (3). For the expert, (3) and (7) dominate all alternatives. In the absence of liability and verifiability, the customer faces a "market for lemons" situation (Dulleck and Kerschbamer, 2006).

All the above conclusions are true if the customer is committed to receiving treatment from the expert who provides the diagnosis. As mentioned above, competition among experts may lead them to post prices that make over-treatment unattractive. However, the drawback of competition among experts may lead to duplication of the cost of diagnosis. We will not discuss competitive models further. Nor will we look at the case of heterogeneous customers. Both issues are treated in Dulleck and Kerschbamer (2006).

Over dinner, Alice and Bob think about possible strategies to prevent the mechanic from cheating. Bob starts: "The positive thing is: He cannot claim to have fixed the car if in fact he hasn't: Without proper repairs, the engine will fail to start." "And if we ask to see the replaced parts, he cannot bill us for anything major if he has only changed a fuse!" Alice adds. Bob is still uneasy: "This may work; this guy may not know that we haven't got a clue about car mechanics. But he could still repair too much just to get a few extra hours of paid work." "Well, I have an idea. Think about this: What would he earn from changing a fuse? $20, I'd guess. What would he earn from changing the whole ignition system? If he keeps 20 per cent of a total bill of $600, that would be $120. That makes a difference of $100. Here comes my idea: Before we get a quote, we must promise a tip of $100 on any bill below, say, $50. Then he has no interest in messing around unnecessarily with our car." "Alice, you are a genius! I'd be happy to part with one hundred dollars, rather than pay six hundred and sit around for a day or two." "You're right, Bob, after all, it's our vacation." "Our first vacation," Bob adds with a confident smile.

16 E c Practical examples

Medical treatment

Dulleck and Kerschbamer (2006) give several examples of practical arrangements that mitigate problems arising from credence goods. One is the Hippocratic oath. Another is the separation of doctors (diagnosis) and pharmacies (treatment). There is empirical evidence for over-treatment by medical doctors. In a study for Switzerland, Domenighetti, Casabianca et al. (1993) derive such evidence from the consumption of medical and surgical services by the most informed consumer in the health care market: the physician-patient. It can be assumed that physicians choose approximately adequate treatments for themselves and for their husbands or wives. Differences in treatment between physician-patients and "ordinary patients" would reveal under- or over-treatment. By means of a questionnaire, Domenighetti, Casabianca et al. (1993) measured the standardized consumption of seven common surgical procedures. They found that: "Except for appendectomy, the age- and sex-standardized consumption for each of the common surgical procedures was always significantly higher in the general population than for the 'gold standard' of physician-patients. The data suggest that (i) contrary to prior research, doctors have much lower rates of surgery than the general population; and (ii) in a fee-for-services health care market without financial barriers to medical care, less-informed patients are greater consumers of common surgical procedures."[109]

Taxi rides

Another example of credence goods are taxi rides. Complaints about over-charging abound. Regulations therefore often require taxis to post their fares and to carry an identification number for possible complaints. In many places, customers are well advised to negotiate the price before the ride. Even if over-charging can be excluded, drivers still have scope for over-treatment (taking long routes). An interesting solution, to be found in many cities, is a two-part tariff, consisting of a fixed fee and a meter charge. The fixed fee gives the driver an incentive to reach the next customer quickly, i.e. to avoid long routes. As Dulleck and Kerschbamer (2006) point out, this incentive works when the taxi operates at a capacity constraint. An idle driver is not discouraged from a detour by a fixed fee he would earn if there were a customer waiting.

The information in a queue

The taxi example illustrates the impact of capacity constraints on the provision of credence goods. From a thorough analysis of the role of capacity constraints, Dulleck and Kerschbamer (2006, p. 36) conclude: "For the meantime, a valuable advice might be to consult an expert featuring a lengthy queue of customers waiting for service." Note that, here, the length of the queue is a signal for the expert's *opportunity cost*. A long queue tells a customer that the expert has no time to waste on an unnecessary treatment. A queue may also be a signal for other customers' *private information* about the quality of the expert. A long queue would then tell a customer that other customers believe that the expert is good. Observing the actions of others is thus a source of information for an individual. Socially, it may not be without risks, as we have tried to show in Chapter 10.

Early next morning, Alice and Bob arrive at the garage. They need to make their offer before the mechanic has a chance to say anything. They are surprised to find him at work already. "Good morning," he smiles, "your car is ready." "What do you mean by ready? We . . ." "Well it was just a fuse, and I have changed it. Actually, I'm glad it was nothing serious. Look around: With all this work waiting, I could hardly have fixed your car within a week."

16 F Conclusions and further reading

Incentive problems are pervasive in the economy and have been studied intensely in the last few years. The principal agent model proved to be a powerful and versatile tool for the analysis of such problems and has yielded a wealth of results, both theoretical and empirical. These results are presented in the great book on the principal–agent model by Laffont and Martimort (2002). The search for optimal incentive contracts in wider areas has led to the study of many real-life arrangements in health or long-term care (Jullien, Salanié and Salanié, 2001; Finkelstein and McGarry, 2006), patent systems (Scotchmer, 2004), or research contests (Che and Gale, 2003).

In the present chapter, we have only presented the most basic model. The model shows how hidden action leads to agency cost, even though parties use optimal contractual arrangements. Presenting the model we tried to simplify the more rigorous presentations in Mas-Colell, Whinston and Green (1995, Ch. 14), Wolfstetter (1999, Ch. 11), and Laffont and Martimort (2002, particularly Ch. 4). We have illustrated the basic model with an application on bank risk (following Blum (2002)) and we have shown how it can be enlarged for credence goods (following Dulleck and Kerschbamer (2006)).

These are but tiny islands in the principal–agent archipelago. There are many issues we have *not* treated or even mentioned. In reality, bilateral situations of one-shot games between one principal and one agent with very few possible actions are rare. But we have to leave it to the literature mentioned above to discuss repeated interaction, competition between several agents or principals (or both), multidimensional action spaces, and many other enrichments. Many real-life situations are burdened with two-sided hidden action. For example, an employer cannot perfectly observe the employee's effort, while the employee is not sure whether the employer will be fair in decisions about salary increases or promotions. Bank depositors may not trust the bank's figures, while the bank is afraid of truthfully reporting a loss from fear that some depositors might opportunistically withdraw their funds. In other situations, hidden action is complicated by hidden information (treated in Chapter 13). A health insurer faces individuals who have both hidden information about their health and opportunities for hidden action like smoking.[110]

Complications of the principal–agent model can very quickly become intractable in reality. Like other models, the principal–agent model has its limits. But there is another limitation to its logic, rooted in the very nature of human beings.

Classical incentive theory builds on the "relative price effect": Offering a higher price for a good or an activity leads to increased production of that good or activity. Offering an agent a higher wage for success leads the agent to put effort into being successful. In many contexts this assumption is warranted, and stronger incentives lead to better performance.

In many contexts, however, it is not. Human beings do not work exclusively for external stimuli like money, titles or degrees. Just as important are internal stimuli like self-perception: We like to experience success or, at least, to perceive ourselves as hard workers or as generous contributors to others' welfare. In recent years economists have come to admit the importance of intrinsic, as distinct from extrinsic, motivation (Frey and Osterloh, 2001). The problem with extrinsic motivation is that it can get in the way of intrinsic motivation. Sometimes we enjoy doing something, provided we do it for fun or for free, or both. Being offered money we can be motivated or demotivated. Demotivation is particularly likely if we consider the monetary reward as paltry or unfair.

Motivation—successfully balancing intrinsic and extrinsic factors—is a science and an art of its own. We cannot further pursue it here. But we will take a closer look at the more general, underlying issue of self-perception. Self-perception, self-management and contracts with oneself are the topic of our final Chapter 17.

Checklist: Concepts introduced in this chapter

- Hidden action
- Principal–agent model
- Incentive contract
- Participation and incentive constraints
- Monotone likelihood ratio property
- Agency cost of contract

- Deposit insurance
- Simultaneous versus sequential game
- Credence goods
- Over-treatment and over-charging

16 G Problem sets: Getting things done

For solutions see: www.alicebob.info.

Box 16.2 "Off-Piste Top Tips"

The *Financial Times* (25/26 February 2006, W22, authored by Belinda Archer) had the following tips:

- Ski with your legs close together and skis flat rather than on edges.
- Stand straighter than you would on-piste.
- Always wear a transceiver.
  Carry a rucksack filled with spare gloves, extra layers, shovel and avalanche probe.
- Never go off-piste without a qualified mountain guide
- Never go with less than three in a group.
- Take a phone that has the local emergency service numbers.
- Check your insurance covers for off-piste with and without a guide.
- Always check avalanche safety levels before heading off (above three is risky).

Problem 16.1 Going off-piste

Which of the tips in Box 16.2 have an element of moral hazard?

Problem 16.2 Sharing a pie

Assume a principal owns a project that will yield an uncertain return π which is positively correlated with the amount of effort an agent invests in the project.

(a) Assume that the effort is *observable*. What kind of contract should the risk-neutral principal offer to a risk-averse agent? What is a real world example of such a contract?
(b) Assume effort is *not observable* to a third party. What kind of contract should the principal offer to an agent if they are both risk-neutral? What is a real world example of such a contract?

Problem 16.3 Making money work

From a rich uncle you inherit $10 million. Having no time to manage your wealth yourself, you mandate a portfolio manager. You draft a contract with the portfolio manager.

(a) What provisions do you include in the contract?
(b) In particular: how would you compensate the manager?

Problem 16.4 Cork!

Some years ago the authors had dinner in a rather pricey restaurant the pride of which was the wine list. On that list, wines with prices above a certain threshold (around $100) were marked with an asterisk translated at the bottom of the page as "We do not guarantee for the quality of the wine."

(a) Assume a guest tastes a bottle of wine and declares "cork!" Should the guest have to pay for the bottle?
(b) What assumptions about the course of events have you made behind your answer to (a)?
(c) What events mentioned under (b) are observable/unobservable?
(d) What assumptions listed under (b) would you have to change to arrive at a different answer to (a)?

Problem 16.5 The pillory

It is sometimes proposed that authorities should make public the results of their examinations of supervised suppliers like physicians, restaurants or banks. List for each of the following type of information the pros and cons of publishing:

(a) Professors' teaching evaluations
(b) Patients' recovery rates for individual physicians or hospitals;
(c) Ratings of restaurants' cleanliness;
(d) Bank supervisory ratings.

Problem 16.6 The price of sweat

You ask a broker to sell your house. The outcome (sales price) is either high (H = $200) or low ($L$ = $100). The probabilities of the two outcomes, Pr(.), depend on the broker's effort, which can be e_1 (small), e_2 (medium), or e_3 (large). The relations between effort and the outcome probabilities are summarized in the table below. The table also shows the cost (in utility units) of the different effort levels.

Effort	Cost	Pr(L)	Pr(H)
e_1 (small)	1	0.75	0.25
e_2 (medium)	2	0.50	0.50
e_3 (large)	4	0.25	0.75

You know that the broker maximizes $u = \sqrt{w} - e$, where w is financial income and e is effort cost. The broker only becomes active if expected utility is at least $\underline{u} = 2$. You are risk neutral.

You cannot observe the broker's effort, but the outcome is publicly observable.

(a) What contract would you offer if effort were observable? What is your expected profit?
(b) Write down your maximization problem (including all constraints) for the case of unobservable effort.
(c) Draw the constraints in a $\sqrt{w_H}$-$\sqrt{w_L}$-graph.
(d) What are the optimal contracts under which you get effort levels (i) e_1, (ii) e_2, and (iii) e_3?
(e) What contract would you offer?
(d) What are the effects of unobservable effort? Does anyone suffer from the information asymmetry?

Problem 16.7 Poor CEOs

Assume that executives have the following utility function:

$$u(x) = \frac{x^{1-\rho}}{1-\rho}$$

An executive is offered a performance bonus of \$100 for achieving a certain target.

(a) Assume that $\rho = 0.5$. What increase in utility would a \$100 bonus represent for an executive with initial wealth of $x = 0$?
(b) Assume that $\rho = 0.5$. What increase in utility would a \$100 bonus represent for an executive with initial wealth of $x = 1000$?
(c) Calculate the coefficient of relative risk aversion ($r_r = xu''(x)/u'(x)$). Is it increasing, decreasing or constant in x? Compare it to absolute risk aversion ($r_a = u''(x)/u'(x)$).
(d) Guay, Core and Larcker (2003, p.38) consider relative performance evaluation: "A widespread concern among both practitioners and academics is that executive portfolios lack relative performance evaluation (RPE) or, equivalently, that stock and stock options gain value not only because the firm performs well, but also because the market rises." Would this be a problem for the CEOs?

Problem 16.8 Deposit insurance

In the Blum (2002) model discussed above, we set $r_0 = 1$ and $a = 1/6$. Compute the levels of return and risk the bank chooses in the three different regimes:

(a) known risk
(b) deposit insurance
(c) hidden risk.

Part IV

The economics of self-knowledge

17 Me, myself, and I

17 A Introduction

Box 17.1 Natascha's pact with herself

In the morning of 2 March 1998, Natascha Kampusch, then a ten year old girl, disappeared on her way to school. The biggest search operation in Austrian police history yielded no tangible results. In fact, Natascha had been abducted, and the kidnapper hid the girl in a makeshift dungeon below his house on the outskirts of Vienna. There she would stay for years to come. Yet, Natascha never lost hope. At the age of twelve, she later reported:

> I swore to myself that I would get older, stronger and more powerful so that one day I could free myself. I made a pact with my later self that I would free that little 12-year-old girl.

On 23 August 2006, aged eighteen and more than eight years after her abduction, Natascha Kampusch finally escaped. A few days later she gave a TV interview from which the above quotation is taken (Transcript from *Times Online*, 6 September 2006, http://www.timesonline.co.uk/article/0,,3-2345787,00.html).

Natascha Kampusch's pact (see Box 17.1) is a striking case of something that is an important part of our everyday life but has been completely ignored, until recently, by economists: contracts people make with themselves.

It is a common experience that we frame decisions as promises to ourselves. The fiction of a present self signing a contract with a future self seems to help us to stick to decisions we have taken. Such contracts seem to bolster our willpower to fight the inner conflicts which lie ahead.

Economists have not overlooked that such conflicts exist. But they have modeled them in a particular way—as a rational choice between an apple and an orange, income or leisure, consuming today or tomorrow. The strong inner battle often involved in subjecting momentary impulses to longer-term reason is not fully captured in the model of rational economic man. "Although Plato compared the human soul to a chariot pulled by the two horses of reason and emotion, modern economics has mostly been a one-horse show. It has been obsessed with reason," *The Economist*[111] wrote.

In recent years, this has changed. Inspired by psychologists, economists have realized that the human mind is not a monolith. Rather, some different "selves" within one person are fighting for control over that person's decisions. This allows a more realistic, but also a more demanding modeling of human behavior, the transition, so to speak, from *homo œconomicus* to *homo sapiens*.

Economists became particularly fascinated by the application of their new approach to intertemporal choice. Splitting the individual into different selves with imperfect communication between them allows for phenomena like the "I couldn't help myself" experience. Humans, in other words, (i) are weak, and (ii) know it. The joint workings of time inconsistency and consciousness are accessible to economic tools, such as the principal–agent model.

Economists have tried to understand why, for example, individuals may ignore demotivating information, why they follow personal rules like saving plans or (sometimes excessive) diets, or why they sign irrevocable life-long bans like the Missouri Voluntary Exclusion Program for problem gamblers (discussed in Box 17.4 below).

The present—final—chapter deals with the information flow or communication with an individual, between the different selves who fight "an intimate contest for self-command" (Thomas Schelling). Unlike the rest of the book it is not about man versus nature or man versus man, but about man's battle against himself.

17 B Main ideas: Contracting with oneself

17 B a From homo œconomicus to homo sapiens

The shortcomings of homo œconomicus

For most of its history, economics was built around the useful abstraction of *homo œconomicus*.[112] *Homo œconomicus* was an easy fellow: He had one set of preferences from which he could derive optimal rational choice, given his budgetary restrictions applicable at a given moment. Homo œconomicus was one single "monolithic" person and he either knew or believed something or he did not.

Unfortunately, rational *homo œconomicus* failed to behave like a real human in a number of ways. As psychologists knew all along, real humans often act irrationally. One source of irrational behavior is cognition (Thaler, 2000): We tend to perceive the outside world as well as ourselves with rose-tinted glasses. There is a long list of well documented cognition biases like the so-called Lake Wobegon Effect (Overconfidence Effect), the human tendency to believe that one is above average.[113]

An important, well-documented form of irrational behavior is time inconsistency. Individuals take decisions today (smoking another cigarette) which they will regret tomorrow (not having stopped smoking). Preferences seem to shift with time. Worse, individuals may even take decisions they *know* they will hate themselves for, like smoking the first cigarette after a three-month break. While one voice within *homo œconomicus* whispers: "Don't do it!", another shouts: "Just this once!"

Behavioral economics

Under the label "behavioral economics", Daniel Kahneman (born 1934), Amos Tversky (1937–1996) and others tried to integrate results from psychology and cognitive science into economics. Their approach, reviewed in Camerer and Loewenstein (2004), proved fertile. For his contribution, Kahneman was rewarded with the 2002 Nobel prize for economics. A research psychologist, he claims to have never taken a single economics course.

Behavioral economics began as a loose collection, mostly from finance, of anomalies not explainable by the standard paradigm of rational economic man. One example is loss aversion, an asymmetric perception of small losses and gains of an equal amount. Later, cognitive psychologists and behavioral economists tried to show that the demise of strict rationality was not just tennis without a net, but that observed non-rational behavior followed some recurring and potentially predictable patterns. Today, behavioral economics seems to be in a third stage, in which researchers try to integrate regularities into a general explanatory framework. For this research program Thaler (2000) also coined the working title "From homo economicus to homo sapiens", which we borrowed as a title for this section.

Homo sapiens

The new *homo* (or *femina*) *sapiens* is a more complicated being than good old *homo œconomicus*. Contrary to what the name might suggest, *homo sapiens* is less intelligent, at least in the narrow sense, than *homo œconomicus*. *Homo sapiens* is simply more human. For the purposes of information economics, two characteristics are important:

- **A split self:** Rather than a monolithic mind, there are several simultaneous instances competing for control over memory, beliefs and decision, some with conscious effort (will), some more automatically (emotion).
- **A serial self:** Rather than a one-shot person, the individual is an intertemporal collection of past, present and future selves. Communication, commitment and contracting between the present and future selves are imperfect.

The intuition of a "partitioned" personality has been confirmed by recent findings in the field of neuroscience. Unlike economics or psychology which try to understand the working of the human brain from observed behavior, neuroscience tries to look into the brain directly, using sophisticated techniques like the imaging of brain activity. The main findings relevant for economics are summarized under the heading "neuroeconomics" in Camerer et al. (2005). These findings suggest that some forms of "irrational" behavior that puzzled economists may indeed be explicable by an allocation of similar tasks (like assessing a gain or a similar loss, or comparing a small with a big gain) to different parts of the brain which follow their own laws.[114]

17 B b The split self

Expressions of everyday language like "my inner voice", "I promised myself", "I couldn't help [whom?]", or "It wasn't me", suggest that we have more than one self. The different selves within a person are often in conflict with each other. Below, we will treat such conflicts in an intertemporal setting. Here we will first focus on another aspect: The fragmentation and selective nature of human memory.

Imperfect memory

Most of the information we receive is forgotten almost immediately. But not all forgotten information is lost forever. Some memories can be recovered by deliberate effort, others pop up by chance. There seems to be a degree of "latent" knowledge. We know something without knowing that we know it. Some Greek philosophers believed that "a man who does not know has in himself true opinions on a subject without having knowledge", as Socrates (470–399 BC) claims.[115] (Note that this violates the *axiom of transparency*, one of the five axioms of knowledge mentioned in Chapter 8.)

Latent knowledge and fact-free learning

Latent knowledge leads to a phenomenon called "fact-free learning", as Aragones, Gilboa et al. (2005) put it. Their example is the econometrician who learns through the discovery of regularities *within* the available database, rather than through collecting additional information. They propose fact-free learning as a complementary paradigm to Bayesian learning (introduced in Chapter 3).

Forgetting and remembering are not purely random, value-free processes. It seems that we often tend to forget (or remember) what we would like to forget (or remember). This becomes relevant whenever we are—technically speaking—not indifferent about our beliefs, as Akerlof and Dickens (1984) point out. Most humans prefer to see themselves as smart, affable individuals. Contradicting evidence has a high chance of being overlooked, forgotten or re-interpreted ("I failed because I am too honest").

Cognitive dissonance

The force at work here is "cognitive dissonance", a phenomenon discovered by psychologists (see Box 17.2) and pervasive to everyday life: Workers tend to be oblivious to the dangers inherent in their jobs. Researchers may be more receptive to evidence consistent with their models than to conflicting theories. Charles Darwin was aware of the problem: "I had during many years followed the Golden Rule, namely, that whenever a published fact, a new observation or thought came across me, which was opposed to my general results, to make a memorandum of it without fail and at once; for I had found by experience that such (contrary and thus unwelcome) facts and thoughts were far more apt to escape from memory than favorable ones" [from: Francis Darwin, *The Life of Charles Darwin* (quoted by Bénabou and Tirole (2002)]).

Selective awareness

Trying to overlook or to forget may seem irrational—a waste of potentially valuable information. However, some authors argue that some strategic "management" of memory can be rational. "Rational" here means that selective awareness helps the right side to win, when there is conflict between different selves within an individual. Given that this is within an intertemporal setting, we should look at the serial nature of the self.

Box 17.2 When the world failed to end

In 1956, a woman from Chicago mysteriously received messages from alien beings about the imminent end of the world in a great flood before 21 December. A group of believers gave up everything on earth and waited to be rescued by a flying saucer.

The psychologist Leon Festinger (1919–1989) forecast that, once the prophecy would have turned out to be wrong, the group would not want to return home in shame. Rather, the "cognitive dissonance" between the beliefs (the imminent end of the world) and the observed facts (the world still turning) would be resolved in favor of beliefs, not of facts. Indeed, the group claimed that, thanks to their prayers, the end of the world had been postponed. They made great efforts at proselytizing to seek social support and lessen the pain of disconfirmation.

Cognitive dissonance (for details, see Harmon-Jones and Mills, 1999, Ch. 1), is a reverse—or perverse—way of Bayesian learning. A rational Bayesian individual takes prior beliefs and new information as inputs to build posterior beliefs. Under cognitive dissonance, fixed posterior beliefs conflict with prior beliefs and additional information. Something then has to give. One example is the "sour grapes" principle. In Aesop's fable *The Fox and the Grapes*, a fox fails to reach the grapes hanging high up on a vine. He retreats, concluding: "The grapes are sour anyway!" Here it is the prior belief (grapes are sweet) that is revised. In the flood example it was the information. George Orwell (1903–1950) in his novel *Nineteen Eighty-Four* uses "doublethink"—the act of holding two contradictory beliefs simultaneously and fervently believing both—as an example of sustained cognitive dissonance.

17 B c The serial self

The 08:25 train

The Beatles song *Yesterday* (1965) has a line "Suddenly, I'm not half the man I used to be". This suggests that one can be a different person today from yesterday, or tomorrow from today. Indeed, from the cradle to the grave we go through great physical and mental changes. Yet, in a puzzling way, we also remain the same person. There must be an identity which persists despite these daily changes. The Swiss linguist Ferdinand de Saussure (1857–1913) pointed out that the identity of a train is not so much its physical existence but the fact of it being the 08:25 train from Geneva to Paris. Likewise one could say that an individual, though different from day to day, keeps an underlying identity, much like the 08:25 train from Geneva to Paris.

Me versus myself

McCloskey (1986) described the human self as an "intertemporal collection of selves". These selves are not just beads on a string: Humans, unlike trains, *know* about their self and about their past and future existences. The present self also has *preferences* about the behavior of its past and future incarnations. It cannot enforce these preferences, however. It can only regret the actions of past selves ("Had I only kept my mouth shut"), but not change them. It can try to influence the actions of future selves, but has little control over them ("I'll not touch a drop of alcohol tonight").

Concepts like time inconsistency and counteracting attempts of self-control start to make sense, once one accepts that some cracks run through our different selves: "The idea of self-control is paradoxical unless it is assumed that the psyche contains more than one energy system, and that these energy systems have some degree of independence from each other." (Donald McIntosh, 1969; cited by (Thaler and Shefrin 1981, pp. 393f.)). Problems of time inconsistency also become accessible to analytical tools. For example, conflicts between the present and future self can be described in the language of the principal–agent model (introduced in Chapter 16), as the present self can be both in this model:

- a child (as an agent) taking decisions the adult (the principal) will regret,
- a far-sighted or clear-minded but imperfect master (the principal) of tomorrow's weak or myopic self (the agent).

The planner and the doer

An interesting solution to deal with intertemporal conflicts between different selves has been proposed by Thaler and Shefrin (1981): The individual is modeled as an organization, consisting (at each point in time) of a *planner* and a *doer*. The planner is far-sighted, i.e. concerned with lifetime utility, while the doer is myopic or selfish and cares only about present consumption. Splitting the self into a planner and a doer reflects both the split, temporary self (a "cross-section" view) and the intertemporal sequence of selves (a "time series" view).

The standard way to model the interaction of a planner and a doer is "hyperbolic discounting", a concept introduced by Laibson (1997). Hyperbolic discounting (to be defined more precisely below) means that a high rate of discount is used between the present and the near future, and a lower rate between the near future and the distant future. For instance, a day before an exam a student would pay a large sum to postpone the exam by just one more day. However, a year ahead of the exam, nobody cares much about its precise date. In a planner-doer formulation, hyperbolic discounting means that the planner formulates an optimal consumption plan, giving due consideration to future periods, while the doer spends an excessive part of present funds on immediate consumption.

17 B d Deals between the selves

Alice comes out of the kitchen with a plate of fresh cookies. "Want to try?" she asks Bob. "Oh, Alice, you know, I promised myself not to have any more sweets until the end of next month." "OK, I'll take them away." "No, let me have just one. Showing myself that I *can* stop after eating one will make me even stronger. And, after all, I deserve one; I've been working so hard these days."

The planner and the doer, or the present self and future selves, have diverging preferences. The planner often loses out to the *present* doer. Yet, the planner has ways to keep the *future* doers in check. This is known as self-management.

Ways of self-management

In the traditional *homo œconomicus* framework there was little room for rational individuals hurting their own long-term interests. One way to allow for such phenomena is the "rational addiction" approach in Becker and Murphy (1988), in which an individual takes into account that smoking a cigarette today commits himself to future smoking. Empirical evidence quoted in Gruber (2002) suggests that particularly young people dramatically underestimate the addictive nature of smoking, thus casting some doubt on the rational addiction approach. Some researchers therefore propose an alternative approach which explicitly allows for intertemporal inconsistency and for the resulting conflicts between planner and doer or today's and tomorrow's selves. Formal models of self-management are provided by Bénabou and Tirole (2002, 2004).

In these models, very much like in the principal–agent model, the planner (the principal) tries to set incentives for the future doer (the agent). There are several instruments to do so:

- Contracts or resolutions,
- Motivation (modifying preferences),
- Rules,
- Beliefs management and identity building.

Contracting would be the standard solution in a principal–agent setting, where principal and agent are different people. Yet, in the form of resolutions, for example, individuals indeed contract with themselves. Unfortunately, when willpower becomes weak, contracts with oneself are difficult to enforce. A court would hardly want to hear a case of me versus myself. There are, however, substitute enforcement mechanisms helping people (i.e. their "planners") to commit themselves (i.e. their future "doers"). Giving away your TV set protects against excessive future TV consumption. Booking a one-month stay in a health farm is effective insurance against over-eating. Signing an exclusion program protects against visiting the casino (see below, Box 17.4).

Motivation, or modification of preferences, is another tool familiar from principal–agent settings, particularly from the labor market. In the context of self-management, motivation would work through habit formation. Getting up early, drinking coffee without sugar, and brushing one's teeth, are activities to which one becomes accustomed and develops a preference for continuing.

Rules or budgets are formulated in order to make violations of a plan transparent. The doer may find it more difficult to indulge in present consumption if the planner has formulated some explicit budget. An example of a rule is thinking in *precedents*, in the sense of "If I eat this tempting dessert, there goes my whole diet" (Bénabou and Tirole 2004, p. 850).

Beliefs management and *identity building* as devices to influence the doer's decisions have been analyzed by Bénabou and Tirole (2004). We mentioned above that overestimating one's abilities may help to get a project started. Self-serving beliefs or identity building in this context would mean that an individual through experience or biased use of information acquires a self-image that suggests potentially excessive abilities. This is an example where self-deception is productive. It may also be harmful, though. A student who has developed excessive confidence may tend to underestimate the difficulty of an upcoming exam and not revise enough. We will have a closer look at these forms of self-management via management of beliefs in Section 17 C b.

17 C Theory: Intertemporal choice and self-management

17 C a Hyperbolic discounting

Tomorrow I will

Many decisions that involve a number of periods are examples of dealing with oneself (see e.g. Laibson, 2003). Planning for her retirement means that Alice implicitly deals with herself at a different stage of the life cycle. She does not really know what she will be like at 65, but to be able to plan her resources adequately, she needs to assume some characteristics of her future self.

The simplest strategy is to assume that her future self will have the same preferences as her current self. This is the traditional assumption made in the theory of life cycle decision-making and in macroeconomics. However, empirical evidence suggests that people think they will be more patient in the future than they are at present. As time goes by, the future becomes the present, and people will be as impatient as they had been in the past, but still cling to the idea that they will be patient in the future. Think of the well-known promise to yourself to start a diet tomorrow, but not today . . .

Such behavior is captured in a concept called "hyperbolic discounting", which is characterized by *time-inconsistent* behavior (see e.g. Laibson, 1997):[116]

1. The individual is myopic in the sense that he overweighs the present with respect to the future, but not the near future with respect to the more distant future;
2. Myopia persists: Once the future becomes the present, the individual is still impatient but believes he will be patient in the future.

If the individual lives for three periods ($t = 0, 1, 2$), his $t = 0$ and $t = 1$ utility functions can be written as:

$$U_0 = u_0 + \beta[\delta u_1 + \delta^2 u_2], \qquad (17.1)$$

$$U_1 = u_1 + \beta[\delta u_2]. \qquad (17.2)$$

where δ captures *time preference* and β captures the *salience of the present*.

Before we discuss the implications of such preferences, let us illustrate the case with Alice and Bob planning their consumption expenditures for the next four years, each having an endowment of \$100 to spread over this period. Their instantaneous utility per year is assumed to be $u(c_t) \equiv \ln(c_t)$, ensuring that Alice's and Bob's consumption is positive in each year.[117] Here, the interest rate on savings is zero.

Box 17.3 Maximizing log utility

Consider the following maximization problem of a logarithmic utility function:

$$U = a\ln(x) + b\ln(y)$$

subject to the resource constraint $p_x x + p_y y = I$. It is illustrative to use the Lagrangian:

$$\mathcal{L} = a\ln(x) + b\ln(y) - \lambda(p_x x + p_y y - I),$$

leading to the first-order conditions:

$$\frac{\partial \mathcal{L}}{\partial x} = \frac{a}{x} - \lambda p_x = 0,$$
$$\frac{\partial \mathcal{L}}{\partial y} = \frac{b}{y} - \lambda p_y = 0.$$

Combining these two conditions yields $p_x x/a = p_y y/b$, meaning that an individual's optimal expenditures on x ($= p_x x$) and y ($= p_y y$) are proportional to the relative utility weights a and b, respectively. By using the resource constraint, it can easily be seen that the fraction of the endowment I spent on x corresponds to the relative weight $a/(a+b)$ of x in the utility function. This simple division rule for logarithmic utility is easily generalized to more than two goods (or time periods in the case of an intertemporal setup).

The economic consequence of this result is that a relative price change does not change the division of available resources on the goods in the utility function. The intuition is as follows: An increase in the relative price of a good has two effects. The first is an income effect that should reduce the consumption of all other goods. The second is a substitution effect that should increase the consumption of all other goods, as these have become cheaper in relative terms. In the logarithmic case, these two effects cancel each other out perfectly.

Non-anticipated future impatience

Bob is an impatient person who values the present twice as much as the future. At the same time he thinks that he will be very patient in the future. So he assumes today that he will value all his consumption expenditures equally from year two onwards. Bob then has the following intertemporal utility:

$$\begin{aligned} U_1 &= u(c_1) + 0.5u(c_2) + 0.5u(c_3) + 0.5u(c_4) \\ &= \ln(c_1) + 0.5\ln(c_2) + 0.5\ln(c_3) + 0.5\ln(c_4) \end{aligned}$$

In Year 1 the value of one unit of utility in Year 2 relative to Year 1 is only one half, meaning that Bob is impatient. On the other hand, the value of one unit of utility in Years 3 or 4 relative to Year 2 is 1, which means that he considers his future self to be patient. In the language of Equations (17.1) and (17.2) Bob has $\beta = 0.5$ and $\delta = 1.0$.

How would Bob allocate his resources over the next four years? Here we use an important result from maximizing logarithmic utility (see Box 17.3): Utility is maximized if the endowment is split according to the relative utility weights. Bob would therefore choose consumption levels of $c_1 = \$40$, and $c_2 = c_3 = c_4 = \$20$. He consumes the \$40 and enters the next year. Unfortunately he is not as patient as he had hoped, but still thinks that he will be patient in the future. He has \$60 left to divide between the remaining three years, and has the following utility function:

$$U_2 = \ln(c_2) + 0.5\ln(c_3) + 0.5\ln(c_4)$$

As before, the relative weight of one util in the immediate future (Year 3) with respect to the present is one half. However, the utility weight of the not so near future (Year 4) relative to the immediate future (Year 3) is one. Bob still thinks he will be more patient in the future. Using the same strategy as before he divides the \$60 according to $c_2 = \$30$, and $c_3 = c_4 = \$15$. Bob thus consumes more than he had planned a year before. As there has been no external reason to change his optimal plan, the overturning of his initial plan is also known as time-inconsistency.

The remaining \$30 are now to be split between the two final years. As the reader might have guessed already, Bob does not keep his promise to himself (to be more patient) and consumes twice as much in the third year ($c_3 = \$20$) than in the fourth year ($c_4 = \$10$). Again, his planning is time-inconsistent. In retrospect, he would have most likely preferred to have been patient and consumed \$25 in every year. Figure 17.1 illustrates how Bob revises his consumption plan over time. The only reason the plan made in Year 3 coincides with the realized consumption plan is that there are no periods and resources left to change the allocation in Year 4.

Anticipated future impatience

Now consider Alice. Like Bob, she is not very patient, albeit not quite as impatient as Bob. She values her immediate future at 70 per cent relative to her present utility. But in contrast to Bob, she does not think she'll be more patient in a year. She always values her consumption in t years at 70 per cent, relative to what utility consumption delivers in $t-1$ years. In formal terms, her intertemporal utility reads as:

$$\begin{aligned} U_1 &= u(c_1) + 0.7u(c_2) + 0.7^2u(c_3) + 0.7^3u(c_4), \\ &= \ln(c_1) + 0.7\ln(c_2) + 0.7^2\ln(c_3) + 0.7^3\ln(c_4), \end{aligned}$$

this means that in the language of Eq. (17.1) Alice has $\beta = 1.0$ and $\delta = 0.7$.

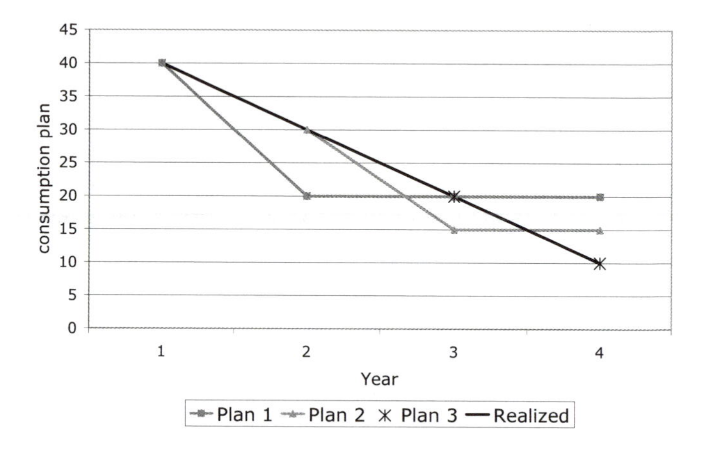

Figure 17.1 Bob's consumption plans under hyperbolic discounting. The anticipated plans do not coincide with the realized consumption levels in the first three years.

Using the same trick as above, it can be shown that Alice consumes $c_1 = \$39.48$ in the first year, $c_2 = \$27.64$ in the second year, $c_3 = \$19.34$ in the third year, and $c_4 = \$13.54$ in the fourth. As Alice has fully anticipated her impatience, she will not revise her optimal consumption plan after the first year.

The unconvinced reader can derive the optimal consumption allocation for Years 2–4 using the leftovers (\$60.52) and the utility function:

$$U_2 = \ln(c_2) + 0.7\ln(c_3) + 0.7\ln(c_4).$$

The optimal consumption plan is $c_2 = 1/(1 + 0.7 + 0.7^2) = \$27.64$ in the second year, and $c_3 = \$19.34$ and $c_2 = \$13.54$ in the third and fourth year, respectively. In other words, Alice's consumption plan is time-consistent.

Figure 17.2 displays Alice's and Bob's realized consumption plan over the four years. Obviously the two profiles do not differ much. In fact, an outside observer, such as an econometrician keen on identifying hyperbolic discounters, would scarcely see any difference. Nonetheless, there is a fundamental difference between Bob and Alice: Alice is happy with how she had divided her initial \$100, whereas Bob constantly revises his plans and would have liked to have been more patient.

Practical examples

Consumption decisions seem to be the "home turf" of hyperbolic preferences. The prospect of acquiring a good within the space of a week is not much more attractive than acquiring it within eight days. However, acquiring it today instead of tomorrow often makes a big difference. This is particularly the case if the present good is in front of our very eyes. In experiments children manage to wait a while to receive two candies rather than one—unless they can *see* that one candy.

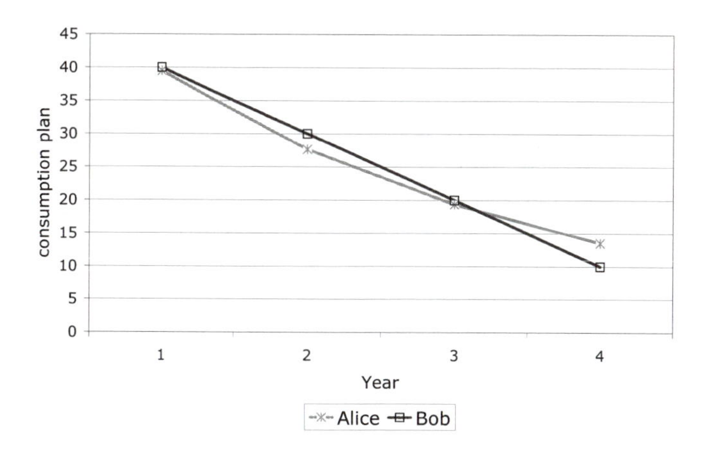

Figure 17.2 Alice's and Bob's consumption profile when Bob uses hyperbolic discounting and Alice uses standard exponential discounting.

Automotive exhibitions, displays of the latest laptop models, or a pair of sneakers in a window may add this hyperbolic twist to adults' preferences.

Hyperbolic discounting may at the same time explain both procrastination and workaholism. On the one hand, a student may decide to work on her thesis until lunch and go swimming thereafter, but work the whole day for the rest of the week. Tomorrow, the situation is the same, and the student ends up swimming again. *Ex post*, she regrets having worked so little.

On the other hand, a manager may decide to work late just one more time, but spend more time with his family the next month. But by then, other pressing problems are on his desk. And one day, the children have grown up and are gone. Consistent with the hypothesis of hyperbolic preferences of fathers, *The Economist* magazine[118] suggests that fathers may repent such a decision: "How many men on their deathbeds say they spent too little time with their boss?"

17 C b *The economics of self-control*

"Oh, don't you like them any more?" Alice asks when she returns after an hour and, to her surprise, finds Bob sitting in front of a still half-full plate of cookies. "You know," she goes on, "they have very little—," "Never mind what's in them," Bob interrupts, "I *need* to believe cookies make me fat, otherwise I could never resist when I see some in front of me."

"Internalities"

If the individual is an intertemporal collection of selves, the decisions by one self can have an effect on all future selves. Such an effect is an externality from the view of one single self. As it takes place within the individual some call it an "internality". There are different channels for such effects:

- Smoking today increases the desire for smoking tomorrow (preferences);
- Smoking today increases the probability of getting cancer tomorrow (endowments);
- Smoking today convinces an individual of her weakness (information).

Under time inconsistency, the effect imposed by one self onto a subsequent self is not internalized: Today's self has different preferences from tomorrow's (and knows tomorrow's self will differ from the day-after-tomorrow's), but cannot negotiate with these. The individual resembles a legislative body who cannot commit to drafting a law in the future.

If the present self cannot commit to its future successors, it can at least *cheat* them. Cheating means withholding information strategically, thus leading a future self to decide as the present self would want it to.

Flattering myself?

"Positive thinking" and self-confidence are key to social and professional success. This is, at least, what a whole industry of self-help literature would like us to believe. Indeed, most of us have experienced first- hand that we succeed when we believe in ourselves and fail on days of lurking self-doubt. Yet, positive thinking is also a form of self-deception and as such may seduce us into taking the wrong decisions.

Is positive thinking, in the sense of strategic overestimation of our abilities (or, symmetrically, underestimation of our tasks), really a good thing? This is the issue discussed in Bénabou and Tirole (2002, p. 872). It involves, as the authors show, two separate questions: (i) "Why should people prefer rosy views of themselves to accurate ones?" (ii) "Can a rational Bayesian individual deceive himself into holding such a rosy view?"

There are three reasons why an individual would want to hold a positive view of himself (Bénabou and Tirole 2002, p. 877):

- **Consumption value:** Self-esteem is a direct source of utility.
- **Signaling value:** A positive view of oneself may also convince others and come back as an "echo" from these.
- **Motivation value:** The productive aspect of a positive self-view is the (potential) improvement of our decisions. A student, for example, is more likely to start a PhD thesis if she overestimates her ability to complete it.

Here we abstract from the consumption value (relevant in the context of cognitive dissonance) and from the signaling value. Instead we focus on the motivation value, i.e. on the productive aspect of self-esteem in the presence of myopia.

The interaction of hyperbolic preferences with information

At the core of the Bénabou and Tirole (2002) model is the interaction between hyperbolic discounting and information. An individual (at least the planner self) knows that a future self is likely to be weak in the face of certain temptations. The present self would like to control the future self, but cannot do so directly. An indirect way is to cheat the future self with regard to the cost and the benefit of some activity. For example, the present self can induce a future self to underestimate the effort required to finish some task the present self would like to see finished.

Overconfidence versus myopia

In the model, an individual starts a project today (in $t = 0$). The project requires no immediate input. Yet, tomorrow (in $t = 1$), the project can be abandoned or continued. Continuing requires a positive effort. The day after tomorrow (in $t = 2$), the project (if continued) yields a return V if successful, and otherwise nothing.

The probability of success depends on the individual's ability. *Ex ante*, the individual is uncertain about his ability. With some probability, though, the individual receives a signal telling him he is of low ability. The individual may try to forget such bad news to allow him to overestimate his ability. Overestimation of ability is equivalent to overestimating the chances of success, i.e. the expected return from continuing.

Overestimating expected returns may improve the continue–abandon decision, because the individual has an innate tendency to underestimate future returns relative to present (in $t = 1$) effort. The individual realizes that, due to his bias towards immediate gratification, he is likely to give up along the way, thereby sacrificing long-run benefits for short-run relief from the effort of completing the task. Managing his beliefs about his true ability, the individual may motivate himself to complete the project (Bénabou and Tirole, 2002). By trying to forget bad news about his ability, the individual balances overconfidence against myopia. This may explain why we have a "psychological immune system" (Bénabou and Tirole 2002, p. 884), shielding us against negative information about our abilities.

The time line

The sequence of events, as represented in Figure 17.3, has an information stage, a decision stage, and a realization stage. In the *information stage* ($t = 0$), the individual receives some information σ about his hidden ability θ to run a project successfully. With probability $1 - q$, the individual receives a bad signal, telling him that his ability is low ($\sigma = L$). With probability q, the individual receives no signal ($\sigma = \emptyset$) which is actually good news.

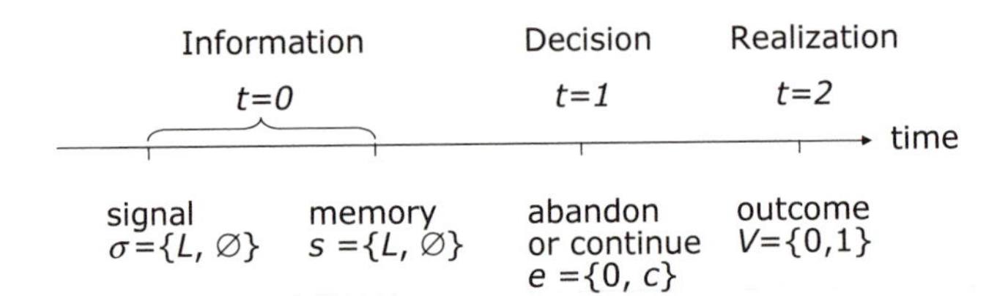

Figure 17.3 At the information stage ($t = 0$), the individual can use selective memory to scramble a received signal about his ability. At the decision stage ($t = 1$), the individual decides whether to abandon or continue a project at effort cost e. At the realization stage ($t = 2$), the project succeeds ($V = 1$) or fails ($V = 0$), the probabilities depending on the individual's ability.

Far from resigning after bad news, the individual can "manage" received information. His memory, as distinct from the "raw" signal, is denoted by s. The individual cannot invent a signal if none has been received, but he can try to forget a received signal: A received (i.e. bad) signal is only remembered with probability λ ($\lambda = Pr[s = L|\sigma = L]$). With corresponding probability $1 - \lambda$, a bad signal is forgotten. The absence of a signal (good news), by contrast, is always remembered.

At the *decision stage* (when $t = 1$), the individual can spend some positive effort to continue the project. Otherwise the project is abandoned (at zero direct cost). The cost of the effort required to continue the project is denoted by c. For simplicity's sake, we work under the assumption (to be relaxed later) that c is (i) known in advance, and (ii) is low enough to make finishing the project efficient even if the individual has low ability:

$$\theta_L \geq c.$$

At the final *realization stage* (in $t = 2$), the project (if not abandoned before), yields an uncertain return V. The project is either a success ($V = 1$) or a failure ($V = 0$). We assume the probability that the project succeeds is identical to the individual's *ability* θ, which can be high or low, with $\theta_H > \theta_L$. As the project yields $V = 1$ in case of success and nothing otherwise, the expected return is equal to an individual's perceived ability θ. The identity of ability, success probability and expected return is very convenient. It is important to keep them in mind in order to avoid confusion.

Hyperbolic discounting

The individual is myopic in the sense that, in every period, he overweighs the then present with respect to the future, but not the near future with respect to the more distant future. Technically speaking, the individual is time-inconsistent in the sense of hyperbolic discounting (see above, Section 17 C a).

The present self, which we will call Self 0, plays the role of the "planner" in the sense of Thaler and Shefrin (1981). Self 0, in its own present, has nothing to overweigh, as there is no income in the pure planning period $t = 0$. Between future income, accruing in $t = 1$ or in $t = 2$, Self 0 is indifferent, unlike tomorrow's self, Self 1. Self 1 (acting as the "doer") will prefer the (then) present ($t = 1$) over the future ($t = 2$). In the eyes of Self 0, Self 1 will overweigh the cost of continuing the project, c, relative to its expected return. While Self 0 would want the project to be continued if:

$$c < \theta, \tag{17.3}$$

Self 1 will only continue the project if:

$$c < \hat{\theta}\beta. \tag{17.4}$$

Self 1. If c happens to fall into the range $\hat{\theta}\beta < c < \theta$, Self 2 inefficiently abandons the project.

Selective memory

This is where selective memory comes into play. By suppressing bad news, Self 0 can make Self 1 overestimate the individual's true ability. This may tilt the balance back in favor of continuing rather than abandoning the project. In technical terms, Self 0 can choose an appropriate quality of the individual's memory λ (the probability of remembering a bad signal). If the individual is of low ability, imperfect memory ($\lambda < 1$) leads Self 1 to overestimate the true ability parameter θ ($\hat{\theta} > \theta$). Self 1's underestimation of future returns (and the resulting temptation to inefficiently abandon the project) is exactly compensated if Self 0 selects a λ that leads to a belief by Self 1 of $\hat{\theta}$, satisfying (from (17.4) and (17.3)):

$$\hat{\theta} \geq \frac{\theta}{\beta}. \tag{17.5}$$

Through "positive thinking", in the sense of overlooking negative evidence today, the individual can prevent himself from myopically abandoning a project tomorrow. It pays the individual to "cheat" himself. This is the productive role of self-serving beliefs, as modeled by Bénabou and Tirole (2002). It is a rare case where scrambling a signal has a positive value (see Chapter 4).

Can self-serving beliefs be rational?

Can an individual really cheat himself? After all, individuals are not naïve. Employees discount a boss who is always full of praise. Similarly, Self 1 should not take an ability assessment by Self 0 at face value. Nevertheless, is it still possible that a rational Bayesian individual holds self-serving beliefs?

Bénabou and Tirole (2002) show that we can. Consider Figure 17.4. The figure is a variation on our probability square used above (in Figure 4.2). The individual receives either no signal (probability q) or a bad signal (probability $1-q$). The bad

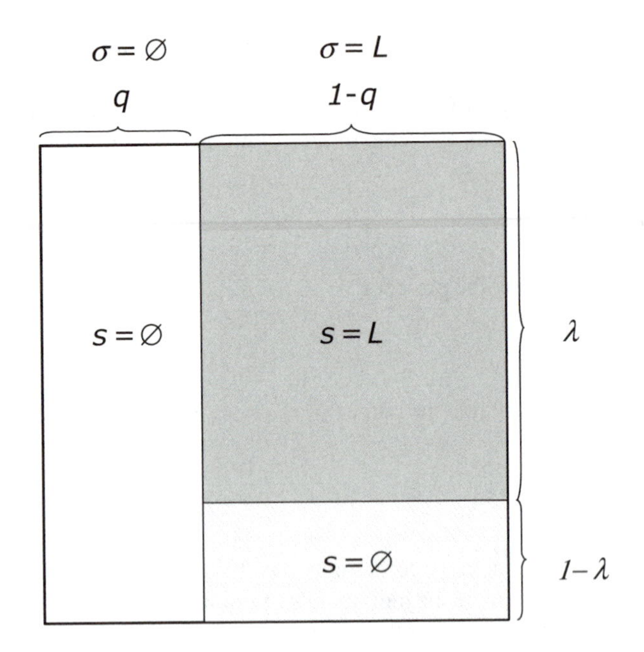

Figure 17.4 With probability q individuals receive no (=good) news ($\sigma = \emptyset$) about their abilities, with probability $1 - q$ they receive bad news ($\sigma = L$). Bad news is remembered with probability λ and forgotten with probability $(1 - \lambda)$. Remembered information is either $s = L$ or $s = \emptyset$.

signal, $\sigma = L$, is remembered with probability λ and forgotten with probability $(1 - \lambda)$. This leaves us with three areas representing the three posterior types of individuals:

- Shaded area: Individuals who have low ability ($\theta = \theta_L$) and who know it ($s = L$). These individuals have a *correct estimate* of their ability.
- Dotted area: Individuals who have low ability (θ_L) but have forgotten their bad signal ($s = \emptyset$). These individuals *overestimate* their ability.
- White area: Individuals who have high ability (θ_H). These individuals have not received any signal. However, they are not sure whether they really have not received a signal ($s = \emptyset$) and have indeed high ability, or whether they forgot a (bad) signal ($\theta = \theta_L$) and have low ability. These individuals *underestimate* their ability.

If Self 1 *remembers* a signal ($s = \sigma = L$), it knows that the individual is of the first type on our list and has low ability (signals cannot be invented). If Self 1 *does not* remember a signal ($s = \emptyset$) the individual may be either of the second type (no signal received), or of the third (signal forgotten). The probability that an individual who does not remember a signal has in fact not received any (and thus has a high ability), r, is called the "reliability" of a memory of "no news" by Bénabou and Tirole (see 2002, p. 889). Using Bayes' Rule, it can be computed as:

$$r \equiv Pr[\sigma = \emptyset | s = \emptyset; \lambda] = \frac{q}{q + (1-q)(1-\lambda)}.$$

In Figure 17.4 r is equal to the white rectangle divided by the L-shaped union of the white and the dotted rectangles.

The perceived ability (and hence the expected project return) of an individual who does not remember any news is:

$$\theta(r) = r\theta_H + (1-r)\theta_L.$$

Bénabou and Tirole (2002) call $\theta(r)$ the "degree of self-confidence". The ranking $\theta_L < \theta(r) < \theta_H$ confirms that an individual who does not remember any news either overestimates his ability (if he is of low ability) or underestimates it (if he is of high ability). Self-serving beliefs are possible in a Bayesian framework, because the overconfidence of one group is compensated by the lack of confidence of another. In other words, strategic optimism (low probability of recollecting a bad signal, λ) is not possible without defensive pessimism (a low credibility of no recollection of a bad signal, $r(\lambda)$). Two extreme examples may illustrate the trade-off. An individual with perfect memory ($\lambda = 1$) cannot forget bad news; to him the non-recollection of a (bad) signal is an infallible indicator of high ability. At the other extreme, an individual with total amnesia ($\lambda = 0$) never recollects bad news, but can never trust its absence either.

The optimal degree of self-confidence ...

The optimal choice of λ as a means to align Self 1's perception with Self 0's views is illustrated in Figure 17.5. The recollection probability λ is shown on the horizontal axis. The vertical axis indicates perceived ability as well as expected returns. (Recall that expected returns, under our assumption of a success return of 1, are equal to success probability θ.) The solid lines describe the perspective of Self 0 and stand for an individual with high ability θ_H (upper line) or with low ability θ_L (lower line), respectively. On the right edge, with $\lambda = 1$ (perfect recall), an individual knows his ability for sure; expected returns (which, by assumption of a success payoff of 1, are equal to abilities or success probabilities) are either θ_H or θ_L. With $\lambda = 0$ (total amnesia), the signal is useless; high and low ability individuals share the prior probability of success q (left edge).

The dotted lines describe the view of Self 1. Self 1 undervalues $t = 2$ returns relative to $t = 1$ cost by a factor of β. The dotted lines therefore lie below the solid lines. Again, the lines connect the extremes of perfect memory (right edge) and no memory (left edge).

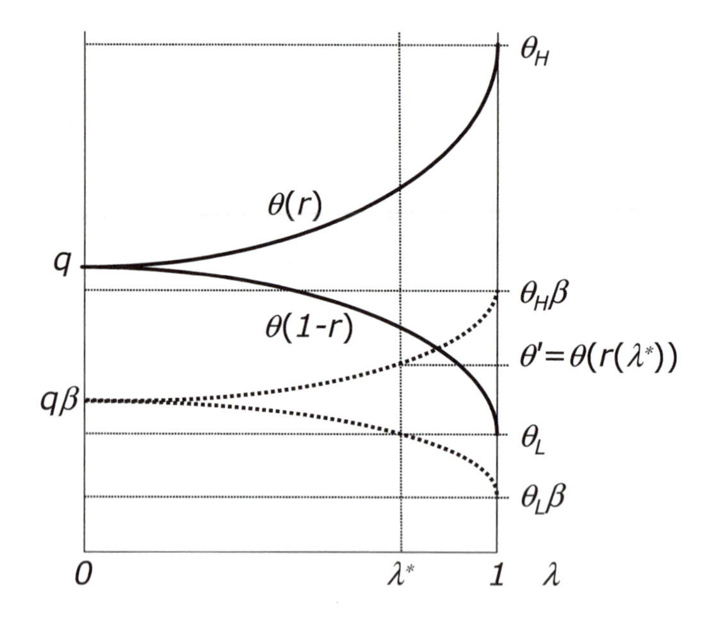

Figure 17.5 Expected returns as a function of λ perceived by Self 0 (solid lines) and Self 1 (dotted lines). The dotted lines for Self 1 correspond to the solid lines multiplied by $\beta < 1$.

. . . with known cost of continuation

To show the logic behind the optimization of λ we start from the special case where $c = \theta_L$. In this case an individual who is sure of being of low ability would find it worthwhile to complete the project. In Figure 17.5 this case is indicated by the horizontal line at level θ_L. Unlike the low-ability individual's Self 0 who prefers to run the project, Self 1 will abandon it, as it only finishes projects with a continuation cost that does not exceed the lower level $\theta_L\beta$.

The maximum λ at which a low-ability Self 1 continues a project with continuation cost $c = \theta_L$ is λ^*. Forgetting bad news with a positive probability $(1 - \lambda^*)$ is necessary for Self 1 to consider continuation worthwhile. Therefore, λ^* would be the optimal recall probability from the perspective of Self 0, the planner (who wants the project to be completed). As defined by equation (17.5) above, λ^* would solve:

$$\hat{\theta}(\lambda^*) \geq \frac{\theta}{\beta}.$$

Needless to say, with λ^* a high-ability individual would also continue the project.

. . . with uncertain cost of continuation

The trade-offs in setting the optimal λ become apparent, once we relax the assumption that c is known and certain. This is analytically difficult. For this reason, we simply try to explain the intuition behind it. For a rigorous treatment we refer interested readers to the original paper Bénabou and Tirole (2004) and its appendix.

Let c be uncertain ($0 < c < \theta_H$) at the time the individual has to fix his λ. Assume that, given the distribution of c, the low-ability individual (represented by Self 0) still finds λ^* optimal. However, once the realization of c is known, λ^* may turn out to be no longer optimal. Either it is too high, leading Self 1 to reject the project, or it is too low, leading Self 1 to continue the project, even though Self 0 would prefer to see it abandoned. If λ is too high, it leads to self-doubts and the inefficient abandoning of a project. If it is too low, it leads to self-overestimation and an unrealistic ambition of finishing overly difficult tasks.

Finally, uncertain continuation cost c also leads to a trade-off between the motivation of high- and low-ability individuals. A higher self-confidence among low-ability individuals, achieved via imperfect memory ($\lambda < 1$), is not free of cost. This cost comes in the form of a lower reliability of the absence of a (bad) signal, that is of the "no news is good news" rule. Such lower reliability of no news is unavoidable: It is the Bayesian correlate to the imperfect recall of bad news. It may, however, undermine the self-confidence of a high-ability individual to a degree that the individual abandons projects that would be worthy of completion.

In Figure 17.5, a λ^* would (in expected terms) lead a high-ability individual to continue the project if its cost turns out to be $c \leq \theta'$. Projects with effort cost in the range $\theta' \leq c \leq \theta_H \beta$, however, would be abandoned (by the individual's Self 1). This is inefficient, since a high-ability individual with unimpaired self-confidence (at $\lambda = 1$) would have continued the project. There is, in other words, an opportunity cost to the building—or better: to the redistribution—of self-confidence. As Bénabou and Tirole (2002, p. 901) put it, the individual has to find a balance between: "The net gain from forgetting bad news" and "the loss from disbelieving good news".

To conclude, an individual in the Bénabou and Tirole (2002) model, choosing the quality of his memory, represented by the recollection probability λ, faces two potential trade-offs:

- The higher self-confidence (overestimation of ability) of a low-ability individual brought about by imperfect memory may lead an individual to complete productive tasks, but also to engage in excessively ambitious tasks.
- A lower λ strengthens the average self-confidence of individuals with low ability, but undermines the self-confidence of individuals with high ability.

Bénabou and Tirole (2002, p. 902) summarize their own findings as follows:

- Self-esteem maintenance ("positive thinking") ($\lambda < 1$) is optimal when a person with weak willpower (high degree of time inconsistency β) faces a relatively demanding task.
- Realism ("accept who you are") ($\lambda = 1$) is optimal when a person with a mild degree of time inconsistency ($\beta \approx 1$) faces a moderately challenging task.

Self-confidence and effort: Complements or substitutes?

In addition, the impact of overconfidence on an individual's utility depends on the relation between ability and effort as inputs to expected return, that is, on whether they are complements or substitutes (Bénabou and Tirole, 2002).

In the model sketched above, the relationship between an individual's ability (θ) and his effort are somewhat *complementary*. Investing the ($t = 1$) effort required to continue the project increases the expected return from zero to θ_H for a high-ability individual, but only to θ_L for a low-ability individual. Effort pays more to a high-ability individual. Under these conditions, overestimating his ability increases the individual's incentive to exert an effort. This is the case of the student who would not start her thesis without a good dose of overconfidence.

Yet, the opposite also occurs. If ability and effort are *substitutes*, over-confidence can decrease the individual's utility. This is the example of the student who thinks he is so intelligent that he does not need to learn very much for the exam. As a result, he (his Self 1) runs an excessive (from the perspective of Self 2) risk of failure. In such cases, "defensive pessimism" pays off, like assuming that the good grades received in the past were influenced by good luck. Note also that defensive pessimism is called for in the case of an inverse sequence of events, i.e. when the returns precede the costs (Bénabou and Tirole 2002, p. 904).

Identity building: Inferring one's preferences from past actions

In the Bénabou and Tirole (2002) model, individuals can try to forget news, especially bad news, but they cannot influence the news itself. In reality, however, people often follow certain actions to *produce* favorable information, such as "If I can do without dessert today, I will be able to stick to my diet tomorrow."

Tomorrow the individual would remember: "I have been strong yesterday; I will certainly remain strong today." This resembles an informational cascade (see Chapter 10), where later individuals (here: incarnations of the same individual) observe and follow the behavior of earlier individuals. The bridge between yesterday and today here is self-reputation, suggesting that restraint today pays off.

Building self-reputation makes sense under two assumptions, as Bénabou and Tirole (2004, p. 850) point out:

- Imperfect willpower (salience of the present), in the form of time-inconsistent behavior;
- Recall of actions but imperfect recall of the motives that led to these actions.

The first assumption is obvious: A time-consistent individual does not need any self-reputation. The second is also easy to understand: An action can only improve self-confidence if it is remembered; it would be fully discounted as cheating if its motive of internal "window dressing" were also remembered. For example, an individual who knows about his tendency to be late and who adjusts his watch to run ten minutes ahead of time quickly learns to subtract ten minutes from the time shown.

Under the above assumptions, past actions can be used for identity building or "self-signaling" (Camerer et al., 2005, p. 38). The individual trades off an instantaneous pleasure against a favorable self-image and, indirectly, against more time-consistent behavior in the longer term. This leads individuals to observation *rules*, like a savings' plan, a diet or the habit of smoking only after meals. Following the rule rather than giving in is each time a signal of willpower. A violation of the rule is a precedent, calling into question the whole rule or the discipline behind it. A rule thus turns decisions into package deals, such as trading in the entire diet for the dessert. Such packaging deprives the myopic self (the doer) of the option to trade-off lapses against momentary violations and strengthens the position of the long-run optimizing self (the planner).

However, the cure can become worse than the disease. Rules may end up dominating an individual. Bénabou and Tirole (2004, p. 850) note that "agents with hyperbolic discounting can actually behave *as though* they overweighed the future". Most at risk are individuals with low confidence in their future self-discipline, as they have a high demand for strict rules. There are "costly forms of self-signaling in which the individual is so afraid of appearing weak to himself that every decision becomes a test of his willpower, even when the stakes are minor or self-restraint is not desirable *ex ante*" (Bénabou and Tirole 2004, p. 851). Within the self-diffident individual there is thus a potential for compulsive behavior in the sense of following potentially self-destructive rules, leading to phenomena like anorexia, parsimoniousness or workaholism.

17 D Application: Soft paternalism

I hate what I want

Traditionally economics builds on revealed preference theory, that is on the assumption that one can infer what people want from what they choose (remember the Ferrari example from Chapter 3). Phenomena like weak self-discipline, procrastination, neglect of retirement provisions, or addiction suggest that there is a difference between what an individual wants and what he likes. Individuals, to put it bluntly, take decisions that violate their own best interests.

So, do they need help? Yes and no. Standard welfare economics claims that if people know and manage to choose what is good for them, government should leave them alone (we abstract from other reasons for government intervention, like externalities). Yet, if individuals are weak, there is a problem. As Camerer et al. (2005, p. 37) state: "If likes and wants diverge, this would pose a fundamental challenge to standard welfare economics". If people hurt themselves there is some reason for governments to step in. Mandatory pensions, for example, correct for insufficient voluntary saving caused by individual myopia. Camerer, Issacharoff et al. (2003) mention usury laws, preventing people from falling into crippling debt, as another example. Hamermesh and Slemrod (2006) finally suggest progressive income taxes to curb workaholism.

The problem with government

The problem is where would intervention stop. "Revealed preference is an attractive political principle because it guards against abuse (albeit quite imperfectly in practice). Once we relax this doctrine, we potentially legitimize government condemnation of almost any chosen lifestyle on the grounds that it is contrary to a 'natural' welfare criterion reflecting an individual's 'true' interests," Bernheim and Rangel (2005, p. 3) warn. Therefore, letting individuals violate their own interests may still be preferable to handing the state a blank cheque for intervention.

Tying myself

Fortunately, there may be a third way. Individuals might choose the degree of external intervention themselves. The present self, in a cool-headed moment (when the planner dominates the doer), could deliberately limit actions by the future self (within which the doer might overwhelm the planner). For example, under the *Missouri Voluntary Exclusion Program* for problem gamblers (Box 17.4) individuals can ban themselves from casinos for the rest of their life. Such programs combine the advantages of both free choice and government intervention. It is not up to a third party to tell people what to do or like. At the same time such plans provide the external coercion necessary to keep future selves "on track".

Such self-commitment works if there is a commitment strategy available. Probably the most famous example is Homer's Odysseus approaching the coast of the Sirens. He knows that the Sirens' song was irresistible and listening to it would lead to a catastrophe. Odysseus asked his crew to tie him to the mast and to stop their ears with beeswax. He admonishes them to keep rowing for the entire time that the Sirens sing, no matter what he says. Thanks to a commitment technology (the mast and a rope), the strategy worked: the planner prevailed over the doer.

A helping hand

Such commitment technologies are not always available when individuals might need them most. Organizations like Alcoholics Anonymous try to create commitment options: Members are organized in groups which make deviations from discipline costly (for example, by requiring individuals to report their experience). Voluntary savings programs introduced in recent years have been very successful: People seem happy, both *ex ante* and *ex post*, about restricting their freedom and committing to a savings' plan. One example of such a scheme in the Philippines is analyzed in Ashraf, Karlan and Yin (2006). Working on this book we voluntarily agreed with the publisher on several fairly artificial deadlines, without which it would never have been finished.

Sometimes, as in these examples, the helping hand is offered by a private partner. Sometimes it is not. Here, the state may have a role to play. Far from telling people what to like or do, the state can offer people the rope with which to lash themselves to the mast, so to speak.

For instance, many smokers seem to advocate bans in public places, like restaurants and bars. Some economists also advocate schemes like taxing cigarettes not per packet but as an up-front lump sum (a license) for, say, 2,500 (tax-free) packets. For a smoker, cigarettes would not become more expensive. But for those who would like to stop, the option of not renewing an expiring license helps them stick to a decision by making short-term deviations very expensive (*The Economist*, 8 April 2006, 69–71).

The provision of such pre-commitment opportunities by government charms conservative free-marketeers and compassionate interventionists equally. Different authors use different terms like "early decisions regulation" (Beshears, Choi, Laibson and Madrian, 2006) or "asymmetric paternalism" (Camerer et al., 2003), but the label in vogue at the moment seems to be"soft paternalism".

If the reasoning behind "soft paternalism" is sound, there is considerable scope for improving existing regulations. In the future, legislators may think harder about offering commitment opportunities for individuals. But before hoping for the helping hand of the state, individuals may find solutions on the market. For example, firms with a non-smoking policy would have an advantage on the labor market if it were really true that smokers (within which the planner momentarily prevails over the doer) appreciate smoking bans. Similarly, some economists have invented successful voluntary savings scheme. Under such schemes, individuals commit to save future, rather than present income (for example wage increases). It turns out that (i) most individuals have a higher propensity to save future compared to present income, and (ii) *ex post*, people tend not to regret their (then past) commitment. For more details see Thaler and Benartzi (2004). We hope that our readers will invent further masts and ropes which will help all of us to withstand the songs of our inner sirens.

Box 17.4 The Missouri Voluntary Exclusion Program

In Missouri, gambling on the riverboat casinos was legalized in 1994. Soon a citizen asked to be banned from the casinos because he found himself unable to control his gambling. In 1996, the Missouri Gaming Commission created a *Voluntary Exclusion Program* for problem gamblers. The program offers a radical method: Individuals can agree to stop visiting casinos for the rest of their life. Such a decision is *irrevocable*. Once a person is placed on the Disassociated Persons List, they can never get off. There is no procedure for removal.

In the first ten years, about 10,000 people chose to sign up to the program. Mental health professionals remained skeptical though, pointing out that in order to achieve long-term recovery, gamblers had to acknowledge that they have a problem and to take personal responsibility for it (including, probably, participation in some overall recovery program). Based on such comments, the Commission revised the rules. Problem gamblers agree that they, not the casinos, bear the responsibility for the success of the program. However, gamblers who are discovered will be arrested for trespassing, and their winnings will be confiscated.

The Voluntary Exclusion Program operates a 24-hour helpline at 1-888-BETSOFF.

17 E Conclusions and further reading

Human beings exhibit many forms of behavior that are hard to reconcile with the view of rational *homo œconomicus*. Yet, in order to understand such behavior it is not necessary to throw the baby out with the bath water. One can reconcile the notion of some logic behind human behavior with seemingly irrational phenomena like addiction, workaholism, self-serving views, etc.

This new logic, the laws of motion of a less rational but more human and richer *homo sapiens*, starts with the distinction of several selves within an individual. A planner and a doer, present and future selves, reason and emotion—all these distinctions try to model a split individual. The assumption of several selves within a time-inconsistent individual allows for many forms of human behavior without giving up rationality altogether, in the sense of some logical optimizing behavior.

The new logic is also consistent with long-known findings from psychology and with more recent advances in neuroscience which have shed some light on the "black box" of the human mind. Economists have taken up these findings and have tried to incorporate them into more familiar settings, such as the principal–agent relationship. The relation between principal and agent, or between different selves within an individual, are characterized by communication, i.e. offers, promises, sending of signals, and the like. This places the economics of *homo sapiens* into the field of information economics.

We have tried to outline the main ideas in this field with an analytical focus on hyperbolic discounting, a particular form of myopia that equally discounts all future consumption compared to present consumption, irrespective of the actual time distance. Hyperbolic discounting formalizes the everyday notion of the "salience of the present".

In a world of imperfect willpower, hyperbolic discounting interacts with information: The present self *knows* about the weakness of the future selves. Yet, communication between the present and the future self—in the form of resolutions, for example—is imperfect. Rather than resort to contracts, a present self must try to keep future selves in check through techniques like self-management (rules that support willpower), or outright cheating in the form of suppression of negative information, as well as identity building.

When self-management ultimately fails, the last resort are commitment opportunities such as a voluntary exclusion program for problem gamblers or a pre-committed savings' plan. Today, the idea of voluntary options for self-commitment is popularly referred to as "soft paternalism".

Even more than in other areas covered in this book, the models discussed in the present chapter are under rapid development. Our account therefore remains preliminary and incomplete. Nor can we yet tell whether, after some time, economists will again give up research on intra-personal relations, or whether, at the other extreme, economics will become part of neuroscience, once the secrets of the human brain are revealed bit by bit.

This is the time to congratulate readers for having read this book to its very end. Your planner, in the language of this book, gloriously prevailed over the doer (there remains one set of problems, alas!). Unfortunately, it is also time to bid farewell to Alice and Bob.

Alice's place. "Your coffee is great, Alice. But I'm sorry, I finished all the cookies tonight." "Don't worry about the cookies, Bob, nor about your coffee consumption. If anything, I think you work too much." "You know, I'm afraid that if I don't work hard on a regular basis, I might become lazy again, as I was in my students days. That's why, Alice." "So you need to go on like that for the rest of your life?" Alice inquires. Somewhat indignant, Bob replies "Well, Alice, first, we're not married. Second, if anything, smoking is more dangerous than working. And third, I actually have the option of taking a part-time job!" Alice is flabbergasted: "You take a part-time job? I find that hard to believe! But, you're right, I shouldn't smoke. Although, it's a matter between me and myself, isn't it?" "That's exactly why you have failed to give up," Bob retorts, "You should try to quit for someone *else*." "Oh, for you, you mean?" "No, Alice, not for me." "Not for you?" "Alice, you used to be much quicker. For whom would you think I'd consider working part-time?" "Bob!! You really . . . —but you'll never have your Ferrari! Are you sure you want . . .?" "Yes, if that's fine with you!" "That's more than fine with me. But right now!"

Checklist: Concepts introduced in this chapter

- Cognition biases
- Behavioral economics
- Homo œconomicus versus homo sapiens
- Learning without information
- Cognitive dissonance
- The split self
- The serial self
- Hyperbolic discounting

- Time inconsistency
- Maximizing log utility
- Strategic ignorance
- Identity building
- Self-management
- Soft paternalism
- Pre-commitment

17 F Problem sets: Tomorrow I will

For solutions see: www.alicebob.info.

Problem 17.1 Fact-free learning
In Chapter 8 we introduced the maxim "Rational individuals cannot agree to disagree." Would you think this still holds under so-called "fact-free learning" in the sense of Aragones, Gilboa et al. (2005)?

Problem 17.2 Procrastination
Bob is supposed to write a report. The work is fairly dull, but it has to be done within the next three days. Bob is sitting at his computer. Should he first finish his online Sudoku puzzle or should he start on his report immediately?

(a) How is Bob likely to split the three days between doing Sudoku puzzles and working, if his planning unit is a day (What should I do today?)
(b) How is Bob likely to split the three days between doing Sudoku puzzles and working, if his planning unit is five minutes (What should I do in the next five minutes?)
(c) How does the length of Bob's planning period influence his tendency towards procrastination?

Problem 17.3 *Protect Me From What I Want*
In a Mediterranean port we once saw an impressive if somewhat mysterious motor yacht with the even more mysterious name PROTECT ME FROM WHAT I WANT. Passers-by wondered what this could mean.

(a) Can you make sense of the name in the light of the present chapter?
(b) Can you find the 1986 background to the title?

Problem 17.4 "Do we now always have to do what we want?"
This paradoxical question was asked by a Russian shortly after the demise of the Soviet Union.

(a) Can you explain the paradox by using concepts from the present chapter?
(b) Does the question raise any issues that go beyond the reach of this book?

Problem 17.5 The road to hell

"The road to hell is paved with good intentions," the saying goes. Can you explain the proverb in economic terms?

Problem 17.6 The planner and the doer

We introduced the Thaler and Shefrin (1981) dichotomy of a "two-self" economic man, consisting of a *planner* and a *doer*. The planner is far-sighted, i.e. concerned with lifetime utility, while the doer is myopic or selfish and cares only about present consumption c_t. Assume an individual lives for T periods and has a lifetime income of Y. Y accrues in a stream of income y_t. Income can be either consumed or saved. There is a perfect capital market in which the individual can freely borrow or lend. The interest rate is zero.

In period 1, the planner perceives utility as the lowest consumption level in all T periods: $U_P = \min c_t$; the doer perceives utility as present consumption: $U_D = c_1$.

(a) What are the planner's and the doer's optimal consumption plans?
(b) What are the planner's and the doer's individual rates of hyperbolic discounting β?
(c) Assume the planner and the doer split the individual's wealth Y. The planner receives fraction f and the doer fraction $1 - f$. At what value of f do they perceive the same level of utility?
(d) Assume the planner and the doer split the individual's wealth Y evenly, i.e. $f = 0.5$. What level of utility does each of the two perceive?

Problem 17.7 Uncle Scrooge

Uncle Scrooge is well known for never spending a penny (unless he receives two back in return).

(a) Is his behavior consistent with hyperbolic discounting?
(b) How would you assess his self-confidence?

Problem 17.8 Hyperbolic discounting

Peter has the following utility function:

$$U_1 = \ln(c_1) + \gamma \ln(c_2) + \gamma \ln(c_3)$$

$$U_2 = \ln(c_2) + \gamma \ln(c_3)$$

$$U_3 = \ln(c_3)$$

where c_t is consumption in period t, and γ is a discount factor equal to 0.5.

Peter's total income is 60, and the interest rate is 0.

(a) Compute the optimal consumption profiles Peter would chose at $t = 1, t = 2$, and $t = 3$.
(b) Are Peter's preferences time-consistent? Why (not)?

Now consider Clara's utility function:

$$U_1 = \gamma_1 \ln(c_1) + \gamma_2 \ln(c_2) + \gamma_3 \ln(c_3)$$

$$U_2 = \gamma_2 \ln(c_2) + \gamma_3 \ln(c_3)$$

$$U_3 = \gamma_3 \ln(c_3)$$

where $\gamma_1 = 1$, $\gamma_2 = \frac{2}{3}$, and $\gamma_3 = \frac{1}{3}$.
Clara's total income is again 60, and the interest rate is 0.

(c) Compute the optimal consumption profiles, Clara chooses at $t = 1$, $t = 2$, and $t = 3$.
(d) Are Clara's preferences time-consistent? Why (not)?
(e) Comment on the consumption paths for Peter and Clara given that the two friends cannot commit (i.e. they re-optimize in each period). In what respect are they alike or different?
(f) Are Peter and Clara equally happy about their realized consumption path? In other words, would it be possible to increase welfare if a benevolent dictator forced the two friends to consume a certain amount at each point in time?

Notes

1 *iMP Magazine*, 22 November 2002.
2 http://www2.sims.berkeley.edu/research/projects/how-much-info-2003/
3 See http://en.wikipedia.org/wiki/Rothschilds or http://www.jewishencyclopedia.com/
4 http://www2.warwick.ac.uk/fac/soc/economics/staff/faculty/wooders/message/
5 The episode is borrowed from the entry for "revealed preference" in: http://www.economist.com/research/Economics/alphabetic.cfm?
6 Finer also means better, i.e. not worse; a utility-maximizing individual is never worse off with the finer information structure.
7 A different definition of precision (the reciprocal of the variance) applies to continuous signals. We will use that definition in Sections 8 C f and 11 C b. A further variation of precision, the correlation with a target variable, will be used in Sections 5 D and and 5 E.
8 The monotone likelihood ratio property is briefly introduced in Section 16 C c. More on both concepts is found in Mas-Colell, Whinston and Green (1995), Laffont (1993), Laffont and Martimort (2002), or Brunnermeier (2001).
9 http://en.wikipedia.org/wiki/Epimenides_paradox.
10 The regression coefficient is $\frac{Cov(s,y)}{Var(y)}$, while the formal definition of the correlation is $\frac{Cov(s,y)}{\sqrt{Var(y)\ Var(s)}}$. As $Var(y) = Var(s) = 1$, the regression coefficient is equal to the correlation of s and y.
11 One version was told by Hal Varian in the *New York Times* (30 June 2005), see: http://www.sims.berkeley.edu/ hal/people/hal/NYTimes/2005-06-30.html
12 The second-order condition for a maximum is satisfied, since the marginal benefit *decreases* in n, while marginal cost is constant.
13 The optimal waiting period *ex ante* and under sequential decision making are the same (see Cukierman, 1980).
14 Marcus Annaeus Lucanus (Lucan) in *The Civil War* (Bellum civile sive Pharsalia, circa 62/65). Isaac Newton took up the quote to acknowledge the fact that his discoveries were greatly facilitated by the work of other scientists: "If I have seen a little further it is by standing on the shoulders of Giants" (letter to Robert Hooke, dated 5 February 1676).
15 Similar parameterizations of the production function of new ideas have been widely used in macroeconomic models of economic growth since the pioneering work by Romer (1990). See Jones (2001, Ch. 5) for an excellent and very accessible exposition of these ideas in growth theory.
16 The second derivative, $-\theta$, is negative, thus ensuring a maximum.
17 Overfishing of the oceans is the corresponding example in our time. For a game-theory discussion of fishing, see Bierman and Fernandez (1995, Ch. 15).

18 The seminal paper of Aghion and Howitt (1992) is rather complicated. An excellent and more accessible reference is their book, Aghion and Howitt (1997).
19 This assumption is similar in spirit to the production function of new ideas in Section 6 C a with $\lambda = 1$. Unlike in equation (6.1), the quantity of new information is fixed, but the probability of finding it is proportional to the number of researchers.
20 The formal analysis to derive the equilibrium allocation by use of dynamic programming is rather demanding.
21 Taking this example literally, the cost-reducing process cannot go on forever. It is relatively easy to find a more realistic example, such as an innovation in the productivity of the production. But this would be more difficult to formalize.
22 As the probability to find a new idea is proportional, there is no "stepping on toes effect" in the outlined creative destruction model.
23 The consumer surplus can be read off as the triangle above the monopolist's surplus rectangle in either market structure.
24 For an overview see Estrella (2000) and White (2001).
25 One problem with this often used argument is that a rating only reduces the borrowing cost if it is better than what the market expected. However, about one third of issuers are unhappy with their ratings. In expected terms, a rating therefore may save bond issuers less money than conventional wisdom holds.
26 See Estrella (2000, Annex I.C).
27 This problem is based on the article of Michael Kremer (1998), "Patent Buyouts: A Mechanism for Encouraging Innovation", *Quarterly Journal of Economics*.
28 Cited from: Kremer (1998), pages 1144–1145. We should add that Nicéphore Niépce and Sir John Herschel have better claims to have invented photography.
29 Without going into technical details, we need at least ω (= number of states) securities to construct as many independent securities as there are states. This is the case if, and only if, the matrix $\mathbf{R}$, defined above, is invertible.
30 Incomplete markets have important effects (Lengwiler, 2004): (i) Equilibrium prices are not uniquely defined; (ii) the equilibrium may not be Pareto efficient; (iii) in a productive economy, agents may not agree on production plans; (iv) ownership may become relevant for efficiency.
31 The securities "one dollar in A" and "one dollar in a", for example, are perfectly (negatively) correlated.
32 In the present context we denote probabilities by π, as the letter p will be used for prices.
33 As probabilities add up to one, one price is redundant.
34 If we can give Bob one more cookie without taking anything away from Alice, we can achieve a Pareto improvement.
35 Hong and Stein (2003) argue that short sale restrictions prevent pessimistic beliefs from flowing into the market during upturns. During downturns, accumulated hidden information comes out when relatively pessimistic investors become marginal buyers. Short sales restriction thus lead to exaggerations during bullish periods and can create some extra negative momentum in a downturn. Real estate markets, in which there are no short sales, thus may be particularly receptive to bubble-like behavior; see also Diether, Malloy and Scherbina (2002).
36 A function $\bar{\mathbf{s}}$ is a sufficient statistic for individual information sets $\mathbf{s}_i$ where knowledge of $\bar{\mathbf{s}}$ leads to the same posterior distribution as the knowledge of the individual information sets $\mathbf{s}_i$.
37 Formally speaking, prices are fully revealing if the mapping of information sets onto the price space is invertible.
38 *The New York Times*, 31 July 2003.
39 See for example http://www.longbets.org/
40 http://plato.stanford.edu/entries/common-knowledge/

41 Though we will encounter "trades" involving the same person in different situations, such as the student who promises himself that he will work more tomorrow, in Chapter 17.

42 Note that this only holds for knowledge, not for beliefs. If Alice believes that Bob knows, Bob may or may not know.

43 Nathan Rothschild—as the legend goes—was a step 2 player, and a lucky one: His guess that the others would think only one step ahead turned out to be correct.

44 O. Henry is the pseudonym for William Sydney Porter.

45 We focus on so-called pure strategy equilibria. There is also a third, mixed strategy equilibrium in which both players decide what to buy by tossing a coin. They may regret the final outcome, but not their action, i.e. their choice of probabilities.

46 Setting $\alpha = 2$ like in Chapter 6 here leads to a degenerate case. If $\theta = 1$, there is an infinite number of equilibria all satisfying $e_1 = e_2$; if $0 \neq \theta \neq 1$, the only equilibrium is $e_1 = e_2 = 0$. Finally if $\theta = 0$, $e_1 = e_2 = \infty$.

47 Or another Unix based system like Mac.

48 There is a third, mixed strategy, equilibrium in which both randomize their choices with 50 per cent.

49 A more difficult focal game arises when there are several equilibria with a high payoff (2,2) and one with a lower payoff (1,1). Here the focal point may support coordination, but on an inferior equilibrium.

50 From a game-theory point of view, it is irrelevant whether the announcing party plays before or after the announcement.

51 While communication may help to coordinate on a particular equilibrium, it does not reduce the *number* of equilibria. The communication game always has one equilibrium in which Alice randomizes over possible announced actions and Bob ignores her announcement. Such a babbling equilibrium leads to the original coordination game without communication, including the possible equilibria.

52 With unobservable private information we are anticipating some subject matter of Chapter 13.

53 We only treat the relatively simple case where R is deterministic. Diamond and Dybvig (1983) also discuss some implications of a stochastic R.

54 The formal proof of this result is as follows: The maximization problem for an individual is to allocate a fraction x to short-term investment, and a fraction of $(1-x)$ to long-term investment, where the latter yields a return L if withdrawn in period $t = 1$, and R in $t = 2$. Assuming no time discounting, utility maximization amounts to

$$\max_x (ku(c_1) + (1-k)u(c_2)) = \max_x (ku(x + (1-x)L) + (1-k)u(x + (1-x)R)),$$

and the corresponding first-order condition reads as

$$-\frac{1-L}{R-1} = -\frac{(1-k)}{k}\frac{u_2'(x+(1-x)R)}{u_1'(x+(1-x)L)} = -\frac{(1-k)}{k}\frac{u_2'(c_2)}{u_1'(c_1)},$$

where $u_1'(c_1)$ and $u_2'(c_2)$ stand for marginal utilities of consumption of an early dier in $t = 1$ and of a late dier in $t = 2$. The first-order conditions are illustrated by the fact that in the optimal point A both line a and the (highest reachable) indifference curve have the same slope.

55 Formally, optimal insurance is the solution to the following expected utility maximization problem:

$$\max_{d,r^*} (ku(d) + (1-k)u(r^*)) = \max_d \left(ku(d) + (1-k)u\left(\frac{1-d \cdot k}{1-k}R\right)\right),$$

where the constraint $r^* = (1 - d \cdot k)R/(1 - k)$ was used to substitute for r^*. The first-order condition then reads as

$$-\frac{1-k}{kR} = -\frac{1-k}{k}\frac{u'(r^*)}{u'(d)}.$$

It follows from the last expression that the promised return to late diers r^* still exceeds d; as $u'(r^*)/u'(d) = 1/R$ implies $u'(r^*) < u'(d)$, r^* must be greater than d for a risk-averse individual with a decreasing marginal utility of consumption.

56 The role of incentives for truth-telling are treated in Chapter 15.
57 The impossibility of contracting on non-observable outcomes is treated in Chapter 14.
58 On the role of truth-telling see Chapter 15.
59 We use "go" rather than "run" or "withdraw" as an homage to the song "Should I stay or should I go?" originally written by The Clash, but given an unsurpassable rendition by The Contractions.
60 This is the definition of a Nash equilibrium.
61 More precisely, there are two equilibria in so-called pure strategies, plus one (economically less relevant) equilibrium in mixed strategies (where individuals randomize their choices).
62 It is important to distinguish between the bank run game and the prisoner's dilemma (see Section 10 D). The former has *multiple* equilibria, while the prisoner's dilemma has one *unique* (but inefficient) equilibrium.
63 For an overview see Freixas and Rochet (1997); a recent contribution is Peck and Shell (2004). There are two crucial assumptions distinguishing Peck and Shell (2004) from earlier contributions: (1) patient and impatient depositors have different utility functions for consumption in their favorite periods (allowing the incentive compatibility constraint to bind at the optimal contract); (2) depositors do not know their place in line when they decide about withdrawing or not.
64 LTCM was founded in 1994 by a group including Myron Scholes and Robert Merton, two Nobel Prize winners in economics.
65 Then under the name of Schweizerische Kreditanstalt.
66 Jung (2000).
67 Global games indeed have unique equilibria. Yet, these equilibria, very much like mixed strategy equilibria, are characterized by heterogeneous actions.
68 In the literature it is sometimes assumed that indifferent individuals toss a coin (Bikhchandani et al., 1998).
69 Note that with prior probabilities of the two events A and B, the precision (the probability that the signal is right) is equal to the probability of receiving the right signal, see Chapter 4.
70 The signal even becomes *common knowledge* (see Chapter 8) but this is not important here.
71 The probabilities indicated in Figure 10.2 may differ from those in Bikhchandani, Hirshleifer and Welch (1998) who assume different prior probabilities as well as a different tie-breaking rule.
72 However, observability of results may lead to an inefficient learning process by an individual in a game against nature. This is the conclusion from the so-called "armed-bandit problem" (Chamley 2004, Ch. 8).
73 On a regular basis, *The Economist* publishes its so-called Big Mac Index, comparing the prices of a Big Mac in several countries. By comparing the prices in local currency and using the market exchange rate, *The Economist* then derives a measure for the deviations of the exchange rates from the rate implied by the purchasing power parity. Surprisingly perhaps, these deviations are high, ranging from an undervaluation vis-à-vis the US-Dollar of approximately 60 per cent for the Chinese Yuan to an overvaluation of the Swiss Franc of nearly 70 per cent (figures for May 2006).

74 See Snowdon and Vane (2005) for an excellent survey of the role of information and expectation formation in macroeconomics.
75 For the US, these are nicely reported in Mankiw, Reis and Wolfers (2003).
76 As a short cut, individuals may (implicitly) agree on a proxy (scapegoat) to make inferences about future values of a variable. See Bacchetta and van Wincoop (2004).
77 Plug $P_{t+1} = \frac{1}{1+r}E_t(P_{t+2} + D_{t+2})$ into the equation, then replace P_{t+2} by a similar equation and so forth. If the discount rate r is constant, one arrives at $P_t = \sum_{s=1}^{\infty} \frac{1}{(1+r)^t} E_t D_{t+s}$ excluding bubbles.
78 The title is taken from Townsend (1983).
79 A zero mean inflation is chosen for simplicity. The analysis would be exactly the same if this was not the case.
80 The precision of a signal was introduced for binary signals in Chapter 3.
81 Central banks that operate under an inflation targeting regime are, among others, New Zealand (inflation targeting adopted in 1990), Canada (1991), United Kingdom (1992), Australia (1993), Finland (1993), Sweden (1993), Spain (1995), the Czech Republic (1997), Israel (1997), Korea (1998), Poland (1999), Brazil (1999), and Switzerland (1999).
82 The tension is similar to the dilemma of the bank supervisor who reacts to market prices thereby disturbing those very prices (see Section 7 E).
83 See the discussion to Morris and Shin (2005) in the same issue of *Brookings Papers on Economic Activity*.
84 It might be worthwhile to review the findings of Capen, Clapp and Campbell (1971) in the light of the huge increase of crude oil prices in the 1970s.
85 Note that by definition of a common value object it is irrelevant from a welfare point of view who wins the auction.
86 http://bear.cba.ufl.edu/ritter/ipolink.htm
87 We are not supposed to disclose the name of the company.
88 We will refer to the repeated nature of interaction and to feedback scores in internet auctions as discussed in Bajari and Hortaçsu (2004) below.
89 More generally speaking, the gains from trade are the sum of producer and consumer surplus.
90 Once more we use π for probabilities instead of p, as we need p for price.
91 There is no premium that would be accepted by the low risks and not by the high risks.
92 The insurance market *ex ante* differs from the market for lemons. In the market for lemons, people want to trade for informational reasons (getting rid of a bad car) as well as for non-informational reasons (acquiring a car). Here, they only want to trade for non-informational reasons (getting insured) (Brunnermeier 2001, pp. 36–37).
93 The *Journal of Political Economy* used this excerpt as a back cover quote entitled “Optimal Risk-Sharing in the Trenches”.
94 At the same time, some pension systems pay additional benefits to widow(er)s and sometimes even to minors. This favors individuals with much younger spouses and old fathers, predominantly at the expense of bachelors.
95 In the language of auctions theory: Informed traders have information with common value, while noise traders have information with private value; see Chapter 12, in particular Box 12.2.
96 This contract strictly dominates its alternatives, once our rigid assumptions are slightly relaxed. It is sufficient to assume that (i) a reliable car has a low probability of breaking down and (ii) the seller is risk-averse.
97 Formally, the slope of iso-profit lines is: $-\frac{\beta(1-p_L)+(1-\beta)(1-p_H)}{\beta p_L+(1-\beta)p_H}$.
98 This is true as long as net assets are positive.
99 Laffont and Martimort (2002) consider messages with several dimensions.
100 This explains a phenomenon that appears to be a violation of the revelation principle. A second price auction (unlike a first price auction) is known to lead to truthful bidding

of valuations. Yet, risk-averse bidders tend to bid higher in a first price auction than in a second price auction. It therefore seems that a mechanism that leads to lying, in terms of the seller's revenue, "beats" a truth-telling mechanism. But this is not true, since no auction in which bidders can only bid in one single dimension can reveal convex preferences.

101 There is a very accessible representation in Lacker (1991).

102 Walter P. Schuetze, former chief accountant at the SEC, as quoted from Brunnermeier (2001).

103 http://www.cockatiels.org/articles/care/frights.html

104 Sappington (1991, p. 45).

105 A random variable $\tilde{X}_2$ first-order stochastically dominates the random variable $\tilde{X}_1$ if it has the higher probability to exceed any arbitrary threshold z, i.e. if $Pr\{\tilde{X}_2 > z\} \geq Pr\{\tilde{X}_1 > z\}$ or, equivalently, if the cumulative distribution function $F(z)$ (the probability that a random variable lies *below* some z) is (slightly) smaller for the *dominating* variable, i.e. if $F_{\tilde{X}_2} \leq F_{\tilde{X}_1}$; see Wolfstetter (1999, Ch. 4), Mas-Colell, Whinston and Green (1995, Ch. 6).

106 Note that the monotone likelihood ratio property is not implied by first-order stochastic dominance (Mas-Colell, Whinston and Green 1995, p. 485).

107 For $ar_0 = 0, p_h$ takes the maximum possible value of $1/2$; for $ar_0 = 0.17, p_h \approx 0.23$.

108 We thank Jürg Blum for this suggestion.

109 The quote is from http://bmj.bmjjournals.com/cgi/content/full/314/7091/1417/a

110 Gardiol, Geoffard and Grandchamp (2005), for example, find that in the Swiss health insurance system 75 per cent of the correlation between coverage and health care expenditures may be attributed to adverse selection (hidden information), and 25 per cent to incentive effects to moral hazard (hidden action).

111 15 January 2005, p. 71.

112 We follow the old-fashioned spelling "*homo œconomicus*" reflecting the Latin and Greek origins of the term. In most of the recent literature it is spelt "*homo economicus*".

113 http://en.wikipedia.org/wiki/List_of_cognitive_biases.

114 Neurological research also suggests that 'mentalizing', the ability to build higher-order beliefs, may have its own "home" in the brain, as mentioned above (see Box 8.7).

115 Quoted by his student Plato in *Meno*.

116 See Angeletos, Laibson et al. (2001) for an accessible introduction of many aspects of hyperbolic discounting in the context of life-cycle decision-making.

117 Technically speaking, their utility function is convex with a finite elasticity of intertemporal substitution. That an individual wants to consume at least a little bit in each year is an implication of the marginal utility of the logarithmic utility function, which is infinite at $c = 0$.

118 8 January 2004.

Bibliography

Aghion, P. and P. Howitt (1992) A Model of Growth through Creative Destruction, *Econometrica*, 60(2), 323–351.

Aghion, P. and P. Howitt (1997) *Endogenous Growth Theory*, MIT Press, Cambridge MA.

Ainley, D. G., G. Ballard, B. J. Karl and K. M. Dugger (2005) Leopard Seal Predation Rates at Penguin Colonies of Different Size, *Antarctic Science*, 17, 323–328.

Akerlof, G. A. (1970) The Market for 'Lemons': Quality Uncertainty and the Market Mechanism, *Quarterly Journal of Economics*, 84(3), 488–500.

Akerlof, G. A. and W. Dickens (1984) The economic consequences of cognitive dissonance, in G. A. Akerlof (ed.), *An Economic Theorist's Book of Tales*, Cambridge University Press, Cambridge UK.

Allen, F. and D. Gale (1999) Innovations in Financial Services Relationships, and Risk Sharing, *Management Science*, 45(9), 1239–1253.

Allen, F., S. Morris and H. S. Shin (2006) Beauty Contests and Iterated Expectations in Asset Markets, *Review of Financial Studies*, 19(3), 719–752.

Angeletos, G.-M., D. I. Laibson, A. Repetto, J. Tobacman and S. Weinberg (2001) The Hyperbolic Consumption Model: Calibration, Simulation, and Empirical Evaluation, *Journal of Economic Perspectives*, 15(3), 47–68.

Aragones, E., I. Gilboa, A. Postlethwaite and D. Schmeidler (2005) Fact-Free Learning, *American Economic Review*, 95, 1355–1368.

Arbatskaya, M. N. (forthcoming) Ordered Search, *Rand Journal of Economics*.

Armendariz de Aghion, B. and J. Morduch (2005) *The Economics of Microfinance*, MIT Press, Cambridge MA (reprint edn).

Arya, A., J. Glover and S. Sunder (2003) Are Unmanaged Earnings Always Better for Shareholders?, *Accounting Horizons*, 17, 111–116.

Ashraf, N., D. Karlan and W. Yin (2006) Tying Odysseus to the Mast: Evidence from a Commitment Savings Product in the Philippines, *Quarterly Journal of Economics*, 121(2), 635–672.

Avins, M. (1999) On the Trail of a Killer, They Discovered Hope, *Los Angeles Times*.

Ayres, I. and B. Nalebuff (1997) Common Knowledge as a Barrier to Negotiation, Yale School of Management Working Papers ysm76, Yale School of Management.

Bacchetta, P. and E. van Wincoop (2004) A Scapegoat Model of Exchange Rate Fluctuations, *American Economic Review*, 94(2), 114–118.

Bacchetta, P. and E. van Wincoop (2006) Can Information Heterogeneity Explain the Exchange Rate Determination Puzzle?, *American Economic Review*, 96(3), 552–576.

Bajari, P. and A. Hortaçsu (2003) The Winner's Curse, Reserve Prices, and Endogenous Entry: Empirical Insights from eBay Auctions, *Rand Journal of Economics*, 34(2), 329–355.

Bajari, P. and A. Hortaçsu (2004) Economic Insights from Internet Auctions, *Journal of Economic Literature*, 42(2), 457–486.

Barro, R. J. (1984) What Survives of the Rational Expectation Revolution? Rational Expectation and Macroeconomics in 1984, *American Economic Review*, 74(2), 179–182.

BCBS (Basel Committee on Banking Supervision) (2002) Basel II: International Convergence of Capital Measurement and Capital Standards: a Revised Framework, June.

Bebchuk, L. A. and J. M. Fried (2003) Executive Compensation as an Agency Problem, *Journal of Economic Perspectives*, 17(3), 71–92.

Becker, G. S. and K. M. Murphy (1988) A Theory of Rational Addiction, *Journal of Political Economy*, 96(4), 675–700.

Bénabou, R. and J. Tirole (2002) Self-Confidence and Personal Motivation, *Quarterly Journal of Economics*, 117(3), 871–915.

Bénabou, R. and J. Tirole (2004) Willpower and Personal Rules, *Journal of Political Economy*, 112(4), 848–886.

Bernanke, B. S. (2003) A Perspective on Inflation Targeting, Remarks by Governor Ben S. Bernanke at the Annual Washington Policy Conference of the National Association of Business Economists, Washington, DC, 25 March.

Bernheim, B. D. and A. Rangel (2005) Behavioral Public Economics: Welfare and Policy Analysis with Non-Standard Decision-Makers, NBER Working Paper 11518, National Bureau of Economic Research, Inc.

Beshears, J., J. J. Choi, D. I. Laibson and B. C. Madrian (2005) Early Decisions: A Regulatory Framework, *Swedish Economic Policy Review*, 12(2), 41–60.

Biais, B., P. Bossaerts and J.-C. Rochet (2002) An Optimal IPO Mechanism, *Review of Economic Studies*, 69(1), 117–146.

Bierman, S. H. and L. Fernandez (1995) *Game Theory with Economic Applications*, Addison-Wesley, Reading MA (reprint edn).

Bikhchandani, S., D. Hirshleifer and I. Welch (1992) A Theory of Fads, Fashion, Custom and Cultural Change as Informational Cascades, *Journal of Political Economy*, 100(5), 992–1026.

Bikhchandani, S., D. Hirshleifer and I. Welch (1998) Learning from the Behavior of Others: Conformity, Fads and Informational Cascades, *Journal of Economic Perspectives*, 12(3), 151–170.

Binmore, K. (1992) *Fun and Games: A Text on Game Theory*, DC Heath, Lexington MA.

Birchler, U. W. (2006) Bankruptcy Priority for Bank Deposits: A Contract Theoretic Explanation, *Review of Financial Studies*, 13(3), 813–840.

Birchler, U. W. and M. Facchinetti (2006) Can Bank Supervisors Rely on Market Data? A Critical Assessment from a Swiss Perspective, Swiss National Bank Working Papers 8, Swiss National Bank.

Birchler, U. W. and D. Hancock (2004) What Does the Yield on Subordinated Bank Debt Measure?, Finance and Economics Discussion Series 2004-19, The Federal Reserve Board.

Blum, J. M. (2002) Subordinated Debt, Market Discipline, and Banks' Risk Taking, *Journal of Banking and Finance*, 26(7), 1427–1441.

Board of Governors of the Federal Reserve System and United States Department of the Treasury (2000) The Feasibility and Desirability of Mandatory Subordinated Debt, Report to the Congress.

Bolton, P. and M. Dewatripont (2005) *Contract Theory*, MIT Press, Cambridge MA.

Bond, P., I. Goldstein and E. S. Prescott (2006) Market-Based Intervention and the Informational Content of Prices, mimeo, University of Pennsylvania and Federal Reserve Bank of Richmond.

Brunnermeier, M. M. (2001) *Asset Pricing under Asymmetric Information*, Oxford University Press, Oxford UK.

Bulow, J. and P. Klemperer (2002) Prices and the Winner's Curse, *Rand Journal of Economics*, 33(1), 1–21.

Camerer, C. F., T.-H. Ho and J.-K. Chong (2004) A Cognitive Hierarchy Model of Games, *Quarterly Journal of Economics*, 109(3), 861–898.

Camerer, C. F., S. Issacharoff, G. Loewenstein, T. O'Donoghue and M. Rabin (2003) Regulation For Conservatives: Behavioral Economics and the Case for "Asymmetric Paternalism", *University of Pennsylvania Law Review*, 151, 1211–1254.

Camerer, C. F. and G. Loewenstein (2004) Behavioral Economics: Past, Present, Future, in: C. F. Camerer, G. Loewenstein and M. Rabin (eds), *Advances in Behavioral Finance*, Princeton University Press, Princeton NJ.

Camerer, C. F., G. Loewenstein and D. Prelec (2005) Neuroeconomics: How Neuroscience can Inform Economics, *Journal of Economic Literature*, 43(1), 9–64.

Capen, E. C., R. V. Clapp and W. M. Campbell (1971) Competitive Bidding in High-Risk Situations, *Journal of Petroleum Technology*, 23, 641–653.

Caplin, A. and J. Leahy (2004) The Supply of Information by a Concerned Expert, *Economic Journal*, 114(497), 487–505.

Chamley, C. P. (2004) *Rational Herds: Economic Models of Social Learning*, Cambridge University Press, Cambridge UK.

Che, Y.-K. and I. Gale (2003) Optimal Design of Research Contests, *American Economic Review*, 93(3), 646–671.

Cooper, R. W. (1999) *Coordination Games: Complementarities and Macroeconomics*, Cambridge University Press, Cambridge UK.

Cooter, R. and T. Ulen (2004) *Law and Economics* (4th edn), Addison-Wesley, Boston MA.

Cukierman, A. (1980) The Effects of Uncertainty on Investment under Risk Neutrality with Endogenous Information, *Journal of Political Economy*, 88(3), 462–475.

Daniels, C. and P. Marlow (2005) Literature Review on the Reporting of Workplace Injury Trends, Research Report 36, Health and Safety Laboratory.

Danielsson, J. and H. S. Shin (2003) Endogenous Risk, in: P. Field (ed.), *Modern Risk Management: A History*, Risk Books, London.

Dasgupta, A., A. Prat and M. Verardo (2006) The Price of Conformism, Working Paper Series, SSRN eLibrary.

Davies, P. (2004) Emergent Biological Principles and the Computational Resources of the Universe, *Complexity*, 10(2), 1.

De Bandt, O. and P. Hartmann (2000) Systematic Risk: A Survey, ECB Working Paper 35, European Central Bank.

Dewatripont, M. and J. Tirole (1994) *The Prudential Regulation of Banks*, MIT Press, Cambridge MA.

DeYoung, R., M. J. Flannery, W. W. Lang and S. Sovescu (2001) The Information Content of Bank Exam Ratings and Subordinated Debt Prices, *Journal of Money, Credit and Banking*, 33(4), 900–925.

Diamond, D. W. (1984) Financial Intermediation and Delegated Monitoring, *Review of Economic Studies*, 51(3), 393–414.

Diamond, D. W. and P. Dybvig (1983) Bank Runs, Deposit Insurance, and Liquidity, *Journal of Political Economy*, 91(3), 401–419.

Diether, K. B., C. J. Malloy and A. Scherbina (2002) Differences of Opinion and the Cross-Section of Stock Returns, *Journal of Finance*, 57(5), 2113–2141.

Dixit, A. K. and R. S. Pindyck (1994) *Investment under Uncertainty*, Princeton University Press, Princeton NJ.

Doherty, N. A. and P. D. Thistle (1996) Adverse Selection with Endogenous Information in Insurance Markets, *Journal of Public Economics*, 63(1), 83–102.

Domenighetti, G., A. Casabianca, F. Gutzwiller and S. Martinoli (1993) Revisiting the Most Informed Consumer of Surgical Services: The Physician-Patient, *International Journal of Technology Assessment in Health Care*, 9(4), 505–513.

Dorus, S., E. Vallender, P. Evans, J. Anderson, S. Gilbert, M. Mahowald, G. Wyckoff, C. Malcolm and B. Lahn (2004) Accelerated Evolution of Nervous System Genes in the Origin of Homo Sapiens, *Cell*, 119(7), 1027–1040.

Dulleck, U. and R. Kerschbamer (2006) On Doctors, Mechanics and Computer Specialists: The Economics of Credence Goods, *Journal of Economic Literature*, 44(1), 5–42.

Estrella, A. (2000) Credit Ratings and Complementary Sources of Credit Quality Information, BIS Working papers 3, Bank for International Settelments.

Evanoff, D. D. and L. D. Wall (2001) Sub-Debt Yield Spreads as Bank Risk Measures, FRB Atlanta Working Paper 2001-11, Federal Reserve Bank of Atlanta.

Fama, E. F. (1970) Efficient Capital Markets: A Review of the Theory and Empirical Work, *Journal of Finance*, 25(2), 383–423.

Farmer, R. E. A. (1990) Rince Preferences, *Quarterly Journal of Economics*, 105(1), 43–60.

Farrell, J. (1987) Information and the Coase Theorem, *Journal of Economic Perspectives*, 1(2), 113–129.

Farrell, J. and M. Rabin (1996) Cheap Talk, *Journal of Economic Perspectives*, 10(3), 103–118.

Feltovich, N. (2002) Information Cascades with Endogenous Signal Precision, Working Paper, University of Houston.

Finkelstein, A. and K. McGarry (2006) Multiple Dimensions of Private Information: Evidence from the Long-Term Care Insurance Market, *American Economic Review*, 96(4), 938–958.

Finkelstein, A. and J. Poterba (2004) Adverse Selection in Insurance Markets: Policyholder Evidence from the U.K. Annuity Market, *Journal of Political Economy*, 112(1), 183–208.

Freixas, X., C. Giannini, G. Hoggarth and F. Soussa (1999) Lender of Last Resort: A Review of Literature, *Financial Stability Review*, November(7), Bank of England, 151–167.

Freixas, X. and J.-C. Rochet (1997) *Microeconomics of Banking*, MIT Press, Cambridge MA.

Frey, B. S. and M. Osterloh (eds) (2001) *Successful Management by Motivation: Balancing Intrinsic and Extrinsic Incentives Series: Organization and Management Innovation*, Springer-Verlag, Berlin.

Friedman, B. M. and M. Warshawsky (1990) The Cost of Annuities: Implications for Saving Behavior and Bequests, *Quarterly Journal of Economics*, 105(1), 135–154.

Fudenberg, D. and D. K. Levine (1998) *The Theory of Learning in Games*, MIT Press, Cambridge MA.

Gale, D. and L. Shapely (1962) College Admissions and the Stability of Marriages, *American Mathematical Monthly*, 69, 9–14.

Gardiol, L., P.-Y. Geoffard and C. Grandchamp (2005) Separating Selection and Incentive Effects in Health Insurance, PSE Working Papers 2005-38, PSE (Ecole normale supérieure).

Gardner, M. (1966) *New Mathematical Diversions from Scientific American*, Simon & Schuster, New York.

Gibbons, R. (1992) *A Primer in Game Theory*, Harvester Wheatsheaf, Brighton, UK.

Gosselin, P., A. Lotz and C. Wyplosz (2006) How much Information should Interest Rate-Setting Central Banks Reveal?, Technical Report No. 5666, C.E.P.R. Discussion Papers.

Graves, R. (1929) *Goodbye to All That*, Anchor, London. (Back cover quote in *The Journal of Political Economy* (111, 2003)).

Green, E. J. and P. Lin (2003) Implementing Efficient Allocations in a Model of Financial Intermediation, *Journal of Economic Theory*, 109, 1–23.

Greenspan, A. (1996) The Challenge of Central Banking in a Democratic Society, Remarks at the Annual Dinner and Francis Boyer Lecture of The American Enterprise Institute for Policy Research, Washington DC.

Greenspan, A. (2001) *Harnessing Market Discipline*, The Region, Banking and Policy Issues Magazine, Federal Reserve Bank of Minneapolis.

Grinblatt, M. S. and S. A. Ross (1985) Market Power in a Securities Market with Endogenous Information, *Quarterly Journal of Economics*, 100(4), 1143–1167.

Grossman, S. J. (1976) On the Efficiency of Competitive Stock Markets Where Trades Have Diverse Information, *Journal of Finance*, 31(2), 573–585.

Grossman, S. J. and J. E. Stiglitz (1980) On the Impossibility of Informationally Efficient Markets, *American Economic Review*, 70(3), 393–408.

Gruber, J. (2002) Smoking's "Internalities", *Regulation*, 25(4), 52–57.

Guay, W. R., J. E. Core and D. F. Larcker (2003) Executive Equity Compensation and Incentives: A Survey?, *Federal Reserve Bank of New York Economic Policy Review*, April, 27–50.

Gürkaynak, R. S., A. Levin and E. T. Swanson (2006) Does Inflation Targeting Anchor Long-Run Inflation Expectations? Evidence from Long-Term Bond Yields in the US, UK and Sweden, C.E.P.R. Discussion Papers 5808, Center for Economic Policy Research.

Hamermesh, D. S. and J. Slemrod (2006) The Economics of Workaholism: We Should not have Worked on this Paper. Available at http://www.eco.utexas.edu/Faculty/Hamermesh/Workaholism.pdf.

Hargreaves, S. P. and Y. Varoufakis (1995) *Game Theory: A Critical Introduction*, Routledge, London and New York.

Harmon-Jones, E. and J. Mills (eds) (1999) *Cognitive Dissonance: Progress on a Pivotal Theory in Social Psychology*, American Psychological Association Books, Washington DC.

Hart, O. (1995) *Firms, Contracts, and Financial Structure*, Oxford University Press, Oxford.

Hausken, K. (2006) Jack Hirshleifer: A Nobel Prize left unbestowed, *European Journal of Political Economy*, 22, 251–276.

Hirshleifer, J. (1971) The Private and Social Value of Information and the Reward to Incentive Activity, *American Economic Review*, 61, 561–574.

Hirshleifer, J. and J. G. Riley (1992) *The Analytics of Uncertainty and Information*, Cambridge University Press, Cambridge UK.

Hong, H. and J. C. Stein (2003) Differences of Opinion, Short-Sales Constraints, and Market Crashes, *Review of Financial Studies*, 16(2), 487–525.

Humphrey, N. K. (1976) The Social Function of Intellect, in P. P. G. Bateson and R. A. Hinde (eds), *Growing Points in Ethology*, Cambridge University Press, Cambridge UK, pp. 303–317.

Jehiel, P. and B. Moldovanu (2003) An Economic Perspective on Auctions, *Economic Policy*, 18, 269–308.

Jensen, M. C. and W. H. Meckling (1976) Theory of the Firm: Managerial Behavior, Agency Costs and Ownership Structure, *Journal of Financial Economics*, 3, 303–360.

Jensen, M. C., K. J. Murphy and E. G. Wruck (2004) Remuneration: Where We've Been, How We Got to Here, What are the Problems, and How to Fix Them, Technical report, Harvard NOM Working Paper No. 04-28; ECGI - Finance Working Paper No. 44/2004.

Jones, C. I. (2001) *Introduction to Economic Growth* (2nd edn), University of California, Berkeley CA.

Jullien, B., F. Salanié and B. Salanié (2001) Screening Risk Averse Agents Under Moral Hazard, CEPR Discussion Papers 3076, C.E.P.R. Discussion Papers.

Jung, J. (2000) *Von der Schweizerischen Kreditanstalt zur Credit Suisse Group*, Neue Zürcher Zeitung.

Keating, E., T. Z. Lys and R. P. Magee (2003) Internet Downturn: Finding Valuation Factors in Spring 2000, *Journal of Accounting and Economics*, 34(1), 189–236.

Keynes, J. M. (1936) *The General Theory of Employment, Interest and Money*, Macmillan, London and Cambridge University Press.

Klemperer, P. (2002) What Really Matters in Auction Design, *Journal of Economic Perspectives*, 16(1), 169–189.

Koessler, F. (2000) Common Knowledge and Interactive Behaviors: A Survey, *European Journal of Economic and Social Systems*, 14(3), 271–308.

Kovári, K. and R. Fechtig (2005) *Historische Alpendurchstiche in der Schweiz* (3rd edn), Gesellschaft für Ingenieurbaukunst, Zürich.

Kremer, M. (1998) Patent Buyouts: A Mechanism for Encouraging Information, *Quarterly Journal of Economics*, 113(4), 1137–1167.

Kreps, D. M. (1990) *A Course of Microeconomic Theory*, Princeton University Press, Princeton NJ.

Krugman, P. (1991) Target Zones and Exchange Rate Dynamics, *Quarterly Journal of Economics*, 106(3), 669–682.

Lacker, J. M. (1991) Why is there Debt?, *Federal Reserve Bank of Richmond Economic Review*, 77(4), 3–18.

Laeven, L., B. Karacaovali and A. Demirgüç-Kunt (2005) Deposit Insurance around the World: A Comprehensive Database, Policy Research Working Paper Series 3628, The World Bank.

Laffont, J.-J. (1993) *The Economics of Uncertainty and Information* (4th edn), MIT Press, Cambridge MA.

Laffont, J.-J. and D. Martimort (2002) *The Theory of Incentives*, Princeton University Press, Princeton NJ.

Laibson, D. (1997) Golden Eggs and Hyperbolic Discounting, *Quarterly Journal of Economics*, 112.

Laibson, D. I. (2003) Intertemporal Decision Making, in L. Nadel (ed.), *Encyclopedia of Cognitive Science*, vol. 1, Nature Publishing Group, London.

Lengwiler, Y. (2004) *Microfoundations of Financial Assets. An Introduction to General Equilibrium Asset Pricing*, Princeton University Press, Princeton NJ and Oxford UK.

LeRoy, S. F. (2004) Rational Exuberance, *Journal of Economic Literature*, 42(3), 783–804.

Lewis, D. K. (1969) *Convention: A Philosophical Study*, Harvard University Press, Cambridge MA.

Lombardelli, C., J. Proudman and J. Talbot (2005) Committees Versus Individuals: An Experimental Analysis of Monetary Policy Decision Making, *International Journal of Central Banking*, 1, 181–205.

Loughran, T. and J. R. Ritter (2002) Why Don't Issuers Get Upset About Leaving Money on the Table in IPOs?, *Review of Financial Studies*, 15(2), 413–444.

Luo, Y. (2005) Consumption Dynamics, Asset Pricing, and Welfare Effects under Information Processing Constraints, 2005 Meeting Papers 345, Society for Economic Dynamics.

McCloskey, D. (1986) *The Rhetoric of Economics*, Harvester Brighton Wheatsheaf, UK.

Maloney, M. and H. J. Mulherin (2003) The Complexity of Price Discovery in an Efficient Market: The Stock Market Reaction to the Challenger Crash, *Journal of Corporate Finance*, 9(4), 453–479.

Mankiw, G. N., R. Reis and J. Wolfers (2003) Disagreement about Inflation Expectations, in M. Gertler and K. Rogoff (eds), *NBER Macroeconomics Annual*, The MIT Press, Cambridge MA, 209–248.

Mas-Colell, A., M. D. Whinston and J. R. Green (1995) *Microeconomic Theory*, Oxford University Press, Oxford UK.

Milgrom, P. (1989) Auctions and Bidding: A Primer, *Journal of Economic Perspectives*, 3(3), 3–22.

Moeller, S. B., F. P. Schlingenmann and R. M. Stulz (2005) Wealth Destruction on a Large Scale? A Study of Acquiring-Firm Returns in the Recent Merger Wave, *Journal of Finance*, 60, 757–782.

Moldovanu, B. and M. Tietzel (1998) Goethe's Second Price Auction, *Journal of Political Economy*, 6, 854–859.

Mookherjee, D. (2006) Decentralization, Hierarchies, and Incentives: A Mechanism Design Perspective, *Journal of Economic Literature*, 44, 367–390.

Morris, S. and H. S. Shin (2002) The Social Value of Public Information, *American Economic Review*, 92(5), 1521–1534.

Morris, S. and H. S. Shin (2005) Central Bank Transparency and the Signal Value of Prices, Technical Report no. 2, Brookings Papers on Economic Activity.

Myatt, D. P., H. S. Shin and C. Wallace (2002) The Assessment: Games and Coordination, *Oxford Review of Economic Policy*, 18(4), 397–417.

Myatt, D. P. and C. Wallace (2002) Equilibrium Selection and Public Good Provision: The Development of Open-Source Software, *Oxford Review of Economic Policy*, 18(4), 446–461.

O'Hara, M. (1995) *Market Microstructure Theory*, Blackwell Publishers, Oxford UK.

Ottaviano, M. and P. Sørensen (2001) Information Aggregation in Debate: Who should Speak First, *Journal of Public Economics*, 81, 393–421.

Paulos, J. A. (2003) *A Mathematician Plays the Stock Market*, Allen Lane (Penguin Press), London, UK.

Peck, J. and K. Shell (2004) Equilibrium Bank Runs, *Journal of Political Economy*, 111(1), 103–123.

Persico, N. (2004) Committee Design with Endogenous Information, *Review of Economic Studies*, 71(1), 165–191.

Prendergast, C. (1999) The Provision of Incentives in Firms, *Journal of Economic Literature*, 37(1), 7–63.

Radner, R. (1993) The Organization of Decentralized Information Processing, *Econometrica*, 61(5), 1109–1146.

Radner, R. and J. E. Stiglitz (1984) A Non-Concavity in the Value of Information, in M. Boyer and R. Kihlstrom (eds), *Bayesian Models in Economic Theory*, Elsevier, Oxford UK, 33–51.

Reis, R. (2006) Inattentive Producers, *Review of Economic Studies*, 73(3), 793–821.

Reis, R. (forthcoming) Inattentive Consumers, *Journal of Monetary Economics*.

Rhode, P. W. and K. S. Strumpf (2004) Historical Presidential Betting Markets, *Journal of Economic Perspectives*, 18(2), 127–142.

Riley, J. G. (2001) Silver Signals: Twenty-Five Years of Screening and Signaling, *Journal of Economic Literature*, 39, 432–478.

Rochet, J.-C. and J. Tirole (2004) Two-Sided Markets: An Overview, Working paper, Institut d'Economie Industrielle.

Romer, P. M. (1990) Endogenous Technological Change, *Journal of Political Economy*, 98(5), Part 2: The Problem of Development: A Conference on the Institute for the Study of Free Enterprise Systems, 71–102.

Rothschild, M. and J. Stiglitz (1976) Equilibrium in Competitive Insurance Markets: An Essay on the Economics of Imperfect Information, *Quarterly Journal of Economics*, 90, 629–649.

Salanié, B. (1997) *The Economics of Contracts: A Primer*, MIT Press, Cambridge MA.

Samuelson, L. (2004) Modelling Knowledge in Economic Analysis, *Journal of Economic Literature*, 42(2), 367–403.

Sappington, D. E. M. (1991) Incentives in Principal–Agent Relationships, *Journal of Economic Perspectives*, 5(2), 45–66.

Schumpeter, J. A. (1939) *Business Cycles*, 1, McGraw-Hill, New York.

Schumpeter, J. A. (1947) *Capitalism, Socialism, and Democracy* (2nd edn), Harper and Brothers, New York.

Schwartz, E. I. (2004) *Juice: The Creative Fuel That Drives World-Class Inventors*, Harvard Business School Press, Boston MA.

Scotchmer, S. (2004) *Innovation and Incentives*, MIT Press, Cambridge MA.

Shannon, C. E. (1948) A Mathematical Theory of Communication, *The Bell System Technical Journal*, 27, 379–423, 623–656.

Shapiro, C. and H. R. Varian (1999) *Information Rules: A Strategic Guide to the Network Economy*, Harvard Business School Press, Boston MA.

Sherman, A. E. (2000) IPO's and Long Term Relationships: An Advantage of Book Building, *Review of Financial Studies*, 13(3), 697–714.

Shiller, R. J. (2000) *Irrational Exuberance*, Princeton University Press, Princeton NJ.

Shiller, R. J. (2003) From Efficient Markets Theory to Behavioral Finance, *Journal of Economic Perspectives*, 17(1), 83–104.

Shleifer, A. (2000) *Inefficient Markets, An Introduction to Behavioral Finance*, Oxford University Press, Oxford UK.

Shy, O. (2001) *The Economics of Network Industries*, Cambridge University Press, Cambridge MA.

Simon, H. A. (1978) Rationality as Process and as Product of Thought, *American Economic Review*, 68(2), 1–16.

Sims, C. A. (2003) Implications of Rational Inattention, *Journal of Monetary Economics*, 50(3), 665–690.

Singer, T. and E. Fehr (2005) The Neuroeconomics of Mind Reading and Empathy, *American Economic Review*, 95, 340–345.

Smith, A. (1759) *The Theory of the Moral Sentiments*, A. Millar, London.

Snowdon, B. and H. R. Vane (2005) *Modern Macroeconomics: Its Origins, Development and Current State*, Edward Elgar, Cheltenham, UK.

Stigler, G. J. (1961) The Economics of Information, *Journal of Political Economy*, 69, 213–225.

Stiglitz, J. E. and A. Weiss (1981) Credit Rationing in Markets with Imperfect Information, *American Economic Review*, 71(3), 393–410.

Sunder, S. (2002) Knowing What Others Know: Common Knowledge, Accounting and Capital Markets, *Accounting Horizons*, 16(4), 305–318.

Temin, P. and P. Klemperer (2001) An Early Example of the "Winner's Curse" in an Auction, *Lagniappe to Journal of Political Economy*, 109(6).

Thaler, R. H. (1992) *The Winner's Curse*, Princeton University Press, Princeton NJ.

Thaler, R. H. (2000) From Homo Economicus to Homo Sapiens, *Journal of Economic Perspectives*, 14(1), 133–141.

Thaler, R. H. and S. Benartzi (2004) Save More Tomorrow: Using Behavioral Economics to Increase Employee Saving, *Journal of Political Economy*, 112, 164–187.

Thaler, R. H. and H. M. Shefrin (1981) An Economic Theory of Self-Control, *Journal of Political Economy*, 89(2), 392–406.

Tirole, J. (1988) *The Theory of Industrial Organization*, MIT Press, Cambridge MA.

Todd, P. M. (1997) Searching for the next best mate, in R. Conte, R. Hegselmann and P. Terna (eds), *Simulating social phenomena*, Springer-Verlag, Berlin, 419–436.

Townsend, R. M. (1983) Forecasting the Forecast of Others, *Journal of Political Economy*, 91(4), 546–588.

Van Rooy, R. (2003) Quality and Quantity of Information Exchange, *Journal of Logic, Language and Information*, 12(4), 423–451.

Vives, X. (2005) Complementarities and Games: New Developments, *Journal of Economic Literature*, 43(2), 437–479.

Von Hayek, F. A. (1945) The Use of Knowledge in Society, *American Economic Review*, 35, 519–530.

Von Krogh, G. and E. von Hippel (2003) Editorial – Special Issue on Open Source Software Development, *Research Policy*, 32, 1149–1157.

Wexler, A. (1996) *Mapping Fate: A Memoir of Family, Risk and Genetic Research*, University of California Press, Berkeley CA.

Wexler, N. S. (1989) The Oracle of DNA, in L. Rowland (ed.), *Molecular Genetics of Neuromuscular Disease*, Oxford University Press, Oxford.

White, L. J. (2001) The Credit Rating Industry: An Industrial Organization Analysis, Working Paper 01-001, New York University Center for Law and Business Research.

Wolfers, J. and E. Zitzewitz (2004) Prediction Markets, *Journal of Economic Perspectives*, 18, 107–126.

Wolfstetter, E. (1999) *Topics in Microeconomics*, Cambridge University Press, Cambridge MA.

Index